Genetically Defined Animal Models of Neurobehavioral Dysfunctions

Genetically Defined Animal Models of Neurobehavioral Dysfunctions

Peter Driscoll
Editor

Birkhäuser
Boston • Basel • Berlin

Peter Driscoll
ETH - Zentrum
Laboratorium für vergleichende Physiologie und Verhaltensbiologie
CH-8092 Zürich
Switzerland

Library of Congress Cataloging-in-Publication Data
Genetically defined animal models of neurobehavioral dysfunctions /
Peter Driscoll, editor.
p. cm.
Includes bibliographical references and index.
ISBN 0-8176-3460-6 (h : alk. paper). — ISBN 3-7643-3460-6 (alk. paper)
1. Mental illness—Genetic aspects—Animal models. 2. Nervous system—Diseases—Genetic aspects—Animal models. I. Driscoll, Peter, 1938–
[DNLM: 1. Central Nervous System Diseases—genetics. 2. Disease Models, Animal. 3. Models, Genetic. WL 300 G3263]
RC455.4.G4G48 1992
616.8′0442—dc20
DNLM/DLC 91-33332
for Library of Congress CIP

Printed on acid-free paper.

ISBN 0-8176-3460-6
ISBN 3-7643-3460-6

Typeset by ATLIS Graphics, Mechanicsburg, PA.
Printed and bound by Quinn Woodbine, Woodbine, NJ.
Printed in the U.S.A.

9 8 7 6 5 4 3 2 1

Contents

List of Contributors

Michel L. Aubert Division de Biologie de la Croissance et de la Reproduction Clinique de Pédiatrie, Hôpital Cantonal Universitaire, 1211 Geneva 4, Switzerland

Christoph N. Berger Department of Pediatrics, University of California, San Francisco, CA 94143, USA

Jeffrey N. Carlson Department of Pharmacology and Toxicology A-136, Albany Medical College, Albany, NY 12208, USA

Allan C. Collins Institute for Behavioral Genetics, Campus Box 447, University of Colorado, Boulder, CO 80309-0447, USA

W. C. Dement Stanford University Sleep Disorders Center, Palo Alto, CA 94304, USA

P. Driscoll Laboratorium für Vergleichende Physiologie und Verhaltensbiologie, ETHZ, Turnerstrasse 1, CH-8092 Zürich, Switzerland

Leonard M. Eisenman Department of Anatomy, Jefferson Medical College, Thomas Jefferson University, Philadelphia, PA 19107, USA

Charles J. Epstein Department of Pediatrics, University of California, San Francisco, CA 94143, USA

S. D. Glick Department of Pharmacology and Toxicology A-136, Albany Medical College, Albany, NY 12208, USA

Dan Goldowitz Department of Anatomy and Neurobiology, University of Tennessee, Memphis, 875 Monroe Avenue, Memphis, TN 38104, USA

C. Grumet Stanford University Sleep Disorders Center, Palo Alto, CA 94304, USA

C. Guilleminault Stanford University, Medical School, Sleep Disorders Center, 701 Welch Road, Suite 2226, Palo Alto, CA 94304, USA

Ting-Ting Huang Department of Pediatrics, University of California, San Francisco, CA 94143, USA

Kalervo Kiianmaa Research Laboratories, State Alcohol Company (Alko Ltd.), P.O. Box 350, SF-00101 Helsinki, Finland

A. D. Lê Biobehavioral Research Department, Addiction Research Foundation, 33 Russell Street, Toronto, Ontario, Canada M5S 2S1

H.-P. Lipp Anatomisches Institut der Universität Zürich-Irchel, Winterthurerstrasse 190, CH-8057 Zürich Switzerland

Wolfgang Löscher Department of Pharmacology, Toxicology, and Pharmacy, School of Veterinary Medicine, Bünteweg 17, 3000 Hannover 71, Germany

M. Marks University of Colorado, Institute for Behavioral Genetics, Campus Box 447, Boulder, CO 80309-0447, USA

Stephen Clark Maxson Biobehavioral Sciences Graduate Degree Program, Department of Psychology, The University of Connecticut, Storrs, CT 06269-4154, USA

Michael J. Meaney Douglas Hospital Research Center, Department of Psychiatry, McGill University, Montréal, Québec, Canada H4H 1R3

E. Mignot Stanford University Medical School, Sleep Disorders Center, 701 Welch Road, Suite 2226, Palo Alto, CA 94143, USA

D. H. Overstreet University of North Carolina, Department of Psychiatry, Center for Alcohol Studies, Chapel Hill, NC 27514, USA

J. Riordan Research Institute, Hospital for Sick Children, 555 University Avenue, Toronto, Ontario, Canada M5G 1X8

Ravina Simons Research Institute, Hospital for Sick Children, 555 University Avenue, Toronto, Ontario, Canada M5G 1X8

John David Sinclair Research Laboratories, State Alcohol Company (Alko Ltd.), P.O. Box 350, SF-00101 Helsinki, Finland

Kinuko Suzuki Department of Pathology, University of North Carolina School of Medicine, Chapel Hill, NC 27599, USA

Kunihiko Suzuki Departments of Neurology and Psychiatry, Brain and Development Research Center, University of North Carolina School of Medicine, Campus Box 7250, Chapel Hill, NC 27599, USA

D. Wahlsten University of Alberta, Department of Psychology, Biological Sciences Building, Edmonton, Alberta, Canada T6G 2E9

C.-D. Walker University of California San Francisco, Department of Physiology, P.O. Box 0444, San Francisco, CA 94143, USA

Introduction

P. DRISCOLL

"A new era of understanding behavior at a genetic and molecular level is on us. . . . Of all the organs of the human body, in terms of sheer numbers and the greatest impact on society, genetic defects of the brain that affect behavior will be the most important. . . . It is the ultimate irony that, as an understanding of our mind and emotions increases, many lay members of our society withdraw in ignorance or fear into an anti-science, anti-intellectual, and pro-mysticism stance. The public needs to be included in the excitement, for they are the ultimate benefactors of one of man's last frontiers." (Comings, 1989).

This book is meant to provide information about genetic animal models for genetic human conditions that affect behavior. As with such widely used genetic animal models as those for spontaneous diabetes, hypertension and many forms of cancer, the models presented here are finding increasing usage by both academic and commercial scientists wherever more realistic and accurate simulations of the human situation are sought. The chapters contained herein cover very diverse subjects, necessitating diverse approaches, thereby resulting in a book that is not only informative but also provides interesting reading.

The first chapter, by Berger et al, covers Down syndrome (DS). Mouse trisomy 16 has been identified as an animal model of DS, based on the results of comparative gene mapping. Midgestation edema, a short and thick neck, flat snout, defect in eyelid closure, hematopoietic and immune system impairments, as well as cardiovascular anomalies (like DS) are seen. As these mice do not survive the perinatal period, chimeras have been developed to extend the lifetime of the trisomy 16 cells. These mice also show altered behavior, but attempts are being made to develop partial trisomy 16 mice which will have even more accurate phenotypes. Transplantation experiments may permit the study of the development of DS and Alzheimer amyloidosis, and methods of generating transgenic mice are described. These mice show abnormalities at neuromuscular junctions in the tongue, as in the "protruding tongue" of DS patients.

Trisomy 16 mice may also help to clarify the molecular mechanisms involved in Alzheimer's disease, as the genes encoding the amyloid precursor protein for the latter are also located on mouse chromosome 16 and human chromosome 21 (Coyle et al, 1988). Genetic factors appear to predominate in the etiology of

Alzheimer's disease (Breitner, 1987), and the pathology of both conditions at middle-age is very similar. (Mann, 1988).

The second chapter, by Suzuki and Suzuki, deals with Krabbe disease (global cell leukodystrophy), which is caused by a deficiency in activity of galactosylceramidase. This condition has been described in several species, of which the most frequently utilized is the twitcher mouse. An extensive degeneration of white matter (severe demyelination) is also seen in the peripheral nerves. The genetically-induced failure to degrade galactosylceramide leads to the accumulation of globoid cells in the brain, and this chapter describes both the clinical pictures and the many directions of treatment. Galactosylceramidase remains the only lyosomal enzyme not to be cloned from any source, thus ruling out transgenic methodology.

The third chapter, by Simons and Riordan, reviews the other myelin disorders. The most common one in humans is multiple sclerosis, which is caused by autoreactive T-cells reacting to myelin antigens. Pelizaeus Merzbacher disease in humans, in which only the CNS is affected, is represented by x-linked mutations such as the jimpy mouse and md rat. Affected humans usually die between 10–30 years of age, and the rodent models within 30 days of birth. Tremors, abnormal gait and nystagmus are among the symptoms. The jimpy mouse results from a mutant myelin proteolipid gene. Autosomal mutations, such as the shiverer and mld mice, both arise from myelin basic protein (MBP) gene defects. Other mutations include twitcher mice (see Chapter 2), quaking and trembler mice.

Shiverer mice show a more severe MBP absence than mld mice. They have both been compared in learning tests (Inagawa et al, 1988), and shiverer mice, along with trembler and quaking mice have been compared with controls in studies investigating the mechanisms involved in the formation of the myelin sheath (Urtizberea et al, 1989). In connection with related problems, much interesting work has been recently executed with models of human motor neuron diseases (Younger et al, 1990), such as the wobbler mouse (Mitsumoto et al, 1990; Pollin et al, 1990) as well as in dogs with inherited spinal muscular atrophy (Cork et al, 1989).

Cerebellar mutations in mice provide the most comprehensive series of mutant loci affecting a single mammalian CNS structure. These are reviewed in Chapter 4 by Goldowitz and Eisenman within the context of cerebellar development, which the mutations facilitate the study of by permitting determination of the genetic and epigenetic instructions that regulate it. Names like weaver, reeler, staggerer, lurcher and nervous indicate the types of phenotypic disorders (ataxias) observable in affected mice. Chimeras have been used to identify the Purkinje cell as a direct target in the pcd, staggerer, lurcher and weaver mutations. All mutants also show a varying loss of granule cells, making possible an analysis of the relationships between the two cell-types. Structural alterations in autistic and schizophrenic patients are also cited.

The authors conclude that more studies are needed which analyze the motor performance in mutants with different cerebellar phenotypes; these would have "non-invasive" advantages over studies employing cerebellar lesions or ablations (an argument reminiscent of that by Lipp et al, 1989, concerning structure/function studies on the hippocampus). Although more behavioral work is required here, there have been some very interesting developments in related fields. For example, immunological abnormalities have been found in staggerer mice (Trenkner and Hoffmann, 1986), and weaver mice have been discussed as a possible genetic model of Parkinsonism (Low et al, 1987). Dystonia musculorum (dt) mutant mice (Sotelo and Guenet, 1988) and gracile axonal dystrophy (GAD) mice (Yamazaki et al, 1991) have also been proposed as genetic models for human spinocerebellar (Friedreich's) ataxia.

A disease which is represented by some excellent animal models, and which is not otherwise covered in this book, is Duchenne muscular dystrophy (DMD). It is the most common and most severe of the muscular dystrophies in man, and is inherited in an X-linked, recessive fashion. This genetic condition has also been described in dogs (Valentine et al, 1988) and in mice. Remarkable recommendations for therapy and/or prevention of DMD have already been made based on studies with the dystrophic mouse models, such as controlled regimens of muscular use and exercise, depending upon the stage of myopathy encountered (Wirtz et al, 1988), gene therapy—using muscle fibrosis and fiber type proportions as indicators (Marshall et al, 1989), detection of heterozygote female carriers (Karpati et al, 1990) and a proposal which has raised great hopes of late in patients and their families, that being myoblast transplantation from normal subjects into the muscle of dystrophic ones (Partridge et al, 1989). Studies with the dog model are expected to make it possible to test the relationship between muscle fiber size and severity of the disease (Worton and Thompson, 1988).

According to Mignot et al (Chapter 5), the presence of a strong association between the DR2 HLA antigen and narcolepsy raises the possibility that the latter could be an autoimmune disorder, but this has not been confirmed as yet. Probably at least two genes are involved, which would be consistent with findings in the canine model. Familial narcolepsy (sustainable over generations) has been obtained with both the doberman and labrador breeds of dogs, transmission being autosomal recessive with complete penetrance. In this drug-oriented chapter (as are the three to follow), it is shown that drugs that increase cataplexy also increase REM sleep. Critical involvement of the amygdala and pontine cholinergic system in narcolepsy are suggested, the former indicating a role for emotional behavior in its precipitation, as well. Linkage studies are now being conducted to identify and isolate the narcolepsy gene(s).

In the sixth chapter, Löscher shows that genetic animal models of epilepsy, either spontaneous or sensory stimulation-induced, offer a means of investigating genetic, neurochemical, neuropathological, neurophysiological and neuropharmacological aspects of epilepsy which no artificially-produced seizure model can provide. The chapter concentrates on the evaluation of anticonvulsant drugs, in particular reference to specific types of human epilepsy. Epilepsy in dogs probably most closely represents the human condition, especially for grand mal epilepsy with spontaneous, recurrent seizures. Rats are described which are models for petit mal epilepsy with spontaneous, recurrent spike-wave discharges. Tottering mutant mice are shown to be a good model for both petit mal and focal seizures. Several genetic animal models of reflex seizures (rare in humans) are described, such as gerbils, baboons with photomyoclonic seizures, audiogenic seizure-susceptible mice and rats, and photosensitive fowl.

Other than the continuing search for new drugs, epilepsy research has lately offered valuable insight into other behavioral dysfunctions. Löscher's group has, for example, proposed renaming the sz mutant hamster dt^{sz}, as it appears to be rather a model for paroxysmal dystonia than for epilepsy (Löscher et al, 1989). In addition, data obtained from clinical and animal research suggest testable hypotheses of how recurrent epileptic seizures may alter neuronal function in ways that might predispose to specific, disruptive interictal behaviors, such as aggression, depression or schizophrenia (Engel et al, 1986).

As for Chapters 7 and 8, the genetics of alcoholism and genetic animal models in alcohol research need no special introduction; they have been amply covered in many reviews and books, several of which are mentioned in both chapters. Sinclair and Kiianmaa add to this store of information in Chapter 7, after a short review, by describing the AT and ANT lines of rats, which differ greatly in motor impairment induced by ethanol. They indicate that the difference between alcohol tolerance and non-tolerance may be found rather in the homeostatic control mechanisms of the lines than in the sensitivity *per se* of some target for ethanol.

Why are alcoholics almost invariably heavy smokers? In Chapter 8, Collins and Marks review the genetic aspects of these behaviors in more detail. They also describe different breeding techniques used in animal studies, including diallel and classical crosses, and inbred and outbred (selectively bred) stocks. Most of the effective genetic work as concerns nicotine has been done with inbred mouse strains, in studies dealing with learning and memory, acute sensitivity to nicotine and/or alcohol, nicotine-induced seizures and acute/chronic tolerance to nicotine (probably the most important area of all). Additional, valuable information has been yielded by the CNS receptor-binding experiments which have accompanied several of these studies.

It follows that the consumption of and addiction to hard drugs, such as heroine and cocaine, may also have a genetic component. Whereas on the clinical side, movement has only recently been initiated toward a systematic investigation of this topic (e.g., Grove et al, 1990), animal researchers have been interested in this subject for years. A recent, comprehensive review has thoroughly covered the area of genetic aspects of opioids and behavior. Individual reactivity to opioids in the C57BL/6 and DBA/2 inbred mouse strains has proved to be a particularly useful model, indicating a genetically determined dissociation of opioid effects on locomotor activity and pain inhibition. Stimulatory locomotor responses in the C57BL/6 reaction type were shown to be linked to a high risk of drug addiction and facilitatory effects on adaptive processes, whereas high analgesic potency in the DBA/2 reaction type was accompanied by a low proneness to drug abuse and amnesic properties of opioids (Frishchknecht et al, 1988). More recently, a selection experiment has demonstrated genetic control of high morphine preference in rats, as well as toxic effects by the third generation (Rönnbäck, 1990). There will undoubtedly be much more work done in this field in the coming years, at all levels.

Data have accumulated over the past 15 years to suggest that morphologic, chemical and behavioral indices of brain asymmetry are present in animals, as well as humans, and that these individual differences in cerebral laterality can predict individual differences in drug responses and/or abuse, spatial behavior and stressor reactivity. Chapter 9, by Carlson and Glick, reviews the proposition that faulty cerebral asymmetry is associated with behavioral dysfunction in both humans and rats, and that these are genetically determined traits. Of special interest are the inherent implications for emotional reactivity and depression which are discussed (both subjects of which are covered from additional aspects in later chapters), their consideration of stimulant sensitization, stressor reactivity and schizophrenia (which involves abnormal brain lateralization), and the deliberations given over to immune system differences and behavioral dysfunction.

Maxson's critical review (Chapter 10) states early that there are three types of genetic models—genic, biological and behavioral. It is impossible at present to consider aggression at the genic level, as individual genes with effects on any type of aggression have not been identified in either mice or humans, even though mice and humans are known to share hundreds of gene homologies to date. But biological (e.g. androgens) and potential behavioral models exist. Four types of the latter—offense, defense, infanticide and predation—are discussed in this chapter.

Another recently published review on mouse behavior, which describes individual coping strategies in aggressive and non-aggressive individuals (Benus et al, 1991), provides an excellent partner for Maxson's chapter. On the human end, evidence is building up that childhood and adult antisocial behavior (defined more broadly than just commission of criminal acts) is heritable (e.g. Grove et al, 1990). A thorough investigation of personality traits, both in humans and animals, may provide a useful key to tying the research done on this subject together (Karli, 1989).

Chapter 11, by Lipp and Wahlsten, summarizes our current knowledge of the corpus callosum and reviews the information available on callosal agenesis in humans and mouse models. Absence of the corpus callosum occurs in a surprisingly large number of human neurological syndromes, many of the most important of which are discussed (although many cases of "asymptomatic" agenesis have also been reported), and in several mouse strains, such as the BALB/c, 129/J, DDN and I/Ln. The latter show several metabolic defects similar to some of those seen in human cases, and are the subjects of most of the behavioral tests described and discussed in this chapter. The chapter concludes by effectively presenting a case for the use of these mouse models in studying callosal agenesis.

Continuing into more nebulous regions of behavioral dysfunctions, the last two chapters in the book discuss potential genetic models for depression/anxiety. The main problems in developing these types of models (attempts at models for autism and attention deficit disorder would be other examples) are actually the uncertainties existing at the syndrome end, i.e., that there are as yet insufficient or inexact definitions of the human conditions, that consistent symptoms for them are difficult to pinpoint, and that it is not yet possible to determine precisely the nature and degree of their heritability. These subjects have, of course, been competently approached on many occasions (e.g. Carey, 1987; Kendler et al, 1987), but the apparent overlapping among the affective disorders, anxiety and, as well, schizophrenia (e.g., Hoff et al, 1990; McGuffin et al, 1987; Squires-Wheeler et al, 1989) makes this a difficult task.

Overstreet's chapter considers not only his selectively bred rats (a cholinergic supersensitivity model for depression), but also discusses the possible suitability of certain inbred mice, as well as of rats which have been selected for "anxiety" traits, within the context of their candidature as models of depression. The main factor to keep in mind is that all of these models are intended to be "trait" models, rather than "state" models, i.e. to be models more representative of the aetiology of the disorder(s) than are the purely environmental-stress models mostly used at present. This differentiation was, in fact, discussed recently in a review which was concerned with models for anxiety disorders (Lister, 1990). The investigation of anxiety in unselected populations of animals was questioned, and it was noted that in humans a drug's efficacy as an anxiolytic is often not apparent in populations of normal volunteers, but is only evident in clinically-anxious populations. It was recommended that future animal research should place a greater emphasis on models with an etiological basis. It might be noted that the usefulness of screening for potential antidepressants by looking at the behavior of unselected populations of animals had been questioned in a previous review (File and Tucker, 1986), once again based upon the crucial differences extant between the general population and clinically-depressed patients.

The last chapter in the book, by Walker et al, deals with genetic aspects of the hypothalamus-pituitary-adrenal axis, focusing on the differences seen in basal and stress-stimulated endocrine responses, on target tissue sensitivity and on the feedback effects of corticosteroids in two lines of rats selected and bred for differences in active, two-way avoidance (which has turned out to reflect differences in anxiety traits—see discussion above). These particular sublines of the Roman high- and low-avoidance lines of rats have been gradually developing into a model which has already established genetically based interactions among emotionality level, neurotransmitter systems, the immunological system and hormonal activity, and additional, vital neurochemical pathways, along with their behavioral representations, are continuing to be studied. Work done on the medial, prefrontal cortex (D'Angio et al, 1988) may provide some impetus for attempting to produce a genetic animal model for schizophrenia, if one considers certain clinical tests (Cohen et al, 1987); and the role of prolactin, which is an important stress hormone both in rats (Minamitani et al, 1987) and humans (Jacobs et al, 1986), has been recently emphasized in these lines of rats as a probable connecting link with the immunological system (Castanon et al, 1992).

Before handing the proceedings over to the more qualified authors of this book's chapters, a few more genetic models and models-to-be should be briefly mentioned, models which will not be further represented here, but which look very promising. Having just entertained the subject of anxiety, we can switch over to nervous pointer dogs, which have been variously presented as a potential model for "pathological anxiety" to panic disorder. Whereas, in response to exposure to humans or novel stimuli the dogs show such interesting behavioral patterns as excessive timidity, freezing, defecation, etc., it has proven difficult to try to pinpoint the biological sources of these behavioral malfunctions, and such complications as sex differences, deafness, etc. have occasionally surfaced (Klein et al 1988a, 1988b, 1990; Newton et al, 1988). There can be no doubt, however, that useful information will continue to come from the use of this model during the course of further studies.

One genetic animal model which has been studied to a considerable extent, but only recently in relation to the CNS and behavior, has been that for systemic lupus erythematosis, the NZB/NZW F_1 mouse. Interest in the immune mediation of neurobehavioral disorders has, at the same time, been steadily increasing, especially in regard to this condition in humans (Hoffman et al, 1988), where the emphasis has been on learning disabilities and coping behavior (Lahita et al, 1988; Engle et al, 1990). Recent studies with the mouse model have certainly been most productive (Kier 1990; Neveu et al, 1989; Sherman et al, 1988), and this trend will undoubtedly continue.

Returning by this indirect route to the source of the quotation which opened this introduction, whereas an animal model for Tourette syndrome (see also Comings and Comings, 1987) may be too much to hope for at present, it would be very worthwhile to keep an eye on the following ones, with which I would like to conclude: the brindled mutant mouse as a model for Menkes disease (Tanaka et al, 1990), the Snell pituitary dwarf mouse and other hormone-deficient mutants (Noguchi, 1988), the senescence accelerated mouse as a model for senile memory deficit and/or spontaneous spongy degeneration associated with aging (Yagi et al 1988, 1989) and hereditary hydrocephalus models in rats and mice (Bruni et al, 1988; Richards et al, 1989). Hoping that no relevant models have been entirely omitted and, at the same time, knowing that some must have been and apologizing profusely for this unavoidable circumstance, I invite the reader to enjoy, and learn from, the following chapters as much as I have while putting this book together. I would also like to take this opportunity to sincerely thank both George Adelman and James Doran of Birkhäuser-Boston for their good-natured assistance and, most of all, patience during that process.

REFERENCES

Benus RF, Bohus B, Koolhaas JM, van Oortmerssen GA (1991): Heritable variation for aggression as a reflection of individual coping strategies. *Experientia* 47:1008–1019

Breitner JCS (1987): Genetic factors in the etiology of Alzheimer's disease. In: *Psychopharmacology: The Third Generation of Progress,* Meltzer HY, ed. New York: Raven Press, 929–939

Bruni JE, Del Bigio MR, Cardoso ER, Persaud TVN (1988): Hereditary hydrocephalus in laboratory animals and humans. *Exp. Pathol.* 35:239–246

Carey G (1987): Big genes, little genes, affective disorder, and anxiety. *Arch. gen. Psychiat.* 44:486–491

Castanon N, Dulluc J, Le Moal M, Mormede P (1992): Neuroendocrine studies exclude the pituitary-adrenocortical system and suggest prolactin as a link between behavioral and immune differences between the Roman rat lines. *Physiol. Behav.* submitted

Cohen RM, Semple WE, Gross M, Nordahl TE, De Lisi LE, Holcomb HH, King AC, Morihisa JM, Pickar D (1987): Dysfunction in a prefrontal substrate of sustained attention in schizophrenia. *Life Sci.* 40:2031–2039

Comings DE (1989): Presidential address: The genetics of human behavior—lessons for two societies. *Amer. J. human Genet.* 44:452–460

Comings DE, Comings BG (1987): A controlled study of Tourette syndrome, I-VII. *Amer. J. human Genet.* 41:701–866

Cork LC, Altschuler RJ, Bruha PJ, Morris JM, Lloyd DG, Loats HL, Griffen JW, Price DL (1989): Changes in neuronal size and neurotransmitter marker in hereditary canine spinal muscular atrophy. *Lab. Invest.* 61:69–76

Coyle JT, Oster-Granite ML, Reeves RH, Gearhart JD (1988): Down syndrome, Alzheimer's disease and the trisomy 16 mouse. *Trends Neurosci.* 11:390–394

D'Angio M, Serrano A, Driscoll P, Scatton B (1988): Stressful environmental stimuli increase extracellular DOPAC levels in the prefrontal cortex of hypoemotional (Roman high avoidance) but not hyperemotional (Roman low avoidance) rats. An in vivo voltammetric study. *Brain Res.* 451:237–247

Engel J Jr, Caldecott-Hazard S, Bandler R (1986): Neurobiology of behavior: anatomic and physiological implications related to epilepsy. *Epilepsia* 27 (Suppl 2):S3–S13

Engle EW, Callahan LF, Pincus T, Hochberg MC (1990): Learned helplessness in systemic lupus erythematosus: analysis using the rheumatology attitudes index. *Arthritis Rheumat.* 33:281–286

File SE, Tucker JC (1986): Behavioral consequences of antidepressant treatment in rodents. *Neurosci. Biobehav. Rev.* 10:123–134

Frischknecht HR, Siegfried B, Waser PG (1988): Opioids and behavior: genetic aspects. *Experientia* 44:473–481

Grove WM, Eckert ED, Heston L, Bouchard TJ Jr, Segal N, Lykken DT (1990): Heritability of substance abuse and antisocial behavior: a study of monozygotic twins reared apart. *Biol. Psychiat.* 27:1293–1304

Hoff AL, Shukla S, Aronson T, Cook B, Ollo C, Baruch S, Jandorf L, Schwartz J (1990): Failure to differentiate bipolar disorder from schizophrenia on measures of neuropsychological function. *Schizophrenia Res.* 3:253–260

Hoffman SA, Narendran A, Shucard DW, Harbeck RJ (1988): Autoantibodies, immune complexes, and behavioral disorders: neuropsychiatric involvement in systemic lupus erythematosis. *Drug Devel. Res.* 15:237–251

Inagawa K, Watanabe S, Tsukada Y, Mikoshiba K (1988): The role of myelination in learning performance observed in two strains of myelin-deficient mutant mice (shiverer and mld). *Behav. neur. Biol.* 50:184–192

Jacobs S, Brown SA, Mason J, Wahby V, Kasl S, Ostfeld A (1986): Psychological distress, depression and prolactin response in stressed persons. *J. human Stress* 12:113–118

Karli P (1989): Is the concept of "personality" relevant to the study of animal aggression? *Eur. J. Personality* 3:139–148

Karpati G, Zubrzycka-Gaarn EE, Carpenter S, Bulman DE, Ray PN, Worton RG (1990): Age-related conversion of dystrophin-negative to -positive fiber segments of skeletal but not cardiac muscle fibers in heterozygote mdx mice. *J. Neuropathol. exp. Neurol.* 49:96–105

Kendler KS, Heath AC, Martin NG, Eaves LJ (1987): Symptoms of anxiety and symptoms of depression: same genes, different environments? *Arch. gen. Psychiat.* 44:451–457

Kier AB (1990): Clinical neurology and brain histopathology in NZB/NZW F_1 Lupus mice. *J. comp. Pathol.* 102:165–177

Klein E, Lenox RH, Uhde TW (1988a): Alpha-2 adrenergic receptor binding in platelets and brains of nervous and normal pointer dogs. J. psychiatr. Res. 22:241–247

Klein E, Steinberg SA, Weiss SRB, Matthews DM, Uhde TW (1988b): The relationship between genetic deafness and fear-related behaviors in nervous pointer dogs. *Physiol. Behav.* 43:307–312

Klein EH, Tomai T, Uhde TW (1990): Hypothalamo-pituitary-adrenal axis activity in nervous and normal pointer dogs. *Biol. Psychiat.* 27:791–794

Lahita RG (1988): Systemic lupus erythematosis: learning disability in the male offspring of female patients and relationship to laterality. *Psychoneuroendocrinology* 13:385–396

Lipp HP, Schwegler H, Crusio WE, Wolfer DP, Leisinger-Trigona MC, Heimrich B, Driscoll P (1989): Using genetically-defined rodent strains for the identification of hippocampal traits relevant for two-way avoidance behavior: a non-invasive approach. *Experientia* 45:845–859

Lister RG (1990): Ethologically-based animal models of anxiety disorders. *Pharmacol. Ther.* 46:321–340

Löscher W, Fisher JE Jr, Schmidt D, Fredow G, Hönack D, Iturrian WB (1989): The *sz* mutant hamster: a genetic model of epilepsy or of paroxysmal dystonia? *Movement Disord.* 4:219–232

Low WC, Triarhou LC, Kaseda Y, Norton J, Ghetti B (1987): Functional innervation of the striatum by ventral mesencephalic grafts in mice with inherited nigrostriatal dopamine deficiency. *Brain Res.* 435:315–321

Mann DMA (1988): The pathological association between Down syndrome and Alzheimer disease. *Mechanisms Ageing Devel.* 43:99–136

Marshall PA, Williams PE, Goldspink G (1989): Accumulation of collagen and altered fiber-type ratios as indicators of abnormal muscle gene expression in the mdx dystrophic mouse. *Muscle Nerve* 12:528–537

McGuffin P, Murray RM, Reveley AM (1987): Genetic influence on the psychoses. *Brit. med. Bull.* 43:531–556

Minamitani N, Minamitani T, Lechan RM, Bollinger-Gruber J, Reichlin S (1987): Paraventricular nucleus mediates prolactin secretory responses to restraint stress, ether stress, and 5-hydroxy-L-tryptophan injection in the rat. *Endocrinology* 120:860–867

Mitsumoto H, McQuarrie JG, Kurahashi K, Sunohara N (1990): Histometric characteristics and regenerative capacity in Wobbler mouse motor neuron disease. *Brain* 113:497–507

Neveu PJ, Betancur C, Barnéoud P, Preud'homme JL, Aucouturier P, Le Moal M, Vitiello S (1989): Functional brain asymmetry and murine systemic lupus erythematosus. *Brain Res.* 498:159–162

Newton JEO, Paige SR, Angel C, Reese WG (1988): Heart rate and activity in response to natural stimuli in nervous and normal pointer dogs. *Biol. Psychiat.* 23:829–833

Noguchi T (1988): Brain development in dwarf mice. *Prog. Neurobiol.* 31:149–170

Partridge TA, Morgan JE, Coulton GR, Hoffman EP, Kunkel LM (1989): Conversion of mdx myofibres from dystrophin-negative to -positive by injection of normal myoblasts. *Nature* 337:176–179

Pollin MM, McHanwell S, Slater SR (1990): Loss of motor neurons from the median nerve motor nucleus of the mutant mouse "Wobbler". *J. Neurocytol.* 19:29–38

Richards HK, Bucknall RM, Jones HC, Pickard JD (1989): The uptake of (^{14}C)deoxyglucose into brain of young rats with inherited hydrocephalus. *Exp. Neurol.* 103:194–198

Rönnbäck L (1990): Is there a genetic control of morphine preference in rat? *Pharmacol. Biochem. Behav.* 35:15–20

Sherman GF, Rosen GD, Galaburda AM (1988): Neocortical anomalies in autoimmune mice: a model for the developmental neuropathology seen in the dyslexic brain. *Drug Devel. Res.* 15:307–314

Sotelo C, Guenet JL (1988): Pathological changes in the CNS of *dystonia musculorum* mutant mouse: an animal model for human spinocerebellar ataxia. *Neuroscience* 27:403–424

Squires-Wheeler E, Skodol AE, Bassett A, Erlenmeyer-Kimling L (1989): DSM-III-R schizotypal personality traits in offspring of schizophrenic disorder, affective disorder, and normal control parents. *J. psychiat. Res.* 23:229–239

Tanaka H, Kasama T, Inomata K, Nasu F (1990): Abnormal movements in brindled mutant mouse heterozygotes: as related to the development of their offspring—biochemical and morphological studies. *Brain Devel.* 12:284–292

Trenkner E, Hoffmann MK (1986): Defective development of the thymus and immunological abnormalities in the neurological mouse mutation "Staggerer". *J. Neurosci.* 6:1733–1737

Urtizberea M, Bourre JM, Debray M, Scherrmann JM (1989): Comparison of colchicine toxicity on different dysmyelinating mutant models. *Neurotoxicology* 10:211–218

Valentine BA, Cooper BJ, de Lahunta A, O'Quinn R, Blue JT (1988): Canine X-linked muscular dystrophy. An animal model of Duchenne muscular dystrophy: clinical studies. *J. neurol. Sci.* 88:69–81

Wirtz P, Loermans HMT, de Haan AFJ, Hendriks JCM (1988): Early immobilization of hindleg muscles of dystrophic mice: short-term and long-term effects. *J. neurol. Sci.* 85:293–307

Worton RG, Thompson MW (1988): Genetics of Duchenne muscular dystrophy. *Annu. Rev. Genet.* 22:601–629

Yagi H, Katoh S, Akiguchi I, Takeda T (1988): Age-related deterioration of ability of acquisition in memory and learning in senescence accelerated mouse: SAM-P/8 as an animal model of disturbances in recent memory. *Brain Res.* 474:86–93

Yagi H, Irino M, Matsushita T, Katoh S, Umezawa M, Tsuboyama T, Hosokawa M, Akiguchi I, Tokunaga R, Takeda T (1989): Spontaneous spongy degeneration of the brain stem in SAM-P/8 mice, a newly developed memory-deficient strain. *J. Neuropathol. exp. Neurol.* 48:577–590

Yamazaki K, Kobayashi A, Kumazawa A, Wakabayashi T, Kikuchi T (1991): Axonal degeneration in the central nervous system of gracile axonal dystrophy (GAD) mice progresses like in human spinocerebellar ataxias. *Biomed. Res.* 12:143–148

Younger DS, Rowland LP, Latov N, Sherman W, Pesce M, Lange DJ, Trojaborg W, Miller JR, Lovelace RE, Haya AP, Kim TS (1990): Motor neuron disease and amyotrophic lateral sclerosis: Relation of high CSF protein content to paraproteinemia and clinical syndromes. *Neurology* 40:595–599

Mouse Models of Human Trisomy 21 (Down Syndrome)

CHRISTOPH N. BERGER, TING-TING HUANG AND CHARLES J. EPSTEIN

Trisomy 21 is one of the most frequent chomosome abnormalities in man. One out of 700 to 1000 newborn babies has this chromosomal imbalance, which arises due to meiotic nondisjunction of chromosome 21. This leads to three copies of chromosome 21 (trisomy 21, Ts21), the consequence of which is the development of characteristic features associated with Down syndrome (DS). Frequently seen in Down syndrome are dysmorphic features (facial, hands), congenital anomalies of the heart, immunodeficiency and moderate mental retardation. Furthermore, individuals with DS develop the pathological changes of Alzheimer's disease, including amyloid plaques and neurofibrillary tangles (Wisniewski et al., 1985; Epstein et al., 1987a).

In mice, trisomy for chromosome 16 (Ts16) results in several features that resemble those found in the phenotype of DS. Besides these phenotypic parallels, there is also a strong similarity on a genetic basis. We shall describe mouse models for the study of human Down syndrome and Alzheimer's disease and summarize recent advances in comparative gene mapping, the characterization of the Ts16 immune system, and the derivation of transgenic mouse strains by the introduction of additional copies of genes encoded by chromosome 21 into the mouse genome.

COMPARATIVE MAPS OF HUMAN CHROMOSOME 21 AND MOUSE CHROMOSOME 16

Standard genetic and molecular mapping procedures have led to the identification of a large conserved region located in the distal portion of human chromosome 21 (HSA21) and murine chromosome 16 (MMU16) (see Fig. 1.1). Three genes (*App, Sod-1,* and *Ets-2*) were initially mapped to this region (Cox et al., 1980; Cox and Epstein, 1985; Watson et al., 1986; Lovett et al., 1987; Reeves et al., 1987; Cheng et al., 1988). The relative positions of these three loci were determined in both species and found to be in the same order. In the mouse, *App, Sod-1,* and *Ets-2* were located distal to the T28H translocation breakpoint, whereas the DNA segments *D21S13* and *D21S52* have been located proximal (MacDonald and Cox, 1989). Additionally, the genes encoding the interferon α/β receptor (*Ifnar*) and

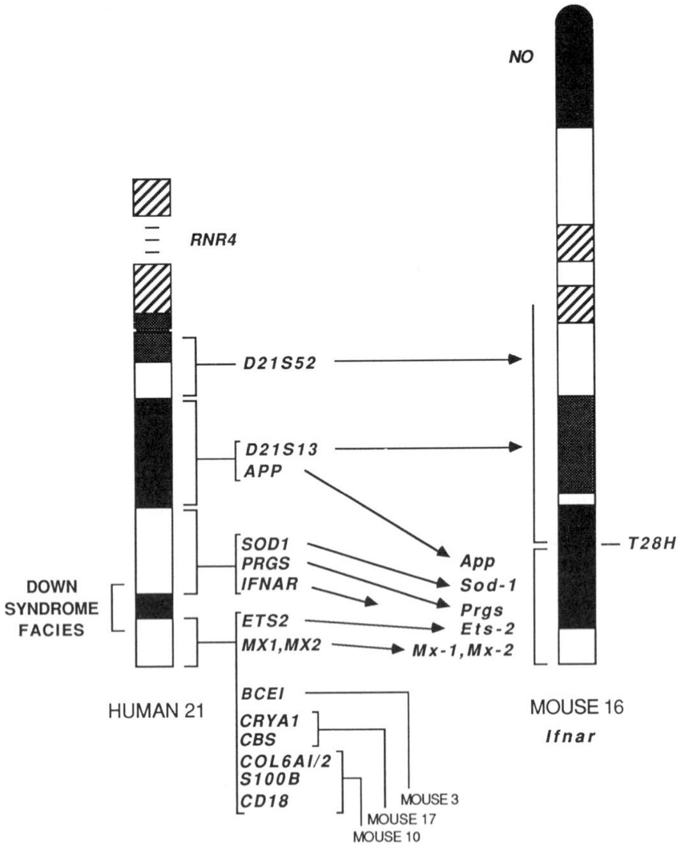

FIGURE 1.1. Localization of human chromosome 21 genes in the mouse genome. *Ifnar* is on both HSA21 and MMU16, but is not further mapped on MMU16. Modified with permission from Epstein (1986), Lyon (1988), Cheng et al. (1988), Reeves et al. 1989), MacDonald et al. (1988, 1989), and Burmeister et al. (1989).

phosphoribosyl glycinamide synthetase *(Prgs)* have been assigned to MMU16 and HSA21 (Lin et al., 1980; Cox et al., 1980, 1981), further expanding the region of homology between the mouse and human genomes. More recently, the neurological mutant, weaver *(wv),* and the loci, *Mx-1* and *Mx-2,* responsible for resistance to orthomyxovirus infection, were found to be linked to this conserved synteny in the mouse (Reeves, et al., 1988; 1989), and the latter are also present on the distal end of HSA 21.

Although aneuploidy of this large conserved region leads to the development of the characteristic features found in DS, the region with the greatest effect on the observable (facial) phenotype is located on the part of chromosome 21 from 21q22.1 to proximal 21q22.3 (Rethoré, 1981; Summitt, 1981; Korenberg et al., 1990). Although it is not known which regions of HSA21 are responsible for other components of the DS phenotype, one would ideally like an animal model to be trisomic for only the critical region that is actually involved. However, Ts16 animals carry three copies of all genes of chromosome 16, not only the ones that have homologs on HSA21. Because of the difference in the relative sizes of the chromosomes as compared to the whole genome (MMU16 comprises about 4% and HSA21 represents only about 1.9% of their respective total genomes), there are genes on MMU16 whose homologs are on human chromosomes different from HSA21: for example, the genes mahoganoid *(md)* and protamine *(Prm)* are found on HSA16, immunoglobulin λ light chain *Igl-1)* is on HSA22, and somatostatin *(Smst)* is on HSA3. Conversely, the region of HSA21 distal to *Mx* contains information that is not on MMU16: for example, the genes for the protein S100β and for CD18 (the β chain of the family of adhesion molecules Mac-1/OKM1, LFA-1 and p150,95) have been mapped to MMU10 (MacDonald et al., 1988). Other HSA 21 genes have been mapped to mouse chromosome 3 and 17 (Cheng et al., 1988; Burmeister et al., 1989). Because of these genetic differences between MMU16 and HSA21, murine Ts16 may develop a number of features that are not represented in DS and, at the same time, other DS features might not be found in Ts16 animals. Nevertheless, the DNA homology between HSA21 and MMU16 comprises the largest segment of conserved synteny between the mouse and the human chromosome complements known up to the present (Searle et al., 1989).

MOUSE TRISOMY 16

Based on the results of comparative gene mapping, mouse Ts16 has been identified as an animal model of DS. Therefore, studies of this mouse model of DS require easy access to Ts16 animals. These can be obtained using Robertsonian translocation-carrying strains of mice, in which a high frequency of meiotic nondisjunction has been observed (Gropp et al., 1975; Lyon et al., 1976). The strain *Rb(16.17)32Lub/Rb(11.16)2H,* in which both chromosomes 16 are fused, one to chromosome 11 and the other to chromosome 17, are used to breed Ts16 animals. If mated to animals of a regular inbred or outbred strains of mice, about 17% of the embryos produced are trisomic for chromosome 16, and about the same proportion are monosomic (Magnuson et al., 1985). Whereas the monosomic embryos die at the implantation stage, the Ts16 fetuses develop to term but do not survive the perinatal period (Epstein, 1985). Ts16 embryos can be recognized by their phenotypic appearance (see below).

The Phenotype of Trisomy 16 Fetuses

Ts16 fetuses can easily be identified by their massive edema in midgestation. The edema is still present to some extent later, but the phenotypic appearance of a short thick neck and flat snout is more prominent. Ts16 fetuses have a defect in eye lid closure, which results in open eyelids at the end of gestation (Miyabara et al.,

TABLE 1.1. Neuroanatomic and neurochemical features of mouse trisomy 16.

Reduced brain weight (to as low as 32–38% of controls)
Reduction of neurons in the basal forebrain at early stages of development (day 16 of gestation)
Retarded development of cerebellar foliation and hippocampal fissure formation
Retarded development of the basicranium and craniofacial apparatus
Decreased catecholaminergic system markers, except for increased DOPA decarboxylase and catechol-O-methyl transferase
Decreased or normal cholinergic markers
Normal GABAergic system marker (glutamate decarboxylase)
Abnormal dorsal root ganglion neuron action potentials

Modified with permission from Epstein et al. (1985a). Data taken from Gearhart et al. (1983), Oster-Granite et al. (1983), Blue et al. (1984), Ozand et al. (1984), Singer et al. (1984), Orozco et al. (1988), and Sweeney et al. (1989).

1982). The edema and the appearance of the face and neck are reminiscent of similar findings in DS (for review, see Epstein, 1986).

NEUROLOGICAL DEFECTS

Moderate to severe mental retardation is a consistent feature of DS (Johnson and Abelson, 1969). The impairments in neurogenesis are reflected in mildly reduced brain weight, a small cerebellum, and a deficit in the number of neurons in certain cortical regions (for review see Epstein, 1986; Coyle et al., 1986). DS patients over the age of 40 years typically show the development of the characteristic pathology of Alzheimer's disease (AD). The major extracellular component of the neuritic plaques characteristic of AD is the cerebrovascular A4− or β-amyloid peptide derived from the amyloid precursor protein (APP) (Masters et al., 1985; Glenner and Wong, 1984). The expression of APP mRNA in normal and aneuploid mice has been studied by *in situ* hybridization and Northern blot analysis. The brains of 15-day-old Ts 16 fetuses had a distribution of APP mRNA similar to that of diploid littermates, but the relative abundance of the transcripts was increased by 55% when measured in whole brain sections (Bendotti et al., 1988). However, the increase was higher in regions of the cerebral cortex. Northern blot analysis showed a clearly elevated level of APP mRNA (2.5 to 3-fold) in the heads of Ts16 fetuses, a value disproportionately greater than that expected from the 1.5-fold gene dosage effect generally found in trisomies (Epstein, 1986). As a further control, brain sections of Ts19 fetuses were hybridized and found to be labeled to the same intensities as the diploid control section (Bendotti et al., 1988).

Detailed neurochemical analyses of the Ts16 fetal brain have been carried out by Gearhart et al. (1983), Singer et al. (1984), and Ozand et al. (1984), and their results are summarized in Table 1.1. Particularly affected, both pre- and postsynaptically, is the cholinergic system, but the catecholaminergic and serotonergic systems are also impaired. More recently, Sweeney et al. (1989) demonstrated that

Ts16 embryos show a reduction (40%) in the number of basal forebrain neurons at early stages of brain development (day 16 of gestation). Also, there is no caudal to rostral gradient of development of acetylcholine esterase-positive mitotic cells in the basal forebrain. This has been interpreted to mean that there might be two different types of progenitor neurons, one of which is missing in the Ts16 brains.

Orozco et al. (1988) and Ault et al. (1989) have investigated the electrophysiological properties of cultured fetal Ts16 dorsal root ganglion (DRG) neurons. Characteristic abnormalities of the action potential, which may be due in part to alterations in sodium and potassium currents, were identified. Of particular interest is the finding of quite similar abnormalities in DRG neurons cultured from human fetuses with Ts21 (Nieminen et al., 1988).

Hematopoietic and Immunological Deficiencies

Many immunological defects, including altered serum immunoglobulin levels and impaired proliferative responses to mitogens and to bacterial or viral antigens (Philip et al., 1985), have been described for human Ts21 (for review, see Epstein, 1986, Murphy and Epstein, 1990). It is therefore of considerable interest that there are significant impairments of the hematopoietic and immune system in Ts16 mice.

Deficiencies in Ts16 Hematopoietic Stem Cells and Myeloid Cells: Fetal hematopoietic stem cells reside in the liver. There they undergo a distinct set of differentiation events and give rise to myeloid and lymphoid cells in the primary and secondary lymphatic organs (Moore and Metcalf, 1970). It has been found that the potency of Ts16 stem cells to repopulate an irradiated host animal and to mature into the different subpopulations of the lymphoid and myeloid cell lineages is very limited (Herbst et al., 1982). Impaired development of the Ts16 lymphoid system has also been found in Ts16 ⟷ 2n chimeric animals (Cox et al., 1984). This implies that the defect in the stem cell population is cell autonomous, rather than being the result of environmental failure. Even in close contact to normal bone marrow or thymic stromal cells, the development of the Ts16 hematopoietic system is impaired.

The deficiencies of Ts16 stem cells have been studied by characterizing the myeloid and lymphoid cell subpopulations. It has been found that the cell numbers for the Ts16 fetal hematopoietic and lymphoid organs (spleen, liver, and thymus) are decreased (Epstein et al., 1985b). In addition, trisomic fetal hematocrits and red cell counts are significantly decreased. These decreases are also reflected in severe decreases in a variety of different hematopoietic stem cell populations. The assays used to quantitate hematopoietic stem cells in the fetal liver are based on culturing the precursor cells either in an *in vivo* adoptive transfer system into an irradiated host animal or in one of several *in vitro* culture systems in which the appearance of differentiated myeloid and lymphoid cells can be enumerated. In the Ts16 fetal liver, a reduced number of colony-forming units in the spleen (CFU-S) or in culture (CFU-c), reflecting granulocyte/macrophage precursor cells, as well as a reduced number of BFU-E and CFU-E (erythroid stem cells), have been observed (Epstein et al., 1985b).

DEVELOPMENT OF Ts16 B LYMPHOCYTES: The Ts16 fetal liver cells were analyzed for the presence of cytoplasmic and surface immunoglobulin μ heavy chains using immunofluorescence. The number of pre-B and B lymphocytes present in the fetal liver in day 17 and 18 embryos were only moderately (10–20%) decreased (Epstein et al., 1985b). To analyze further the cells in the fetal liver, the expression of several surface markers has been examined with flow cytometry (Berger and Epstein, 1990). Ts16 cells were found to stain more intensely than diploid controls for B220 antigen (Kincade et al., 1981), as well as with an antibody against the transferrin receptor (Trowbridge et al., 1982). B220 has been described as an antigen on B-progenitor cells. Higher intensities of staining could imply that the surface molecules might be overexpressed due to a gene dosage effect, or because of some abnormality in development. By contrast, Ts16 cells stained less intensely with J11d (Bruce et al., 1981), a marker found on subsets of T and B cells, neutrophils, and erythrocytes.

Since human Down syndrome patients have an increased susceptibility to the development of leukemia (Kaneko et al., 1981), the sensitivities of Ts16 and diploid fetal liver cells to transformation by Abelson murine leukemia virus were compared. This retrovirus preferentially transforms precursors of the B lymphocyte differentiation pathway. There was a profound (94–96%) decrease in the number of colonies obtained after transformation of the trisomic liver cells (Epstein et al., 1985b). This decrease in transformation sensitivity was far in excess of the diminution in numbers of pre-B and B cells, as measured by the presence of cytoplasmic and surface immunoglobulin heavy chains. This might indicate that the target cell population for the Abelson virus is very severely diminished, or alternatively, that the expansion of the retroviral transformed cell clones is very limited for Ts16 fetal liver cells.

DEVELOPMENT OF Ts16 THYMOCYTES: The thymus of the Ts16 mouse is profoundly hypoplastic throughout fetal development, with a five-to-ten-fold reduction in the number of thymocytes (Epstein et al., 1985b). However, when Ts16 and diploid control thymuses were cultured *in vitro* for a week, the relative rates of proliferation of thymocytes in the trisomic and diploid thymic lobes were found to be identical. Taken together, these results suggest that the numerical defect results from a deficiency of prothymocytes very early in thymic development. Thymocyte depletion and reconstitution (by fetal liver stem cells) experiments gave further support for the defect being cell autonomous in nature (Epstein et al., 1987c), rather than being caused by a failure of interaction with the thymic epithelial cells. The question of altered migration properties of Ts16 precursor cells from the fetal liver to the thymic rudiment still needs to be examined.

In addition to the quantitative deficiencies, there is also an abnormality in the functional maturation of trisomic thymocytes *in vitro*. When placed into culture at 17 days of gestation, the development of peak responses to mitogenic (ConA)

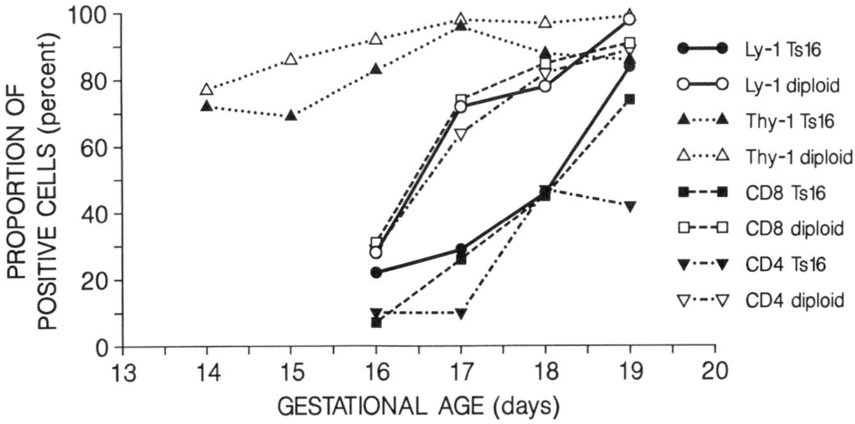

FIGURE 1.2. Ontogeny of Thy-1, Ly-1, CD4, and CD8 on Ts16 and diploid thymocytes at various stages of gestation.

stimulation, as measured by ^{3}H-thymidine incorporation and production of interleukin-2 (IL-2), was delayed by approximately 1 week, although a normal degree of responsiveness was ultimately attained (Epstein et al., 1985b).

ONTOGENY OF SURFACE ANTIGENS ON TS16 THYMOCYTES: T lymphocyte subpopulations and stage of development can be defined using molecules expressed on the cell surface. Some of these surface antigens are expressed, during the ontogeny of T lymphocytes, very early in gestation (as early as day 13), whereas others appear at later stages of development. Furthermore, some antigens are downregulated in the course of differentiation, whereas others are not. The ontogeny of the surface antigens. Thy-1, CD4, CD8, and Ly-2 in Ts16 and control thymocytes is shown in Figure 1.2. Although all of these molecules are found on Ts16 thymocytes, the development of the thymocytes is severely retarded (Berger and Epstein, 1989). At the end of gestation, the delay in development is about 2 days, a quite significant period of time in view of the fact that there are only 15 days between implantation and birth in the mouse. There is a high proportion of single CD8 positive cells, which probably represents the differentiation stage before thymocytes become CD4/CD8 double positive. In *in vitro* organ cultures, the expression of TCR vβ and CD3 has also been shown. Similar to the results obtained from the ConA stimulation experiments, the differences between trisomic and diploid cells decrease upon continued culturing (Berger and Epstein, 1989).

The fetal development and ontogeny of surface antigens of human trisomy 21 thymocytes has not been extensively studied because of the difficulty of obtaining human Ts21 fetal thymuses. However, the thymocytes of young DS patients undergoing heart surgery show an abnormal pattern of expression for the CD3-T cell receptor complex (Murphy and Epstein, 1990).

In diploid thymocytes, the surface IL-2 receptor is downregulated in midgestation. By contrast, IL-2 receptor positive cells constitute a much larger fraction at the end of gestation in trisomic thymuses than in normal animals (Berger and Epstein, 1990). The retardation of the development of the Ts16 thymocytes is therefore manifested not only by the delayed expression of newly synthesized surface antigens, but also by the delayed onset of the downregulation of the normal production of membrane molecules.

Defects in the Heart and Skeleton

Congenital heart disease is nearly an invariant (96%) feature of mouse Ts16 (Pexieder et al., 1981; Miyabara et al., 1982; Bacchus et al., 1987). Abnormalities of the great vessels represent the majority of the cardiovascular anomalies, but more than half of trisomy 16 fetuses having cardiovascular anomalies also have endocardial cushion defect. The latter lesion is frequently (~40%) found in Down syndrome (Greenwood and Naclas, 1976), but conotruncal abnormalities are also known to occur. Bacchus et al. (1987) have also described malformations of the sternum, ribs, and thoracic and lumbar spine affecting about a quarter of Ts16 fetuses.

The Use of Trisomy 16↔ Diploid Chimeras as an Experimental System

Many aspects of DS cannot be studied in Ts16 mice because they do not survive beyond the perinatal period. One way to extend the time during which Ts16 cells can be studied is to make use of Ts16↔ 2n chimeras. Chimeras can be obtained by aggregation of diploid 4- or 8-cell stage embryos with Ts16 embryos of the same developmental stage, followed by transfer into the uteri of pseudopregnant females. This has been successfully done by Cox et al. (1984) and Gearhart et al. (1986). Alternatively, embryonic stem cell lines could replace one of the partners used to form the chimeras. With a trisomic embryonic stem cell line, not only would the efficiency of obtaining chimeras be expected to be higher than can be obtained by aggregating 8-cell stage embryos, but there would also be the possibility of exploring methods for altering the development of Ts16 cells by recombinant DNA techniques.

A possible disadvantage to experiments with chimeric animals is that the diploid cells might complement some of the effects caused by Ts16, especially if humoral factors or hormones are involved. Nevertheless, chimeric animals certainly offer advantages over complete Ts16 animals in extending the lifetime of Ts16 cells.

The Nervous System of Ts16 Chimeras

As has already been noted, the experiments by Cox et al. (1984) showed a marked deficiency to Ts16 cells in the blood, spleen, thymus, and bone marrow of life-born Ts16↔ 2n chimeras. Other organs, however, contained a significant portion of trisomic cells. Neurochemical analyses of chimeras revealed changes in the activities of the cholinergic and catecholinergic synthesizing enzymes, choline acetyltransferase, and tyrosine hydroxylase in the brains of Ts16↔ 2n chimeras (Gearhart et al., 1986). The behavior of such chimeras, as measured by distances moved and the number of stereotypic movements, was also altered, as compared to 2n ↔ 2n controls (Gearhart et al., 1986).

SCRAPIE: Patients with AD, as well as scrapie-infected mice, both show changes in the expression of brain-specific genes. Both exhibit increased abundance of mRNA transcripts coding for sulfated glycoprotein 2 and glial fibrillary acidic protein (GFAP) (Wietgrefe et al., 1985; Duguid et al., 1989). Furthermore, brains of scrapie-infected mice show senile plaque depositions (Prusiner, 1984), although the composition of the plaques is different from those found in AD. It is conceivable that the etiology of Alzheimer's disease might involve an unrecognized exogenous (infectious) agent, and it was thought that the Ts16 ↔ 2n chimera would provide an interesting test system to determine whether the trisomic state can result in an increased susceptibility to infection with an agent such as the scrapie prion. Accordingly, Ts16 ↔ 2n and control 2n ↔ 2n chimeras were prepared and inoculated intracerebrally with the scrapie prion. It was found that the disease proceeds more rapidly in the Ts16 chimeras than in the control animals and that the time of survival of the inoculated chimeras was significantly reduced (Epstein et al., 1991). The mechanism by which the more rapid and fulminant course of scrapie is produced remains to be elucidated.

THE MARKING OF COMPONENTS OF CHIMERAS

Working with chimeric animals requires that the cells of the two genotypes within the chimeras develop in a reasonably equivalent manner. For this to occur, it is important to work with mouse strains with minimal differences in their genetic background. A combination of congenic strains of mice is favorable. For further analysis, the cells of interest have to be identified within the chimera, and there are several marker systems that can be applied. The most commonly used markers are the electrophoretic variants of the enzyme glucosephosphate isomerase (GPI) (Cox et al., 1984; Gearhart et al., 1986). GPI is easily detectable, but it is applicable only to large groups of cells or to whole organs. It is difficult to resolve chimeric tissue on a single cell level with such a marker.

Other marker systems are provided by allelic differences of surface molecules, such as Thy-1.1/Thy-1.2, or the many major histocompatibility complex (H-2) variants, which are detectable with specific monoclonal antibodies. Using flow cytometry, such markers can be employed to distinguish cells at a single cell level and also to quantitate the contribution of each genotype to the various tissues of the chimera. A further possibility is the use, as one partner for the chimera formation, of transgenic mice that have integrated several copies of a specific DNA sequence into their genome (Lo et al., 1987). These transgenic sequences can be detected by *in situ* hybridization with radioactively labelled or biotinylated probes and, by this means, the trisomic cells of Ts ↔ 2n chimeras could be identified.

Transplantation of Ts16 Fetal Brain

To circumvent the perinatal lethality of Ts16 mice, fetal trisomic tissues can be grafted into adult recipients. Richard et al. (1989) transplanted cortical and hippocampal regions of 16-to-18-day-old Ts16 fetuses into adult hosts, and the transplanted animals were analyzed after 4 to 6 months. Palmgren stain revealed occasional extraneuronal tangles of fibers and pyramidal neurons with abnormal fibrils within the proximal dendrites. In the grafted tissues, as well as in the immediately adjacent host brain, numerous densely stained cells were detected by immunocytochemical staining with an antiamyloid antibody that has been raised against synthetic A4 peptide. More recently, fetal Ts16 and control basal forebrain neurons have been transplanted into the hippocampus of young adult mice (Holtzman et al. 1992). The transplanted neurons survived and grew neurites in all grafts, but over time there was a selective atrophy of cholinergic neurons in the Ts16 grafts. Denervation of the hippocampus produced a significant increase in the size of these neurons, presumably through the mediation of hippocampus-derived neurotrophic factors. A4 containing plaques were not seen.

The Use of Partial Trisomy 16 as an Experimental System

The region of homology between HSA21 and MMU16 lies at the distal end of chromosome 16 (see above), and it has been proposed that this region constitutes a critical region for many of the phenotypic alterations found in Down syndrome. Therefore, a mouse with partial Ts16 for just this region might constitute a better animal model for DS than does complete Ts16. The contribution of aneuploid loci outside the critical region to the development of the phenotypic appearance would then be minimized. If the appropriate breeding stocks were available, partial Ts16 animals could be obtained by using mouse strains carrying reciprocal translocations of chromosome 16 or duplications of a part of chromosome 16. Attempts are now being made to develop the necessary stocks.

TRANSGENIC ANIMALS

To understand fully the pathogenic mechanisms underlying the phenotype of Down syndrome, it is necessary to "dissect" HSA 21 to determine which loci are involved with which components of the phenotype. Although mouse Ts16 and Ts16 ⟷ 2n chimeras serve as useful genetic models of human Ts21 and Ts21/2n mosaics (Cox et al., 1984), respectively, a very large number of genes are unbalanced simultaneously. It is therefore impossible to attribute, definitively, a particular phenotypic effect to a specific gene or set of genes. Furthermore, since the Ts16 fetus does not live to term, the mouse Ts16 model does not permit the observation of long-term effects of chromosome imbalance on the brain or other organs. The use of transgenic mice, on the other hand, would allow the dose of individual genes of interest and the amounts of their gene products to be increased to the level found in the usual trisomic state, and would thereby facilitate the analysis of the effect of imbalance of single loci.

It has been observed that the phenotypic features of individual aneuploid states can often be added together to generate phenotypes of combined aneuploidies

(Epstein, 1986; 1990). In principle, then, it should be possible to decompose the Down syndrome phenotype into individual components that can then be assigned to specific loci on chromosome 21. If so, it should ultimately be possible (once we have generated transgenic mice with different transgenes obtained from human chromosome 21 and/or mouse chromosome 16, and have determined the effects that result) to regenerate sequentially the trisomic phenotype by breeding different transgenic strains together, to create transgenic animals simultaneously unbalanced for several human chromosome 21 and/or mouse chromosome 16 loci.

Methods of Generating Transgenic Mice

The pioneering work in mouse embryo manipulation, recombinant DNA technology, and gene transfer techniques in the 1970s has led to the development of methods for generating transgenic mice. Two systems have been developed and widely used to generate transgenic mice. One is by means of microinjection of DNA fragments into fertilized mouse eggs (Gordon and Ruddle, 1981; 1983; Brinster and Palmiter, 1986), and the other is by use of embryo-derived stem (ES) cells (Robertson, 1987).

Microinjection

Successful microinjection of DNA into pronuclei of fertilized mouse eggs was first carried out independently in several laboratories during 1980 and 1981 (Gordon and Ruddle, 1981). The injected DNAs were shown to become integrated into the mouse genome, and expression of the injected DNA was detected. These early observations and subsequent experiments clearly demonstrated the possibility of generating transgenic mice carrying and expressing foreign genes. Furthermore, by fusing transgenes with specific promoter/regulator regions, such as the metallothionin promoter (Durman et al., 1984; Karin and Richards, 1982), expression of the injected genes can be controlled and studied throughout different developmental stages and in different tissues.

Embryo-Derived Stem Cells

In recent years, successes in the cultivation of embryo-derived stem (ES) cells has provided an alternative approach to generating transgenic mice. As primary stem cell lines, ES cells have a high capacity for differentiation both *in vivo* and *in vitro*. The majority of cultured cells can stably retain the euploid chromosome constitution, and ES cells can generate chimeras with high efficiency when returned back to embryonic environment (Robertson et al., 1986; Robertson, 1987). Since ES cells divide rapidly, a large number of cells can be obtained. As a result, purified DNA fragments coupled to selectable markers can be transferred into ES cells using conventional DNA transfection techniques, such as electroporation. Cell colonies that have retained, and expressed, the input DNA can then be preselected before injection into host blastocysts to form chimeras. Germ line transmission of the transgene can usually be achieved by subsequent breedings of the chimeras.

Human CuZn Superoxide Dismutase Transgenic Mice

To date, the most thoroughly studied transgenic mouse strains generated for the study of gene dose imbalance of Down syndrome are those carrying the human CuZn-superoxide dismutase (hCuZnSOD) transgene (Levanon et al., 1985). Human CuZnSOD, a key enzyme in the metabolism of oxygen free radicals, is encoded by the gene *SOD1* in band 21q22.1 (Touati, 1988). A 1.5-fold increase of CuZnSOD activity has been detected in many different cells and tissues from persons with Down syndrome (see Epstein, 1986, for review). Overexpression of the CuZnSOD gene as a result of increased gene dose may disturb the steady-state equilibrium of active oxygen species within the cell, resulting in oxidative damage to biologically important molecules. High amounts of superoxide dismutase have indeed been shown to cause increased lipid peroxidation in different experimental cell lines and in the brain tissue of Down syndrome fetuses (Balazs and Brooksbank, 1985; Elroy-Stein et al., 1986).

Preliminary Studies of Transfected Cells

To permit the systematic study of the possible involvement of CuZnSOD overproduction in the etiology of Down syndrome, the human *SOD1* gene and its cDNA were cloned and introduced first into human HeLa cells, mouse L cells, and rat PC12 cells (Elroy-Stein et al., 1986). Stably transfected cell clones expressing an elevated level of hCuZnSOD activity were isolated. Increased lipid peroxidation and higher resistance to the toxic effects of paraquat, a superoxide generator, were demonstrated in the transfected HeLa cells and L cells (Elroy-Stein et al., 1986). In addition, impaired neurotransmitter (dopamine and norepinephrine) uptake was shown in the transfected PC12 cells (Elroy-Stein and Groner, 1988), and the lesion was localized to the chromaffin granule transport mechanism. A diminished transmembrane proton gradient (pH gradient, ΔpH), the driving force for the uptake of biogenic amines such as dopamine and norepinephrine has been identified as the defect of the chromaffin granule transport system in *SOD1* transfected PC12 cells. This finding is consistent with the observations that the content of these neurotransmitters in nerve terminals of DS and AD brains appears to be substantially less than that of normal brains (Yates et al., 1983). The above observations with *SOD1*-transfected PC12 cells, HeLa cells, and L cells suggested that transgenic mice overexpressing CuZnSOD would provide an animal model for critically testing the hypothesis that the increased CuZnSOD activity present in Down syndrome has adverse phenotypic effects.

The method employed in generating the *SOD1* transgenic mice, and the studies conducted on these transgenics, are briefly described below. A linear 14.5-kb fragment of human genomic DNA (Elroy-Stein et al., 1986) containing the entire hCuZnSOD gene, including the sequences required for expression in transfected cells (Levanon et al., 1985), was prepared, and approximately 500 copies of the purified fragment were microinjected into the male pronuclei of fertilized eggs (Epstein et al., 1987b). Founder mice containing 1 to 12 copies of the hCuZnSOD gene were obtained, and transmission of the transgenes to the second generation was between 18% and 50%, indicating the possibility of germ-line mosaicism. However, transmission in second generation crosses was approximately 50% for all transgenic strains. Southern blot analysis of genomic DNA isolated from transgenic mice revealed the integration of the injected DNA, and the transmission of the transgenes, to their progeny. Expression of the transgenes in transgenic founder mice and their progeny was demonstrated by both Northern blot RNA hybridization and assay for CuZnSOD enzymatic activity. Northern blot hybridization of total RNA from transgenic animals revealed that the hCuZnSOD gene is expressed in a fashion similar to its expression in humans, with two RNA transcripts of 0.9 kb and 0.7 kb in a ratio of 1:4. Although increased CuZnSOD activity was detected in nearly all tissues of transgenic animals, the ratios of transgenic to normal CuZnSOD activity varied considerably from tissue to tissue within a given line.

Transgenic Neurons

A study of the toxicity of hydrogen peroxide to neurons obtained from normal and transgenic mice was conducted, and the amount of lactate dehydrogenase released into culture supernatant was used as a measurement of the degree of cell injury. At 10 and 50 mM hydrogen peroxide concentrations, the transgenic neurons were more resistant to the hydrogen peroxide treatment, indicating a protective effect of higher amounts of CuZnSOD activity (Chan et al., 1989). This protective effect is probably due to a feedback control of glutathione peroxidase concentration by the amount of superoxide. Primary cortical neurons derived from 15-day transgenic fetuses were also partially protected against the toxic effects of exposure to 0.5 mM glutamate for 5 minutes (Chan et al. 1990).

Transgenic Brain

The activity of CuZnSOD in the brains of transgenic mice has been increased between 1.5- and 5-fold by the insertion of the CuZnSOD transgenes into the mouse genome. Although physically normal, these transgenic mice exhibit several differences from control mice when exposed to known or presumed forms of oxidative stress. After a 30-second cold injury to the cerebral cortex, transgenic mice with approximately 3-times normal CuZnSOD activity have less cerebral edema, a smaller increase in vascular permeability, and a smaller area of infarction than do the control animals (Chan et al. 1991). These transgenic animals are also protected to a significant degree against the effects of transient focal cerebral ischemia in the middle cerebral artery distribution, with sparing particularly of the penumbra region surrounding the ischemic brain tissue (Kinouchi et al. 1991). Furthermore, in contrast to control animals, the injection of 1-methyl-4-phenyl-1,2,3,6-tetrahydropyridine (MPTP) into transgenic mice does not induce any depletion in striatal levels of dopamine and its metabolites nor decreases in dopamine uptake sites (Przedborski et al. 1992). MPTP is an agent which causes changes in the dopaminergic system similar to those observed in Parkinson's disease.

In addition to the protection afforded against the several types of acute stress just enumerated, increased CuZnSOD activity in the brain may also have long-term effects, the functional consequences of which are unknown. Such effects have been detected with regard to the numbers of neurotensin binding sites in the brains of aged transgenic animals (Cadet et al., submitted). In certain regions, the transgenic animals display age-dependent reductions in binding which were not observed in control animals, whereas in other regions the transgenic animals did not have the reductions that the control animals had. Similar results, although involving other regions, were also obtained for the binding of μ opioid receptors (Kujirai et al. 1991).

Transgenic Fibroblasts

Although, as has already been demonstrated, CuZnSOD has been shown to reduce oxidative damage, the quantitative relationship between the degree of protection and CuZnSOD activity has not been well investigated. Therefore, the ability of cells to tolerate superoxide toxicity was assessed as a function of endogenous CuZnSOD activity in a series of five transgenic fetal fibroblasts lines with levels of CuZnSOD activity ranging from 1.7 to 7.1-times the nontransgenic level (Huang et al. 1992). A direct relationship was observed between the LD_{50} to paraquat, a strong generator of superoxide radicals, and enzyme activity. Furthermore, a protective effect could be shown at the DNA level, with a reduction in the generation of thymine glycols in transgenic cells.

Neuromuscular Junctions in Transgenic Animals

In many persons with Down syndrome, a protruding tongue is a striking clinical feature. Studies on the tongue muscles of Down syndrome patients obtained after partial glossectomy have revealed defects in the neuromuscular junctions (Yarom et al., 1987). Analysis of the tongue muscles of the 2-to-4-month-old *SOD1* transgenic mice also revealed significant pathological abnormalities in the neuromuscular junctions (Avraham et al., 1988). Namely, there are a decreased number of terminal axons, a decreased ratio of axon area to the length of the postsynaptic membrane, and an increase in degenerated membranes, vacuoles, confluent vesicles, and filaments. These pathological changes in *SOD1* transgenic mice were similar to those observed in tongue muscles of aging mice and rats, as well as of Down syndrome patients, suggesting the involvement of elevated CuZnSOD activity in these neuropathological changes of Down syndrome.

Taking together the results of studies on *SOD1* transgenic mice, an increased level of CuZnSOD activity may be protective, deleterious, or have no effect at all, depending on the cell types examined and the agent used to perturb oxygen metabolism. Perhaps the most interesting observations are those of the neuronal tissues. The findings of impaired neurotransmitter uptake in *SOD1*-containing PC12 cells, pathological abnormalities in the neuromuscular junctions of *SOD1* transgenic tongue muscles, and altered binding of neurotransmitters all indicate an impact of an elevated level of CuZnSOD on the development of neuronal system in Down syndrome patients.

Beta-Amyloid Transgenic Mice

Another human chromosome 21 gene that has evoked wide interest is the APP gene. As has already been mentioned, by the fourth decade of life most Down syndrome patients develop Alzheimer's disease with a characteristic pathology of amyloid deposition in neuritic plaques and along the walls of cerebral blood vessels. Cerebrovascular amyloid is a 4.2- to 4.5-kD hydrophobic (A4) peptide derived from a larger precursor protein (Kang et al., 1987). Given the extra copy of chromosome 21 in persons with Down syndrome, a gene dosage effect for APP products has been considered as a possible cause of the early onset of Alzheimer's disease among Down syndrome patients.

The promoter region of the human APP gene (Salbaum et al., 1988) has the characteristics of housekeeping gene promoters (i.e., lack of a typical TATA box), shows a high GC content of 72%, and has multiple transcription initiation sites. Sequences that are similar to some transcription factor consensus binding sites, including Hox1.3, AP-1, and heat shock control element binding protein, have been located upstream of the RNA start sites. As was mentioned earlier, transplantation of the brain tissue from mouse Ts16 embryos to the brains of normal adult mice show a deposition of amyloid peptide in the Ts16 grafts (Richards et al., 1991). Furthermore, a peptide derived from the amyloid precursor, including the entire amyloid sequence and the remaining carboxyl terminal portion of APP, has been shown to be neurotoxic (Yankner et al., 1989).

Attempts are now in progress to develop strains of transgenic mice carrying transgenes either for the entire precursor protein, for its secreted isoforms, or for just the A4 peptide that accumulates in amyloid fibrils. It is expected that studies of such mice, expressing high levels of amyloid peptide in the brain tissue, either alone or in combination with *SOD1* or other transgenic strains, will provide important information toward the understanding of the development of Alzheimer's disease in persons with Down syndrome. Transgenic mice with the human form of APP (the APP gene) under the control of the human metallothionein-IIa promoter have been made. They overproduced APP in neurons, but amyloid production and other pathology of Alzheimer's disease were not seen (Beer et al. 1990). However, when human β-APP 751 cDNA under control of a rat neural-specific enolase was used, both diffuse and compact extracellular β-amyloid immunoreactive deposits were observed in the hippocampus and cortex of transgenic animals (Quon et al. 1991).

FUTURE APPROACHES TO MAKING MODELS OF DOWN SYNDROME

Improved Cloning Methods

With the conventional cloning approaches, the size of DNA fragments that can be obtained is limited to less than 50 kb. In many cases, especially in eukaryotic systems, the control elements for the regulation of gene expression may not all be included in the initial cloning of a gene. This possibility is indicated by the observations of insertional site-dependent expression of the transgenes in most transgenic mice. As in the case of human *SOD1* transgenic mice (Epstein et al., 1987b), the level of hCuZnSOD expression is not directly related to its copy number and is presumed to be dependent on its site of integration into the genome. This phenomenon suggests that there may be some *SOD1* control elements that, due to their remote positions from the structural gene itself, are being excluded. As a consequence, transgenic mice so generated may not permit the complete developmental regulation of the transgene. The recently reconstructed "minilocus" of the human β-globin gene (Grosveld and van Assendelft, 1987; Lores et al., 1988), which is expressed in a tissue-specific manner at a level directly related to its copy number, and yet independent of its position of integration, is one good example of important regulatory elements that are not included in the initial cloning of the gene, being located far upstream and downstream from the structural gene. In order to overcome this problem, it is essential to have effective means of obtaining large DNA fragments.

Although several genes on human chromosome 21 have already been identified, they represent only a small portion of the total numbers of genes present on chromosome 21. To speed up the process of cloning and identifying genes that are responsible for the development of Down syndrome, alternative cloning avenues need to be taken. Several systems for obtaining large DNA fragments, from several hundred kilobases up to several megabases in length, have been developed. These include yeast artificial chromosome (YAC) cloning (Burke et al., 1987), reconstruction of large DNA fragments from smaller pieces via homologous recombination (O'Connor et al., 1989), and dissection of chromosome fragments from human metaphase spreads for microinjection into mouse pronuclei (Richa and Lo, 1989).

Gene Deletion or Inactivation and Controlled Gene Expression

In contrast to the addition of extra copies of genes into the mouse genome, the removal or inactivation of specific genes of interest with regard to Down syndrome is another approach to understanding the effects on the pathogenic effect, or effects, contributed by a single locus. The targeted mutation of specific genes by homologous recombination and insertional inactivation in mouse ES cells (Doetschmann et al., 1988, Capecchi, 1989; Johnson et al., 1989; Zimmer and Gruss, 1989) has been recently, and successfully, carried out. In addition, the use of antisense DNA or RNA to inhibit gene expression is another way to reduce specifically the expression of certain genes (Bevilacqua et al., 1988; Cabrera et al., 1987; Katsuki et al., 1988; McGarry and Lindquist, 1986; van der Krol et al., 1988). With inducible promoters, such as the metallothionein I and II promoters (Karin and Richards, 1982), the heat shock hsp70 promoter (Li, 1983; Pelham, 1982), and mouse mammary tumor virus promoter (Lee et al., 1981), and appropriate inducing methods, antisense DNA/RNA constructs should be able to provide different levels of controlled gene expression in cultured tissue cells as well as in transgenic mice.

Acknowledgments. Preparation of this chapter was supported by National Institutes of Health grants HD-17001 and AG-08938. C.N.B. was supported by fellowship 83.662.0.88 from Swiss National Fund, and T.T.H. was supported by grants from the Weingart Foundation and the American Cancer Society (CN-2).

REFERENCES

Ault B, Caviedes P, Rapoport SI (1989): Neurophysiological abnormalities in cultured dorsal root ganglion neurons from the trisomy 16 mouse fetus, a model for Down syndrome. *Brain Res* 485:165–170

Avraham KB, Schickler M, Groner Y (1988): Down's syndrome: abnormal neuromuscular junction in tongue of transgenic mice with elevated levels of human Cu/Zn-superoxide dismutase. *Cell 54:823–829*

Bacchus C, Sterz H, Buselmaier W, Sahai S, Winking H (1987): Genesis and systematization of cardiovascular anomalies and analysis of skeletal malformations in murine trisomy 16 and 19. *Hum Genet* 77:12–22

Balazs R, Brooksbank BWL (1985): Neurochemical approaches to pathogenesis of Down's syndrome. *J Ment Defic Res* 29:1–14

Bendotti C, Forloni GL, Morgan RA, O'Hara BF, Oster-Granite ML, Reeves RH, Gearhart JD, Coyle JT (1988): Neuroanatomical localization and quantification of amyloid precursor protein mRNA by *in situ* hybridization in the brains of normal, aneuploid, and lesioned mice. *Proc Natl Acad Sci USA* 85:3628–3632

Beer J, Salbaum JM, Schlichtmann, Hoppe P, Earley S, Masters CL, Carlson GA, Beyreuther K (1990): Transgenic mice and Alzheimer's disease (abstract). *Neurobiol Aging* 11:327

Berger CN, Epstein CJ (1989): Delayed thymocyte maturation in the trisomy 16 mouse fetus. *J Immunol* 143:389–396

Berger CN, Epstein CJ (1990): Expression of the IL-2 receptor and J11d antigen during thymocyte and stem cell development in the trisomy 16 mouse fetus. In: *Advances in Gene Technology: The Molecular Basis of Immune Disease*. Oxford: IRL Press, pp 216—217.

Bevilacqua A, Erickson RP, Hieber V (1988): Antisense RNA inhibits endogenous gene expression in mouse preimplantation embryos: lack of double-stranded RNA "melting" activity. *Proc Natl Acad Sci USA* 85:831–835

Blue ME, Molliver ME, Gearhart JD, Coyle JT (1984): Abnormalities of cortical development in murine trisomy 16, an animal model for Down's syndrome. *Soc Neurosci Abstr* 10:953

Brinster RL, Palmiter RD (1986): Introduction of genes into the germ line of animals. *The Harvey Lectures* Series 80:1–38

Bruce J, Symington FW, McKearn TJ, Sprent JJ (1981): A monoclonal antibody discriminating between subsets of T and B cells. *J Immunol* 127:2496–2501

Burke DT, Carle GF, Olson MV (1987): Cloning of large segments of exogeneous DNA into yeast by means of artificial chromosome vectors. *Science* 236:806–812

Burmeister M, Kim SW, de Lange T, Tantravahi U, Frazier K, Cox DR (1989): The fine structure map of the distal long arm of chromosome 21: hot spots of recombination and homology to several mouse chromosomes. *Am J Hum Genet* 45:A133

Cabrera CV, Alonso MC, Johnston P, Phillips RG, Lawrence PA (1987): Phenocopies induced with antisense RNA identify the *wingless* gene. *Cell* 50:659–663

Cadet JL, Kujirai K, Jackson-Lewis V, Carlson E, Epstein CJ (submitted) Quantitative autoradiographic distribution of [^{3}H]neurotensin binding sites in the brains of superoxide dismutase transgenic mice: age effects.

Capecchi MR (1989): Altering the genome by homologous recombination. *Science* 244:1288–1292

Chan PH, Chu L, Chen SF, Carlson EJ, Epstein CJ (1990): Reduced neurotoxicity in transgenic mice overexpressing human copper-zinc-superoxide dismutase. *Stroke* 21 (suppl III):III-80–III-82

Chan PH, Yu ACH, Chen S, Epstein CJ (1989): Oxidative stress exacebrates cellular damage in primary cultures of astrocytes from human *SOD1* transgenic mice. *J Cell Biol* 107:726a

Cheng SV, Nadeau JH, Tanzi RE, Watkins PC, Jagadesh J, Taylor BA, Haines JL, Sacchi N, Gusella JF (1988); Comparative mapping of DNA markers from the familial Alzheimer disease and Down syndrome regions of human chromosome 21 to mouse chromosome 16 and 17. *Proc Natl Acad Sci USA* 85:6032–6036

Cox DR, Epstein CJ (1985): Comparative gene mapping of human chromosome 21 and mouse chromosome 16. *Ann NY Acad Sci* 450:169–177

Cox DR, Epstein LB, Epstein CJ (1980): Genes coding for sensitivity to interferon (*IfRec*) and soluble superoxide dismutase (*SOD-1*) are linked in mouse and man and map to mouse chromosome 16. *Proc Natl Acad Sci USA* 77:2168–2172

Cox DR, Goldblatt D, Epstein CJ (1981): Chromosomal assignment of mouse PRGS: further evidence for homology between mouse chromosome 16 and human chromosome 21. *Am J Hum Genet* 33:145A

Cox DR, Smith SA, Epstein LB, Epstein CJ (1984): Mouse trisomy 16 as an animal model of human trisomy 21 (Down syndrome): production of viable trisomy 16$\leftrightarrow$diploid mouse chimeras. *Dev Biol* 101:416–424

Coyle JT, Oster-Granite ML, Gearhart JD (1986): The neurobiologic consequences of Down syndrome. *Brain Res Bull* 16:773–787

Doetschman T, Maeda N, Smithies O (1988): Targeted mutation of the *Hprt* gene in mouse embryonic stem cells. *Proc Natl Acad Sci USA* 85:8583–8587

Duguid JR, Bohmont CW, Liu N, Tourtellotte WW (1989): Changes in brain gene expression shared by scrapie and Alzheimer disease. *Proc Natl Acad Sci USA* 86:7260–7264

Durman DM, Hoffaman JS, Quaife CJ, Benditt EP, Chen HY, Brinster PL, Palmiter RD (1984): Induction of mouse metallothionein-I mRNA by bacterial endotoxin is independent of metals and glucocorticoid hormones. *Proc Natl Acad Sci USA* 81:1053–1056

Elroy-Stein O, Berstein Y, Groner Y (1986): Overproduction of human Cu/Zn superoxide dismutase in transfected cells: extenuation of paraquat-mediated cytotoxicity and enhancement of lipid peroxidation. *EMBO J* 5:615–622

Elroy-Stein O, Groner Y (1988): Impaired neurotransmitter uptake in PC12 cells overexpressing human Cu/Zn-superoxide dismutase—implication for gene dosage effects in Down syndrome. *Cell* 52:259–267

Epstein CJ (1985): The mouse trisomies: experimental systems for the study of aneuploidy. In: *Issues and Reviews in Teratology,* Vol. 3. Kalter H, ed. New York: Plenum, pp 171–217

Epstein CJ (1986): *The Consequences of Chromosome Imbalance. Principles, Mechanisms, and Models.* New York: Cambridge University Press

Epstein CJ (1990): The consequences of chromosome imbalance. *Am J Med Genet, Suppl* 7:91–97

Epstein CJ, Anneren KG, Foster D, Groner Y, Prusiner SB, Smith SA (1987a): Pathogenetic relationships between Down's syndrome and Alzheimer's disease: Studies with animal models. *Banbury Report.* 27:339–355

Epstein CJ, Avraham KB, Lovett M, Smith S, Elroy-Stein O, Rotman G, Bry C, Groner Y (1987b): Transgenic mice with increased Cu/Zn superoxide dismutase activity: Animal model of dosage effects in Down syndrome. *Proc Acad Natl Sci USA* 84:8044–8048

Epstein CJ, Cox DR, Epstein LB (1985a): Mouse trisomy 16: An animal model of human trisomy 21 (Down syndrome). *Ann NY Acad Sci* 450:157–168

Epstein CJ, Foster DB, DeArmond SJ, Prusiner SB (1991) Acceleration of scrapie in trisomy 16 ↔ diploid aggregation chimeras. *Ann Neurol* 29:95–97

Epstein CJ, Hofmeister BG, Yee D, Smith SA, Philip R, Cox DR, Epstein LB (1985b): Stem cell deficiencies and thymic abnormalities in fetal mouse trisomy 16. *J Exp Med* 162:695–712

Epstein CJ, Weil J, Epstein LB (1987c): Abnormalities in the interferon response and immune systems in Down Syndrome: studies in human trisomy 21 and mouse trisomy 16. In: *Oncology and Immunology of Down Syndrome,* McCoy EE, Epstein CJ, eds. New York: Alan R. Liss, pp 191–208.

Gearhart JD, Singer HS, Moran TH, Tiemeyer MC, Oster-Granite ML, Coyle JT (1986): Mouse chimeras composed of trisomy 16 and normal (2n) cells: preliminary studies. *Brain Res Bull* 16:815–824

Gearhart J, Singer HS, Tiemeyer MC, Coyle JT (1983): Neurochemical studies of the trisomy 16 mouse. *Am J Hum Genet* 35:164A

Glenner GG, Wong CW (1984): Alzheimer disease and Down syndrome: sharing of a unique cerebrovascular amyloid fibril protein. *Biochem Biophys Res Commun* 122:1131–1135

Gordon JW, Ruddle FH (1981): Integration and stable germ line transmission of genes injected into mouse pronuclei. *Science* 214:1244–1246

Gordon JW, Ruddle FH (1983): Gene transfer into mouse embryos: production of transgenic mice by pronuclear injection. *Meth Enzymol* 101:411–433

Greenwood RD, Naclas AS (1976): The clinical course of cardiac disease in Down syndrome. *Pediatrics* 58:893–902

Gropp A, Kolbus U, Giers D (1975): Systematic approach to the study of trisomy in the mouse. II. *Cytogenet Cell Genet* 14:42–62

Grosveld F, van Assendelft GB (1987): Position-independent, high-level expression of the human β-globin gene in transgenic mice. *Cell* 51:975–985

Herbst EW, Gropp A, Nielson K, Hoppe H, Freyman M, Pluznik DH (1982): Reduced ability of mouse trisomy 16 stem cells to restore hematopoiesis in lethally irradiated animals. In: *Experimental Hematology Today,* Baum SJ, Ledney GD, Thierfelder S, eds. Basel: S. Karger, pp 119–126

Holtzman DM, Li YW, DeArmond SJ, McKinley MP, Gage FH, Epstein CJ, Mobley WC (1992): A mouse model of neurodegeneration: Atrophy of basal forebrain cholinergic neurons in Ts16 transplants. *Proc Natl Acad Sci USA* (in press)

Huang TT, Carlson EJ, Leadon SA, Epstein CJ (1992): Relationship of resistance to oxygen free radicals to CuZn-superoxide dismutase activity in transgenic, transfected, and trisomic cells. *FASEB J* (in press)

Johnson RC, Abelson RB (1969): The behavioral competence of mongoloid and non-mongoloid retardates. *Am J Ment Defic* 73:856–857

Johnson RS, Sheng M, Greenberg ME, Kolodner RD, Papaioannou VE, Spiegelman BM (1989): Targeting of nonexpressed genes in embryonic stem cells via homologous recombination. *Science* 245:1234–1236

Kaneko Y, Rowley JD, Variakojis D, Chilcote RR, Moohr JW, Patel D (1981): Chromosome abnormalities in Down syndrome patients with acute leukemia. *Blood* 58:459–466

Kang J, Lemaire HG, Unterbeck A, Salbaum JM, Masters CL, Grzeschik KH, Multhaup G, Beyreuther K, Muller-Hill B (1987): The precursor of Alzheimer's disease amyloid A4 protein resembles a cell-surface receptor. *Nature* 325:733–736

Karin M, Richards RI (1982): Human metallothionin genes—primary structure of the metallothionin II gene and a related processed gene. *Nature* 299:797–802

Katsuki M, Sato M, Kimura M, Yokoyama M, Kobayashi K, Nomura T (1988): Conversion of normal behavior to shiverer by myelin basic protein antisense cDNA in transgenic mice. *Science* 241:593–595

Kincade PW, Lee G, Watanabe FT, Sun L, Scheid M (1981): Antigens displayed on murine B lymphocyte precursors. *J Immunol* 127:2262–2268

Kinouchi H, Epstein CJ, Mizui T, Carlson E, Chen SF, Chan PH (1991): Attenuation of focal ischemic injury in transgenic mice overexpressing CuZn-superoxide dismutase. *Proc Natl Acad Sci USA* 88:11158-11162

Korenberg JR, Kawashima H, Pulst S-M, Ikeuchi T, Ogasawara N, Yamamoto K, Schonberg S, West R, Allen L, Magenis E, Ikawa K, Taniguchi N, Epstein CJ (1990): Molecular definition of a region of chromosome 21 that causes features of the Down syndrome phenotype. *Am J Hum Genet* 47:236–246

Kujirai K, Jackson-Lewis V, Fahn S, Carlson E, Epstein CJ, Cadet JL (1991) Comparative quantitative autoradiographic distribution of μ opioid receptors in the brains of superoxide dismutase-transgenic (SOD-Tg) mice and of their non-transgenic (Non-Tg) littermates: age effects (abstract). *Soc Neurosci Abstr* 17:812

Lee F, Mulligan R, Berg P, Ringold G (1981): Glucocorticoids regulate expression of dihydrofolate reductase cDNA in mouse mammary tumor virus chimaeric plasmids. *Nature* 294:228–232

Levanon D, Lieman-Hurwitz J, Dafni N, Wigderson M, Sherman L, Berstein Y, Laver-Rudich Z, Danciger E, Stein O, Groner Y (1985): Architecture and anatomy of the chromosomal locus in human chromosome 21 encoding the Cu/Zn superoxide dismutase. *EMBO J* 4:77–84

Li GC (1983): Induction of thermotolerance and enhanced heat shock protein synthesis in Chinese hamster fibroblasts by sodium arsenite and by ethanol. *J Cell Physiol* 115:116–122

Lin PF, Slate DL, Lawyer FC, Ruddle FH (1980): Assignment of the murine interferon sensitivity and cytoplasmic superoxide dismutase to chromosome 16. *Science* 209:285–287

Lo CW, Coulling M, Kirby C (1987): Tracking of mouse cell lineage using microinjected DNA sequences: analyses using genomic Southern blotting and tissue-section in situ hybridizations. *Differentiation* 35:37–44

Lores P, Besnard C, Jami J (1988): A mouse minialbumin gene is specifically expressed in differentiated hepatoma cells but not in transgenic mice. *Differentiation* 38:35–41

Lovett M, Goldgaber D, Ashley P, Cox DR, Gajdusek DC, Epstein CJ (1987): The mouse homolog of the human amyloid beta protein (AD-AP) gene is located on the distal end of mouse chromosome 16: further extension of the homology between human chromosome 21 and mouse chromosome 16. *Biochem Biophys Res Commun* 144:1069–1075

Lyon MF (1988): Mouse chromosome atlas. *Mouse News Lett* 81:20–41

Lyon MF, Ward HC, Simpson GM (1976): A genetic method for measuring non-disjunction in mice with Robertsonian translocations. *Genet Res* 26:283–295

MacDonald GP, Cox DR (1989): The mouse T28H translocation break point occurs in a region of mouse chromosome 16 homologous to human chromosome 21, separating the sequences D21S13 and D21S52 from App, Sod-1, and Ets-2. *Am J Hum Genet* 45:A149

MacDonald GP, Price ER, Chu ML, Timpl R, Allore R, Marks A, Dunn R, Cox DR (1988): Assignment of four human chromosome 21 genes to mouse chromosome 10: Implications for mouse models of Down syndrome. *Am J Hum Genet* 43:151A

Magnuson T, Debrot S, Dimpfl J, Zamora T, Epstein CJ (1985): The early lethality of autosomal monosomy in the mouse. *J Embryol Exp Morph* 69:223–236

Masters CL, Simms G, Weinman NA, Multhaup G, McDonald BL, Beyreuther K (1985): Amyloid plaque core protein in Alzheimer disease and Down syndrome. *Proc Natl Acad Sci USA* 82:4245–4249

McGarry TJ, Lindquist S (1986): Inhibition of heat shock protein synthesis by heat-inducible antisense RNA. *Proc Acad Natl Sci USA* 83:399–403

Miyabara S, Gropp A, Winking H (1982): Trisomy 16 in the mouse fetus associated with generalized edema, cardiovascular and urinary tract anomalies. *Teratology* 25:369–380

Moore MA, Metcalf D (1970): Ontogeny of the haemopoietic system: yolk sac origin of in vivo and in vitro colony forming cells in the developing mouse embryo. *Br J Haematol* 18:279–296

Murphy M, Epstein LB (1990): Down syndrome thymuses have a decreased proportion of cells expressing high levels of TCR alpha, beta, and CD3. *Clin Immunol Immunopathol* 55:453–467

Nieminen K, Suarez-Isla BA, Rapoport SI (1988): Electrical properties of cultured dorsal root ganglion from normal and trisomy 21 human fetal tissues. *Brain Res* 474:246–254

O'Connor M, Peifer M, Bender W (1989): Construction of large DNA fragments in *Escherichia coli*. *Science* 244:1307–1312

Orozco CB, Epstein CJ, Rapoport SI (1988): Voltage activated sodium conductances in cultured normal and trisomy 16 dorsal root ganglion neurons from fetal mouse. *Dev Brain Res* 38:265–274

Oster-Granite ML, Baker C, Ozand PT (1983): Neuroanatomic, ocular and audiovestibular malformations in trisomy 16 mice. *Pediatr Res* 17:300A

Ozand PT, Hawkins RL, Collins Jr RM, Reed WD, Baab PJ, Oster-Granite ML (1984): Neurochemical changes in murine trisomy 16: delay in cholinergic and catecholaminergic systems. *J Neurochem* 43:401–408

Pelham HRB (1982): A regulatory upstream promoter element in the Drosophila *hsp*70 heat-shock gene. *Cell* 30:517–528

Pexieder T, Miyabara S, Gropp A (1981): Congenital heart disease in experimental (fetal) mouse trisomies: Incidence. In: *Perspectives in Cardiovascular Research,* Vol 5: Mechanism of Cardiac Morphogenesis and Teratogenesis, Pexieder T, ed. New York: Raven Press pp 389–399

Philip R, Berger AC, McManus NH, Warner NH, Peacock MA, Epstein LB (1985): Abnormalities of the in vitro cellular and humoral responses to tetanus and influenza antigens with concomitant numerical alterations in lymphocyte subsets in Down syndrome (Trisomy 21). *J Immunol* 136:1661–1667

Prusiner SB (1984): Some speculations about prions, amyloid, and Alzheimer's disease. *N Engl J Med* 310:661–663

Przedborski S, Kostic V, Jackson-Lewis V, Naini AB, Fahn S, Carlson E, Epstein CJ, Cadet JL (1992) Increased CuZn-Superoxide dismutase activity confers resistance to MPTP-induced neurotoxicity in transgenic mice. *J. Neurosci,* in press

Quon D, Wong Y, Catalano R, Marian Scardina J, Murakami K, Cordell B (1991): Formation of β-amyloid protein deposits in brains of transgenic mice, *Nature* 352:239–341

Reeves RH, Crowley MR, Lorenzon N, Pavan WJ, Smeyne RJ, Goldowitz D (1989): The mouse neurological mutant weaver maps within the region of chromosome 16 which is homologous to human chromosome 21. *Genomics,* 5:522–526

Reeves RH, Gallahan D, O'Hara BF, Callahan R, Gearhart JD (1987): Genetic mapping of Prm-1, Igl-1, Smst, Mtv-6, Sod-1 and Ets-2 and localization of the Down syndrome region on mouse chromosome 16. *Cytogenet Cell Genet* 44:76–81

Reeves RH, O'Hara BF, Pavan WJ, Gearhart JD, Haller O (1988): Genetic mapping of Mx within the region of mouse chromosome 16 that is homologous to human chromosome 21. *J Virol* 62:4372–4375

Rethoré M-O (1981): Structural variation of chromosome 21 and symptoms of Down's syndrome. In: *Trisomy 21. An International Symposium,* Burgio GR, Fraccaro M, Tiepolo L, Wolf U, eds. Berlin: Springer–Verlag, pp 173–182

Richa J, Lo CW (1989): Introduction of human DNA into mouse eggs by injection of dissected chromosome fragments. *Science* 245:175–177

Richards SJ, Waters JJ, Beyreuther K, Masters CL, Wischik CM, Sparkman DR, White CC III, Abraham CR, Dunnett SB (1991): Transplants of mouse trisomy 16 hippocampus provide a model of Alzheimer's disease neuropathology. *EMBO J* 10:297-303

Robertson EJ (1987): *Teratocarcinomas and Embryonic Stem Cells: A Practical Approach.* Oxford, Washington, DC: IRL Press.

Robertson E, Bradley A, Kuehn M, Evans M (1986): Germ-line transmission of genes introduced into cultured pluripotential cells by retroviral vector. *Nature* 323:445–448

Salbaum JM, Weidmann A, Lemaire HG, Masters CL, Beyreuther K (1988): The promoter of Alzheimer's disease amyloid A4 procursor gene. *EMBO J* 7:2807–2813

Searle AG, Peters J, Lyon MF, Hall JG, Evans EP, Edwards JH, Buckle VJ (1989): Chromosome maps of man and mouse. IV *Ann Hum Genet* 53:89–140

Singer HS, Tiemeyer M, Hedreen JD, Gearhart J, Coyle JT (1984): Morphologic and neurochemical studies of embryonic brain development in murine trisomy 16. *Dev Brain Res* 5:155–166

Summitt RL (1981): Chromosome specific segments that cause the phenotype of Down syndrome. In: *Trisomy 21 (Down syndrome). Research Perspectives,* de la Cruz FF, Gerald PS, eds. Baltimore: University Park Press, pp 225–235

Sweeney JE, Hohmann CF, Oster-Granite ML, Coyle JT (1989): Neurogenesis of the basal forebrain in normal and trisomy 16 mice: An animal model for developmental disorders in Down syndrome. *Neuroscience* 31:413–425

Touati D (1988): Molecular genetics of superoxide dismutases. *Free Radical Biol Med* 5:393–402

Trowbridge IS, Lesley J, Schulte R (1982): Murine cell surface transferrin receptor: studies with an anti-receptor monoclonal antibody. *J Cell Physiol* 112:403–410

van der Krol AR, Mol JNM, Stuitje AR (1988): Modulation of eukaryotic gene expression by complementary RNA or DNA sequences. *BioTechniques* 6:958–976

Watson DK, Smith MJ, Kozak C, Reeves R, Gearhart J, Nunn MF, Nash W, Fowle JR, Duesberg P, Papas TS (1986): Conserved chromosomal positions of dual domains of the *ets* protooncogene in cats, mice and humans. *Proc Natl Acad Sci USA* 83:1792–1796

Wietgrefe S, Zupancic M, Haase A, Chesbro B, Race R, Frey W, Rustan T, Friedman RL (1985): Cloning of a gene whose expression is increased in scrapie and in senile plaques in human brain. *Science* 230:1177–1179

Wisniewski KE, Wisniewski HM, Wen GY (1985): Occurence of neuopathological changes and dementia of Alzheimer's disease in Down's syndrome. *Ann Neurol* 17:278–282

Yankner BA, Dawes LR, Fisher S, Villa-Komaroff L, Oster-Granite ML, Neve RL (1989): Neurotoxicity of a fragment of the amyloid precursor associated with Alzheimer's disease. *Science* 245:417–420

Yarom R, Sherman Y, Sagher U, Peled IJ, Wexler MR (1987): Elevated concentrations of elements and abnormalities of neuromuscular junctions in tongue muscles of Down's syndrome. *J Neurol Sci* 79:315–326

Yates CM, Simpson J, Gordon A, Maloney AFJ, Allison Y, Ritchie IM, Urguhart A (1983): Catecholamine and cholinergic enzymes in presenile and senile Alzheimer-type dementia and Down's syndrome. *Brain Res* 280:119–126

Zimmer A, Gruss P (1989): Production of chimaeric mice containing embryonic stem (ES) cells carrying a homoebox Hox1.1 allele mutated by homologous recombination. *Nature* 338:150–156

Genetic Animal Models of Krabbe Disease

KINUKO SUZUKI AND KUNIHIKO SUZUKI

Krabbe disease or globoid cell leukodystrophy (GLD) in humans is one of the classical genetic leukodystrophies, mainly affecting infants (Krabbe, 1916). Its prevalence rate is approximately 1 per 50,000 in Sweden but is far lower in other countries. The mode of inheritance is autosomal recessive. The underlying cause of the disease is a genetic deficiency of the activity of a lysosomal hydrolase, galactosylceramidase (Suzuki and Suzuki, 1970).

Globoid cell leukodystrophy has been described in several other mammalian species: the Cairn and Westhighland terrier dogs (Fankhauser et al., 1963; Fletcher et al., 1966; Jortner and Jones, 1968; Yajima et al., 1977a), the miniature poodle (Zaki and Kay, 1973), the blue-tick hound dog (Boysen et al., 1974), the beagle dog (Johnson et al., 1975), the sheep (Pritchard et al., 1980), the cat (Johnson, 1970), the mouse (Duchen et al., 1980; Kobayashi et al., 1980), and most recently in the rhesus monkey (Baskin et al., 1989). All models showed clinical and neuropathological features typical of Krabbe disease, and the genetic deficiency of galactosylceramidase has been confirmed in all species except for the feline model and the sporadic cases of dogs.

The canine model occurring among the Cairn and Westhighland terriers was maintained for many years for research purposes at the School of Veterinary Medicine, University of Minnesota. Unfortunately, the colony was dissolved some years ago due to lack of funds. The murine model was discovered at the Jackson Laboratory, Bar Harbor, Maine, and named the twitcher. Since 1980 when it was recognized as an enzymatically authentic model, the twitcher mouse has become a popular tool for investigation of the pathogenesis and therapeutic approaches on Krabbe disease. Twitcher mice are commercially available from the Jackson Laboratory. Colonies of this murine mutant are being maintained in many laboratories throughout the world. Availability of such authentic animal models is essential for studies of rare genetic diseases, such as globoid cell leukodystrophy. The disease in other species has been all sporadic and, to our knowledge, is not readily available for research purposes. In this chapter we review clinical, pathological, biochemical, and enzymatic aspects of the animal models of Krabbe disease and discuss their usefulness as the tool for the therapeutic approach to the disease. This updates an earlier review on the same subject several years ago (Suzuki and Suzuki, 1985).

CLINICAL FEATURES

The disease in all species is transmitted as an autosomal recessive trait. In all species clinical manifestations of the disease are largely limited to those reflecting pathology in the central and peripheral nervous systems.

Human

The classical form of the human disease has an onset in infancy. Affected infants develop first clinical signs and symptoms of disease at around 3 to 6 months of age. The disease is steadily and rapidly progressive with fatal outcome usually within 1 to 2 years from the onset. Hagberg et al. (1969) divided the clinical course into three main stages. The first stage is characterized by irritability, fever of unknown origin, stiffness of limbs, and elevated spinal fluid protein. In the second stage, the patients show marked hypertonicity with the characteristic opisthotonic posture and diminished or hypoactive deep tendon reflexes. The third stage is that of a "burn out" stage. In recent years, however, many atypical cases occurring in late infantile, childhood, or even in the adult period have been reported (for detailed clinical description, see Suzuki and Suzuki, 1989). Most of the atypical cases in the past 20 years have been confirmed enzymatically. Although these atypical cases are likely to be allelic to the classical infantile form, they probably reflect different mutations.

Dog

The canine disease was the first genetically authentic animal model of human GLD. The disease occurs among two breeds of dogs (Cairn terriers and West Highland terriers) (Fankhauser et al., 1963; Fletcher et al., 1966; Hirth and Nielsen, 1967; Jortner and Jones, 1968; McGrath et al., 1969). Typically, clinical signs and symptoms develop as early as the second and as late as the seventh month of life with rapid progression. Affected dogs usually die within 2 to 3 months. Major clinical symptoms are severe pelvic limb paresis leading to complete paralysis, head tremor, ataxia, and other cerebellar signs. There are also sporadic clinicopathological case reports of GLD in the literature; in a 2-year-old miniature poodle whose clinical signs were ataxia, forelimb dysmetria, and mental retardation starting around 2 months of age (Zaki and Kay, 1973), in the blue-tick hound dog (Boysen et al., 1974), and in a beagle (Johnson et al., 1975). Enzymatic confirmation of the diagnosis is lacking in these cases.

Cat

Two domestic short-haired kittens have been reported with clinical and pathological features of GLD (Johnson, 1970). The feline disease, however, lacks enzymatic confirmation. Dysmetria of hindlimbs and incoordination were the initial signs recognized at 5 and 6 weeks after birth. Both developed generalized tremors. Since they were euthanatized at 6½ and 8½ weeks, the natural course of this feline model is not known.

Sheep

There is a report of GLD in two Polled Dorset sheet from Australia (4 and 18 months of age). Progressive hindlimb ataxia and incoordination ascending to quadriplegia and tremor of the head and neck were the major clinical symptoms. Again they were euthanatized and thus the natural course of the illness was not established (Pritchard et al., 1980).

Monkey

A recent meeting abstract is available that describes enzymatically confirmed GLD in an infant rhesus monkey (Baskin et al., 1989). No clinical details are available at this time.

Mouse (C57BL/6J/twi)

This mutant was first discovered in the colony of the CE/J strain at Jackson Laboratory in Bar Harbor, Maine. The twi mutant allele was later transferred to and maintained in a C57BL/6J background (Duchen et al., 1980). The affected mice cannot be clinically differentiated from other littermates at birth. Affected mice show early general retardation of body growth, but neurological signs of unsteadiness with hindlimb weakness and generalized tremors do not become obvious until about 20 days. They become less alert and inactive, lose weight, and usually die before reaching 40 days. Twitcher mice in earlier reports (Duchen et al., 1980) appeared to have developed neurological signs and symptoms later and lived up to 3 months, but those in our current colony do not survive beyond 40 days. Before the onset of clinical symptoms, significant reduction of motor conduction velocity in the tibial nerve (Toyoshima et al., 1986) and retardation of neurological and locomotor development (Olmstead, 1987) can be documented.

PATHOLOGY

Significant pathological lesions are limited to the nervous system and are similar in all species. The pathology of GLD has been studied most extensively in three species: human, dog, and mouse.

Central Nervous System

Characteristic pathological changes in the central nervous system (CNS) of patients with Krabbe disease are extensive degeneration of the white matter, consisting of severe demyelination, reactive astrocytic gliosis, and an accumulation of multinucleated macrophages (globoid cells) clustered around blood vessels (Fig. 2.1). The term "globoid cell leukodystrophy" is derived from the presence of these abnormal cells. Unlike in other demyelinating diseases, these macrophages do not contain sudanophilic materials. They are strongly positive for Sudan black, Sudan IV, and periodic acid-Schiff (PAS) stains. These cells possess numerous pseudopods and contain inclusions characteristic of the disease (Schochet et al., 1969; Yunis and Lee, 1969; Suzuki and Grover, 1970; Andrews and Cancilla, 1970; Andrews et al., 1971; Shaw and Carlson, 1970; Liu, 1970). Ultrastructurally, the inclusions in the macrophages in the human, canine, and murine forms of GLD are essentially identical (Yunis and Lee, 1976; Yajima et al., 1977b; Duchen et al., 1980) (Fig. 2.2). Macrophages with similar morphological characteristics can be induced experimentally by intracerebral injection of galactocerebroside but of no similar or related lipids (Austin and Lehfeldt, 1965; Suzuki, 1970; Andrews and Menkes, 1970). The subplasmalemmal linear density, a unique feature in the cells of mononuclear phagocytic origin (Kawanami et al., 1980), was observed in these macrophages in the human, canine, and murine diseases as well as in the experimentally induced globoid cells (Yajima et al., 1977b; Takahashi and Suzuki, 1982). More recently, the globoid cells in twitcher mice were shown to

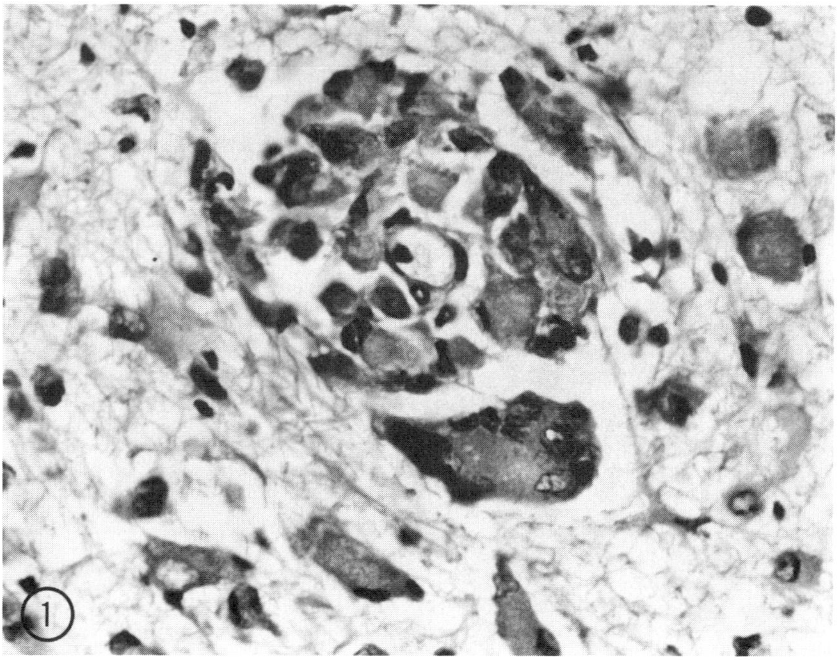

FIGURE 2.1. Gliotic cerebral white matter of a 7-month-old Cairn terrier affected by GLD. Many globoid cells are clustered around a blood vessel. Six-micron thick paraffin section stained with hematoxylin-eosin. Magnification: ×600.

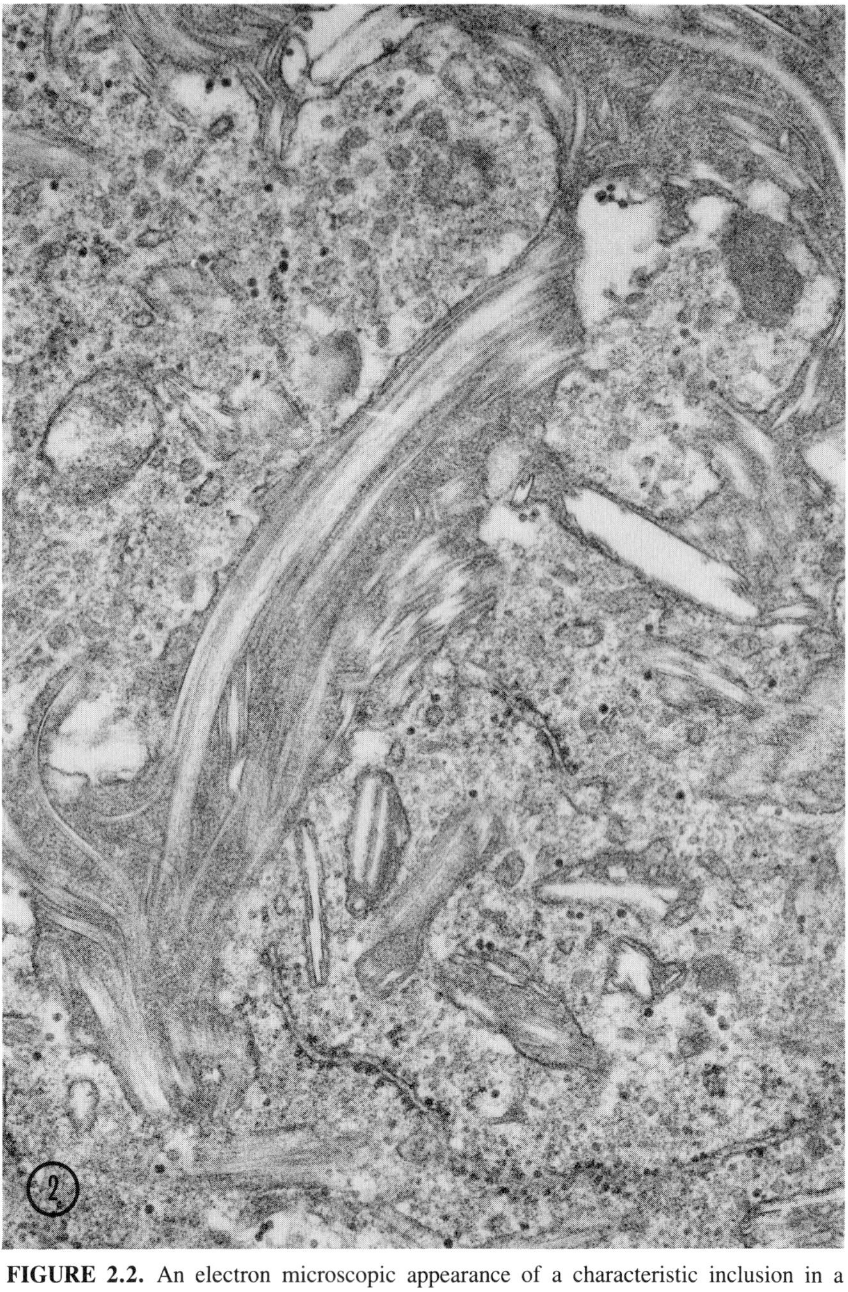

FIGURE 2.2. An electron microscopic appearance of a characteristic inclusion in a macrophage (globoid cell) in the cerebral white matter of an affected dog. Magnification: ×68,000. Reprinted with permission of Elsevier Science Publishers BV from Yajima K, Fletcher JF, and Suzukik (1977a): Canine globoid cell leukodystrophy I. Further ultrastructural study of the typical lesion. *J Neurol Sci* 33:179–197.

express a surface antigen, Mac-1, which is intimately associated with the type three complement receptor on mononuclear phagocytic cells (Kobayashi et al., 1985a). D'Agostino et al. (1963) indicated that these cells were more numerous in the area where myelin was still preserved and that they decreased as demyelination and gliosis progressed. The axonal loss becomes conspicuous in later stages of the illness. In a study of fraternal twin fetuses at 23 weeks' gestation, one affected and the other heterozygous, the degree of myelination in the spinal cord was indistinguishable between the two, indicating that early myelination proceeds normally in this disease (Okeda et al., 1979). Normal progression of early myelination was also demonstrated with morphometric analyses of the CNS of twitcher mice (Nagara et al., 1982; Tanaka et al., 1989).

Peripheral Nervous System

The peripheral nerves are always involved in human, canine, and murine GLD. No information as to the peripheral nervous system (PNS) pathology is available for the feline and sheep models. The murine model appears to have the most severe PNS involvement. Grossly affected nerves are abnormally white, firm, and enlarged with prominent endoneurial fibrosis and edema. The main pathological findings are segmental demyelination with preserved axons and infiltration of macrophages (Fig. 2.3), which are morphologically similar to those of globoid cells in the CNS (Sourander and Olsson, 1968; Hogan et al., 1969; Fletcher et al., 1971; Yunis and Lee, 1976; Duchen et al., 1980; Scaravilli et al., 1980; Powell et al., 1983). Characteristic inclusions are also present in the Schwann cells at the ultrastructural level. In a study of twitcher sciatic nerves, the characteristic inclusions in the Schwann cells and infiltration of macrophages appeared as early as 5 days and hypomyelination became apparent at 15 days before development of demyelination (Jacobs et al., 1982; Tanaka et al., 1988).

Other Organs

In human GLD, pathology is essentially limited to the nervous system although some pathological changes in other organs have occasionally been described (For references, see Suzuki and Suzuki, 1989). The characteristic inclusions are abundant in the lymphnodes and renal tubular epithelial cells in the twitcher mice (Takahashi et al., 1983, 1984) and, to a lesser degree, in the renal epithelial cells in the canine disease (Suzuki, 1986). Search for similar abnormalities in the kidney of human GLD patients failed.

BIOCHEMISTRY

The genetic defect of human Krabbe disease was identified in 1970 as a deficiency of activity of a lysosomal hydrolase, galactosylceramidase (Suzuki and Suzuki, 1970, 1989). In all animal models of GLD, except for the feline model, deficiency

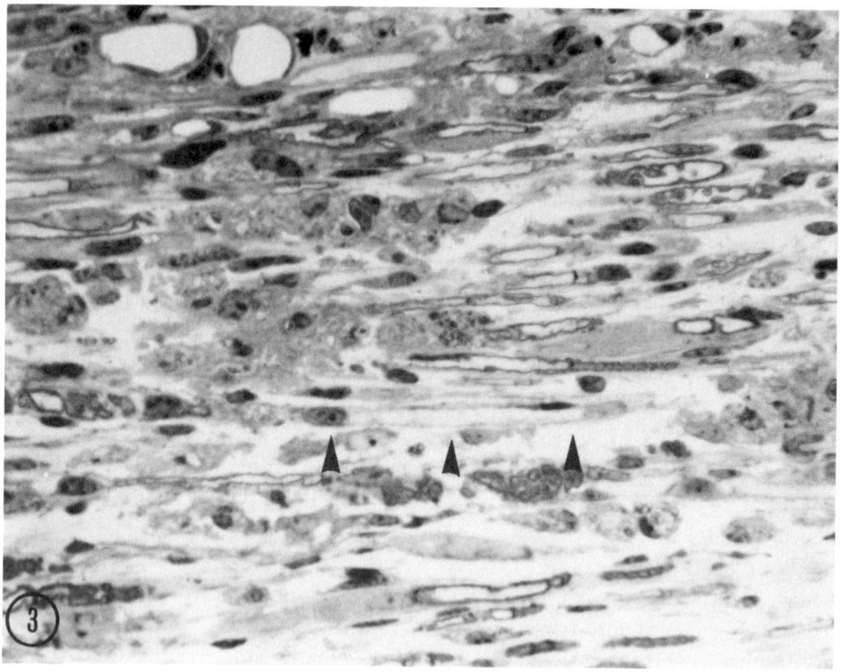

FIGURE 2.3. Sciatic nerve of a 45-day-old twitcher mouse. There are numerous infiltrating macrophages. Significant interstitial edema is present. Arrows indicate a demyelinated segment of a nerve fiber. One-micron thick Epon-embedded section stained with toluidine blue. Magnification: ×400.

of this enzyme has been confirmed (Suzuki et al., 1970; Kobayashi et al., 1980, 1982; Pritchard et al., 1980; Baskin et al., 1989). The gene coding for galactosylceramidase has been localized to chromosome 12 in mouse (Sweet, 1986). More recently, the human galactosylceramidase gene was mapped to chromosome 14, which is homologous to murine chromosome 12 (Zlotogora et al., 1990). At least in *in vivo* assay systems, the enzyme has a relatively broad substrate specificity in that, under suitable conditions, it can hydrolyze not only galactosylceramide but also galactosylsphingosine and lactosylceramide (Miyatake and Suzuki, 1974; Tanaka and Suzuki, 1975). On the other hand, the other lysosomal β-galactosidase, GM1-ganglioside β-galactosidase, has been shown to have catalytic activity toward galactosylceramide under specific assay condition (Kobayashi et al., 1985b). It is important to keep in mind that these observations on the substrate specificity are all based on *in vitro* assays in the presence of detergents. The substrate specificities of these enzymes *in vivo* may or may not be parallel to those under the artificial *in vitro* conditions. Galactosylceramidase is exceedingly difficult to purify. No preparation from any of the mammalian sources pure enough for either antibody production or amino acid sequencing is available. This difficulty is keeping GLD from entering the inevitable molecular genetics era at this time.

Globoid cell leukodystrophy is conceptually a "storage disorder" in that the natural substrate of the defective catabolic enzyme would be expected to accumulate abnormally in the tissues. Characteristically and perhaps uniquely, the natural substrate, galactosylceramide, does not accumulate beyond the normal level in the CNS in all species examined (Vanier and Svennerholm, 1974; Suzuki and Suzuki, 1989; Costantino-Ceccarini et al., 1976; Igisu et al., 1983a). In contrast, levels of galactosylceramide are highly variable in the kidney in different species, ranging from no abnormal accumulation in human patients (Suzuki, 1971), a two-to-three-fold increase in the dog model (unpublished), and up to a 50-fold increase in the twitcher mutant (Ida et al., 1982; Igisu et al., 1983b). These analytical findings in the brain and kidney indicate that the absence of abnormal accumulation is more closely related to the overall metabolic characteristic of the organ than to substrate specificities of the two β-galactosidases. In the CNS, an early cessation of normal process of myelination due to dysfunction and then destruction of the oligodendroglia appears to provide the most plausible explanation for the lack of galactosylceramide accumulation because galactosylceramide is highly concentrated in the myelin sheath (Suzuki and Suzuki, 1970).

Pathogenetic Mechanism (Psychosine Hypothesis)

Although there is no gross accumulation of galactosylceramide in the nervous system of patients or animals affected by GLD, the unique pathological feature of the disease, the globoid cells, is clearly the consequence of the genetic failure to degrade galactosylceramide. Free galactosylceramide has long been known to elicit infiltration of macrophages specifically into the brain (Austin and Lehfeldt, 1965; Suzuki, 1970). Once in the brain, they are transformed to multinucleated globoid cells. The characteristic inclusions in the globoid cells have ultrastructural characteristics of galactosylceramide itself (Yunis and Lee, 1970). No other glycolipids or phospholipids have such a capacity, including glucosylceramide, sulfatide, lactosylceramide, and gangliosides. On the other hand, the devastating early destruction of the myelin-forming cells, the oligodendroglia in the CNS and the Schwann cells in the PNS, is difficult to explain in the absence of accumulation of galactosylceramide. A hypothesis, known as the psychosine hypothesis, was first proposed for the human disease (Miyatake and Suzuki, 1972) in order to explain the pathogenetic mechanism of the disease. The hypothesis postulates that in GLD the toxic metabolite of galactosylceramide, galactosylsphingosine (psychosine), also cannot be degraded because of the galactosylceramidase deficiency and causes destruction of the myelin-forming cells. Inability of tissues of human patients and of affected dogs to degrade psychosine was demonstrated early (Miyatake and Suzuki, 1972; Suzuki et al., 1974). Actual accumulation of psychosine in the brain of human patients was then demonstrated analytically several years later (Svennerholm et al., 1980; Kobayashi et al., 1987, 1988). Its accumulation in the CNS and PNS of the dog and murine models was then demonstrated (Igisu and Suzuki, 1984; Shinoda et al., 1987; Tanaka et al., 1988, 1989). The only known metabolic pathway to generate psychosine is by galactosylation of sphingosine presumably by the same synthetic enzyme that galactosylates ceramide to galactosylceramide, UDP-galactose:ceramide (sphingosine) galactosyltransferase. This is the most likely source of psychosine that accumulates in GLD (Mitsuo et al., 1989). An enzymatic pathway to remove fatty acid from galactosylceramide has never been convincingly demonstrated. Nevertheless, contribution from an alternate route of galactosylceramide breakdown, either enzymatic or chemical, has not been completely ruled out. For details of the psychosine hypothesis, refer to a recent review (Suzuki and Suzuki, 1989).

USE OF AUTHENTIC ANIMAL MODELS FOR RARE GENETIC DISORDERS

Genetic defects underlying the inborn lysosomal diseases have been largely clarified. The field is now moving in two different directions; toward specific mutations in the genes responsible for the abnormal or absence of the gene products, and toward the pathogenetic mechanisms that lead from the genetic defects to the eventual biological dysfunctions. Animal models of rare human genetic diseases are expected to provide invaluable tools for research into both of these directions in the near future. Recent uses of the twitcher mutant for treatment of the disease well illustrate this potential.

Approaches to Treatment

Although any of the animal models can be useful for experimental exploration of treatment for this rare genetic disorder, the twitcher mutant is the most often used model because of its small size and ready availability. When the cDNA and the gene coding for galactosylceramidase become available, the twitcher mutant would be expected to provide an unusual opportunity for experimental trials for treatment of this devastating genetic disorder of the nervous system, because experimental procedures for transfer and expression of the normal gene through transgenes, vector-mediated transfer, and other means are already being extensively and intensively explored in this species. As of this writing, galactosylceramidase remains the only lysosomal enzyme involved in human sphingolipidoses that has not yet been cloned from any source. Therefore, therapeutic approaches so far conducted are conceptually all in the category of the enzyme replacement.

In earlier studies, the sciatic nerve of affected mice was grafted to the sciatic nerve of either normal or a trembler mutant mice (Scaravilli and Jacobs, 1981, 1982). When twitcher sciatic nerves were grafted to the sciatic nerve of normal littermates, no sign of demyelination was detected 1 month after the operation. However, at 2 months, endoneurial edema became pronounced and features of demyelination appeared. Even at 5 months after the operation, demyelinated axons were still present and the typical GLD inclusions were found in the Schwann cells. At 6 months, many nerve fibers were myelinated but myelin was far thinner for the size of the axons (Scaravilli and Jacobs, 1981). Galactosylceramidase activities of the grafted twitcher sciatic nerves were indistinguishable from those in the host nerves at 4, 5, and 9 months (Scaravilli and Suzuki, 1983).

More recently, two groups of investigators have explored transplantation of normal bone marrow tissue to twitcher mice as a potential enzyme source (Yeager et al., 1984; Hoogerbrugge et al., 1988a). Following determination of genetic status by assay of galactosylceramidase activity in clipped tails (Kobayashi et al., 1982) and sublethal whole body irradiation, normal bone marrow tissue was transplanted to affected mice before the onset of the clinical disease. With bone marrow transplantation (BMT), the life span of the recipient mice could be extended with a 36% probability of survival at 100 days in contrast to untreated mice, which rarely survive beyond 35 to 40 days. Hindlimb weakness was significantly improved but tremulousness persisted. Nevertheless, the body weight of the treated mice remained much below that of normal mice throughout their lifespan. After BMT, galactosylceramidase activity in the hematopoietic organs (spleen and bone marrow) increased to the donor level. Also, the enzyme activity was significantly increased in the liver, lung, and heart. In the CNS, the enzyme activity gradually increased to 15% of the control level in 100-day-old treated mice (Hoogerbrugge et al., 1988a). On the other hand, Ichioka and coworkers found the normal level of galactosylceramidase activity in the twitcher mouse brain by 90 days after BMT (Ichioka et al., 1987). Perhaps more importantly, psychosine decreased in the CNS following BMT and remained stable (Ichioka et al., 1987; Hoogerbrugge et al., 1988b). However, a slight additional increase of psychosine level was noted after 100 days of age. In the PNS, the psychosine level was relatively low in the sciatic nerves of 50-day-old treated mice, but in 100-day-old mice, it reached the same high level as observed in the nerve of 40-day-old untreated twitcher mice (Ichioka et al., 1987). Light microscopic examination indicated some remyelination in the PNS with decreased perineurial edema by 70 to 80 days after BMT, although mononuclear cells were still abundant (Yeager et al., 1984). The motor nerve conduction velocity improved in the sciatic nerve following BMT (Toyoshima et al., 1986). In the CNS, improvement of the pathology was not evident in the treated mice examined at 71 days or younger but was clearly observed in 100-day or older treated mice. In the latter, many nerve fibers in the cerebellum, brain stem, and spinal cord were myelinated and myelin degeneration was rare in these regions, although thickness of the myelin sheaths was far less than normally expected for the size of axons, indicating hypomyelination or remyelination. In these regions of improvement, numerous foamy macrophages replaced globoid cells. Similar foamy cells were also present in the subarachnoid space. Their distribution was almost identical to that of globoid cells in untreated twitcher mice, and thus they are likely to be infiltrating donor macrophages in response to CNS pathology (Suzuki et al., 1988). Unlike in untreated mice, these macrophages can degrade galactosylceramide and thus are not transformed to globoid cells. With allogeneic transplantation using C3H strain of mice [H-2K^k allele of the major histocompatibility complex (MHC)] as the bone marrow donor to twitcher mice that are of C57BL strain (H-2K^b allele of MHC complex), the donor origin of the foamy macrophages was demonstrated with immunocytochemical techniques (Hoogerbrugge et al., 1988b). Remyelinating oligodendrocytes appeared normal at the light microscope level with abundant perikaryal cytoplasm. However, examination on the ultrastructural level indicated presence of the GLD inclusions in their perikarya as well as in the inner or outer tongue processes (Suzuki et al., 1988). Apparent recovery from demyelination in the regions where foamy cells were conspicuously present suggested that these cells were responsible for the improvement of the CNS pathology by somehow supplementing the defective galactosylceramidase with the normal enzyme. In agreement with the previous light microscope observation (Yeager et al., 1984), improved pathology was also well recognized in the PNS on the ultrastructural level. However, remyelinating Schwann cells contained the GLD inclusions in their perikarya and the degree of remyelination was limited (Kondo et al., 1988). Similarly, a limited degree of remyelination was also observed following nerve graft and nerve crush as described above (Scaravilli et al., 1980; Scaravilli and Jacobs, 1982).

Acknowledgment. Many young investigators who worked in our laboratories in the past 20 years have contributed to many aspects of the studies summarized in this chapter. The number is too large to acknowledge them individually but their names appear in the references cited. The work from the authors' laboratories has been supported in part by research grants NS-24453, NS-24928, HD-03110, and ES-01104 from the National Institutes of Health, USA.

REFERENCES

Andrews JM, Cancilla PA (1970): Cytoplasmic inclusions in globoid cell leukodystrophy. *Arch Pathol* 89:53–55

Andrews JM, Cancilla PA, Grippo J, Menkes JH (1971): Globoid cell leukodystrophy (Krabbe's disease). Morphological and biochemical studies. *Neurology* 21:337–353

Andrews JM, Menkes JH (1970): Ultrastructure of experimentally produced globoid cells in the rat. *Exp Neurol* 29:483–493

Austin JH, Lehfeldt D (1965): Studies in globoid (krabbe) leukodystrophy III. Significance of experimentally-produced globoid-like elements in rat white matter and spleen. *J Neuropathol Exp Neurol* 24:265–289

Baskin G, Alroy J, Li Y-T, Dayal Y, Raghavan S, Shaver L (1989): Galactosylceramide-lipidosis in Rhesus monkeys. *Lab Invest* 60:7A

Boysen GB, Tryphonas L, Harries NW (1974): Globoid cell leukodystrophy in the blue-tick hound dog. I. Clinical manifestations. *Can Vet J* 15:303–308

Costantino-Ceccarini E, Fletcher TF, Suzuki K (1976): Glycolipid metabolism in the canine form of globoid cell leukodystrophy. In: *Current Trends in Sphingolipidoses and Allied Disorders,* BW Volk, Schneck L, eds. New York: Plenum, pp 127–145

D'Agostino AN, Sayre GP Hagles AB (1963): Krabbe's disease. *Arch Neurol* 8:82–96

Duchen LW, Eicher EM, Jacobs JM, Scaravilli F, Teixeira F (1980): Hereditary leucodystrophy in the mouse: the new mutant twitcher. *Brain* 103:695–710

Fankhauser R, Luginbühl H, Hartley WJ (1963): Leukodystrophie vom Typus Krabbe beim Hund. *Arch Tierheilk* 105:198–207

Fletcher TF, Kurtz HJ, Low DG (1966): Globoid cell leukodystrophy (Krabbe type) in the dog. *J Am Vet Med Assoc* 149:165–172

Fletcher TF, Kurtz HJ, Stadlan EM (1971): Experimental Wallerian degeneration in peripheral nerves of dogs with globoid cell leukodystrophy. *J Neuropathol Exp Neurol* 30:593–602

Hagberg B, Kollberg H, Sourander P, Akesson HO (1969): Infantile globoid cell leukodystrophy (Krabbe disease). A clinical and genetic study of 32 Swedish cases. 1953–1967. *Neuropädiatrie* 1:74–88

Hirth RS, Nielsen SW (1967): A familial canine globoid cell leukodystrophy (Krabbe type). *J Small Anim Pract* 8:569–575

Hogan GR, Gutmann L, Chou SM (1969): The peripheral neuropathy of Krabbe's (globoid) leukodystrophy. *Neurology* 19:1094–1100

Hoogerbrugge PM, Poorthuis BJHM Romme A, van de Kamp JJP Wagemaker G, van Bekkum DW (1988a): Bone marrow transplantation on enzyme levels and clinical course in the neurologically affected twitcher mouse. *J Clin Invest* 81:1790–1794

Hoogerbrugge PM, Suzuki K, Suzuki K, Poorthuis BJHM, Kobayashi T, Wagemaker G, van Bekkum DW (1988b): Donor derived cells in the central nervous system of twitcher mice after bone marrow transplantation. *Science* 239:1035–1038

Ichioka T, Kishimoto Y, Brennan S, Santos GW, Yeager AM (1987): Hematopoietic cell transplantation in murine globoid cell leukodystrophy (the twitcher mouse): effect of level of galactosylceramidase, psychosine and galactocerebrosides. *Proc Natl Acad Sci USA* 84:4259–4263

Ida H, Umezawa F, Kasai E, Eto Y, Maekawa K (1982): An accumulation of galactocerebroside in kidney from mouse globoid cell leukodystrophy (twitcher). *Biochem Biophys Res Commun* 109:634–638

Igisu H, Shimomura K, Kishimoto Y, Suzuki K (1983a): Lipids of developing brain of twitcher mouse, an authentic murine model of human Krabbe disease. *Brain* 106:405–417

Igisu H, Suzuki K (1984): Progressive accumulation of toxic metabolite in genetic leukodystrophy. *Science* 224:753–755

Igisu H, Takahashi H, Suzuki K, Suzuki K (1983b): Abnormal accumulation of galactosylceramide in the kidney of twitcher mouse. *Biochem Biophys Res Commun* 110:940–944

Jacobs JM, Scaravilli F, de Aranda FT (1982): The pathogenesis of globoid cell leukodystrophy in peripheral nerve of the mouse mutant twitcher. *J Neurol Sci* 55:285–304

Johnson GR, Oliver JE, Jr, Selcer R (1975): Globoid cell leukodystrophy in a beagle. *Am J Vet Med Assoc* 167:380–384

Johnson KH (1970): Globoid leukodystrophy in the cat. *J Am Vet Med Assoc* 157:2057–2064

Jortner BS, Jones AM (1968): The neuropathology of globoid cell leucodystrophy in the dog. *Acta Neuropathol* 10:171–182

Kawanami O, Ferrans VI, Crystall RG (1980): Subplasmalemmal linear densities in cells of the mononuclear phagocytes system in lung. *Am J Pathol* 100:131–150

Kobayashi T, Goto I, Yamanaka T, Suzuki Y, Nakano T, Suzuki K (1988): Infantile and fetal globoid cell leukodystrophy: analysis of galactosylceramide and galactosylsphingosine. *Ann Neurol* 24:517–522

Kobayashi S, Katayama M, Bourque EA, Suzuki K, Suzuki K (1985a): The twitcher mouse: positive immunohistochemical staining of globoid cells with monoclonal antibody against Mac-1 antigen. *Dev Brain Res* 20:49–54

Kobayashi T, Nagara H, Suzuki K, Suzuki K (1982): The twitcher mouse: determination of genetic status by galactosylceramidase assay on clipped tail. *Biochem Med 27:*8–14

Kobayashi T, Shinnoh N, Goto I, Kuroiwa Y (1985b): Hydrolysis of galactosylceramide is catalyzed by two genetically distinct acid β-galactosidases. *J Biol chem* 250:14982–14987

Kobayashi T, Shinoda H, Goto I, Yamanaka T, Suzuki Y (1987): Globoid cell leukodystrophy is a generalized galactosylsphingosine (psychosine) storage disease. *Biochem Biophys Res Commun* 144:41–46

Kobayashi T, Yamanaka T, Jacobs JM, Teixeira F, Suzuki K (1980): The twitcher mouse: an enzymatically authentic model of human globoid cell leukodystrophy (Krabbe disease). *Brain Res* 202:479–483

Kondo A, Hoogerbrugge PM, Suzuki K, Poorthuis BJHM, Van Bekkum DW, Suzuki K (1988): Pathology of the peripheral nerve in the twitcher mouse following bone marrow transplantation. *Brain res* 460:178–183

Krabbe K (1916): A new familial, infantile form of diffuse brain sclerosis. *Brain* 39:74–114

Lui HM (1970): Ultrastructure of globoid leukodystrophy (Krabbe's disease) with reference to the origin of globoid cells. *J Neuropathol Exp Neurol* 29:441–462

McGrath J, Schutta H, Yaseen A, Steinberg S (1969): A morphological and biochemical study of canine globoid leukodystrophy. *J Neuropathol Exp Neurol* 28:171

Mitsuo K, Kobayashi T, Shinnoh N, Goto I (1989): Biosynthesis of galactosylsphingosine (psychosine) in the twitcher mouse. *Neurochem Res* 14:899–903

Miyatake T, Suzuki K (1972): Globoid cell leukodystrophy: additional deficiency of psychosine galactosidase. *Biochem Biophys Res Commun* 48:538–543

Miyatake T, Suzuki K (1974): Galactosylsphingosine galactosyl hydrolase in rat brain: probable identity with galactosylceramide galactosyl hydrolase. *J Neurochem* 22:231–237

Nagara H, Kobayashi T, Suzuki K, Suzuki K (1982): The twitcher mouse: normal pattern of early myelination in the spinal cord. *Brain Res* 244:289–294

Okeda R, Suzuki Y, Horiguchi S, Fujii (1979): Fetal globoid cell leukodystrophy in one of twins. *Acta Neuropathol* 47:151–154

Olmstead CE (1987): Neurological and neurobehavioral development of the mutant twitcher mouse. *Behav Brain Res* 25:143–153

Powell HC, Knobler RL, Meyers RR (1983): Peripheral neuropathy in the twitcher mutant. A new experimental model of endoneurial edema. *Lab Invest* 49:19–25

Pritchard DH, Napthine DV, Sinclair AJ (1980): Globoid cell leukodystrophy in Polled Dorset sheep. *Vet Path* 17:399–405

Scaravilli F, Jacobs JM (1981): Peripheral nerve grafts in hereditary leukodystrophic mutant mice (twitcher). *Nature* 290:56–58

Scaravilli F, Jacobs JM (1982): Improved myelination in nerve graft from the leucodystrophic twitcher into trembler mice: evidence for enzyme replacement. *Brain Res* 237:163–172

Scaravilli F, Jacobs JM, Teixeira F (1980): Quantitative and experimental studies on the twitcher mouse. In: *Neurological Mutations Affecting Myelination,* Baumann N, ed. Amsterdam: Elsevier/North-Holland Biomedical Press, pp 115–122

Scaravilli F, Suzuki K (1983): Enzyme replacement in grafted nerve of twitcher mouse. *Nature* 305:713–715

Schochet SS Jr, Hardman JM, Lampert PW, Earle KM (1969): Krabbe's disease (globoid leukodystrophy). Electron microscopic observations. *Arch Pathol* 88:305–313

Shaw CM, Carlson CB (1970): Crystalline structure in globoid epithelioid cells. *J Neuropathol Exp Neurol* 29:306–319

Shinoda H, Kobayashi T, Katayama M, Goto I, Nagara H (1987) Accumulation of galactosylsphingosine (psychosine) in the twitcher mouse: determination by HPLC. *J Neurochem* 49:92–99

Sourander P, Olsson Y (1968): Peripheral neuropathy in globoid cell leukodystrophy (Morbus Krabbe). *Acta Neuropathol* 11:69–81

Suzuki K (1970): Ultrastructural study of experimental globoid cells. *Lab Invest* 23:612–619

Suzuki K (1971): Renal cerebroside in globoid cell leukodystrophy (Krabbe's disease). *Lipids* 6:433–436

Suzuki K (1986): Characteristic inclusions in the kidney of canine globoid cell leukodystrophy. *Acta Neuropathol* 69:33–37

Suzuki K, Grover WD (1970): Krabbe's leukodystrophy (globoid cell leukodystrophy): an ultrastructural study. *Arch Neurol* 22:385–396

Suzuki K, Hoogerbrugge PM, Poorthuis BJHM, van Bekkum DW, Suzuki K (1988): The twitcher mouse: central nervous system pathology after bone marrow transplantation. *Lab Invest* 58:302–309

Suzuki K, Suzuki Y (1970): Globoid cell leucodystrophy (Krabbe disease): deficiency of galactocerebroside-β-galactosidase. *Proc Natl Acad Sci USA* 66:302–309

Suzuki K, Suzuki K (1985): Genetic galactosylceramidase deficiency (globoid cell leukodystrophy, Krabbe disease) in different mammalian species. *Neurochem Pathol* 3:53–68

Suzuki K, Suzuki Y (1989): Galactosylceramide lipidosis: globoid cell leukodystrophy (Krabbe disease). In: *The Metabolic Basis of Inherited Disease,* 6th ed, Scriver CK, Beaudet AL, Sly WS, Valle D, eds. New York: McGraw Hill, pp 1699–1720

Suzuki Y, Austin J, Armstrong D, Suzuki K, Schlenker J, Fletcher T (1970): Studies in globoid leukodystrophy: enzymatic and lipid findings in the canine form. *Exp Neurol* 29:65–75

Suzuki Y, Miyatake T, Fletcher TF, Suzuki K (1974): Glycosphingolipid β-galactosidases III. Canine form of globoid cell leukodystrophy: comparison with the human disease. *J Biol Chem* 249:2109–2112

Svennerholm L, Vanier MT, Månsson JE (1980): Krabbe disease, a galactosylsphingosine (psychosine) lipidosis. *J Lipid Res* 21:53–64

Sweet H (1986): Twitcher is on Ch12. *Mouse News Let* 75:30

Takahashi H, Igisu H, Suzuki K, Suzuki K (1983): Murine globoid cell leukodystrophy (the twitcher mouse). The presence of characteristic inclusions in the kidney and lymph nodes. *Am J Pathol* 112:147–154

Takahashi H, Igisu H, Suzuki K, Suzuki K (1984): Murine globoid cell leukodystrophy: the twitcher mouse. An ultrastructural study of the kidney. *Lab Invest* 50:42–50

Takahashi H, Suzuki K (1982): Globoid cell leukodystrophy: specialized contact of globoid cell with astrocyte in the brain of twitcher mouse. *Acta Neuropathol* 58:237–242

Tanaka K, Nagara H, Kobayashi T, Goto I (1988): The twitcher mouse: accumulation of galactosylsphingosine and pathology of the sciatic nerve. *Brain Res* 454:340–346

Tanaka K, Nagara H, Kobayashi T, Goto I (1989): The twitcher mouse: accumulations of galactosylsphingosine and pathology of the central nervous system. *Brain Res* 482: 347–350

Tanaka H, Suzuki K (1975): Lactosylceramide β-galactosidase in human sphingolipidoses: evidence for two genetically distinct enzymes. *J Biol Chem* 250:2324–2332

Toyoshima E, Yeager AM, Brennan S, Santos GW, Moser MW, Mayer RF (1986): Nerve conduction studies in the twitcher mouse (murine globoid cell leukodystrophy). *J Neurol Sci* 74:307–318

Vanier MT, Svennerholm L (1974) Chemical pathology of Krabbe's disease. I. Lipid composition and fatty acid patterns of phosphoglycerides in brain. *Acta Paediatr Scand* 63:494–500

Yajima K, Fletcher JF, Suzuki K (1977a): Canine globoid cell leukodystrophy I. Further ultrastructural study of the typical lesion. *J Neurol Sci* 33:179–197

Yajima K, Fletcher TF, Suzuki K (1977b): Sub-plasmalemmal linear density: a common structure in globoid cell and mesenchymal cells. *Acta Neuropathol* 39:195–200

Yeager AM, Brennan S, Tiffany C, Moser HW, Santos GW (1984): Prolonged survival and remyelination after hematopoietic cell transplantation in the twitcher mouse. *Science*225:1052–1054

Yunis EJ, Lee RE (1969): The ultrastructure of globoid (Krabbe) leukodystrophy. *Lab Invest* 21:415–419

Yunis EJ, Lee RE (1970): Tubules of globoid leukodystrophy: a right-handed helix. *Science* 169:64–66

Yunis EJ, Lee RE (1976): The morphologic similarities of human and canine globoid leukodystrophy: thin section and freeze fracture studies. *Am J Pathol* 85:99–114

Zaki FA, Kay WJ (1973): Globoid cell leukodystrophy in a miniature poodle. *J Am Vet Med Assoc* 163:248–250

Zlstogora J, Chakraborty S, Knowlton RG, Wenger DA (1990): Krabbe disease locus mapped to chromosome 14 by genetic linkage. *Am J Human Genet* 47:37–44

Genetically Defined Myelin Disorders

RAVINA SIMONS AND JOHN R. RIORDAN

Myelin is a multilayered membranous sheath characteristic of nervous tissue. It insulates the axons of selected neurons along their entire length in discrete segments, increasing the conduction velocity of electrical nerve impulses traveling down these axons. The myelin is interrupted by bare patches of axon called nodes. Each myelinated segment is known as an internode. Myelin occurs in vertebrate nervous systems in two main forms, central and peripheral. It also occurs in some invertebrates in a form that resembles the myelin found in the peripheral nervous system (PNS) of vertebrates (Raine, 1984a).

The myelin sheath arises from oligodendrocytes in the central nervous system (CNS) and Schwann cells in the PNS. Each Schwann cell myelinates only one patch of axon, whereas each oligodendrocyte may myelinate as many as 30 to 50 internodes on neighboring axons. These myelin-forming cells send out membrane-bound cytoplasmic processes toward axons, each process engulfing one patch of axon destined to become an internode. The multilayered structure of the sheath is a consequence of the spiral infolding of each process around the segment of axon that it is myelinating. The multiple layers become compacted and lose most of their cytoplasmic content during the course of normal myelin development. A tightly wound membranous sheath is formed around the axon, with the lipid bilayers closely apposed to each other at the intracellular as well as the extracellular faces. The cytoplasmic space between apposed membranes forms an electron-dense band known as the major dense line. The extracellular space between successive layers of membrane forms a less dense band known as the intraperiod line (Kirschner and Ganser, 1984).

The unique morphology of myelin is likely to be brought about by interactions between the major myelin proteins. Two classes of proteins, the myelin proteolipid proteins (PLP) and the myelin basic proteins (MBP) make up 60% to 80% of the total protein content of CNS myelin (Lees, 1984). Myelin basic protein is the major extramembranous protein of both the CNS and the PNS and is thought to be involved in interactions between the intracellular faces of apposed membranes. Proteolipid protein is a membrane-bound protein specific to the CNS. A large portion of the protein is extramembranous and accordingly is probably involved in interactions between adjacent membrane layers. The major membrane-bound protein in the PNS is P_0 glycoprotein, which likely plays a role analogous to PLP (Raine, 1984a; Lees, 1984). Since the primary role of myelin is to provide high-resistance insulation around axons, it is to be expected that the most abundant proteins of myelin should be devoted to organizing the unique structure of myelin that gives it the necessary electrical properties.

Myelin is clearly not essential for axonal conduction of nerve impulses since most invertebrate nervous systems function without myelination, and not all vertebrate axons are myelinated. However, it does greatly improve the efficiency of conduction, and the myelination of selected axons is essential for the proper functioning of vertebrate nervous systems. This is demonstrated by the debilitating and often fatal consequences of genetically defined as well as acquired disorders of myelination observed in several vertebrate species.

In humans, the most commonly occurring myelin disorder is multiple sclerosis. It is a chronic demyelinating disease affecting the CNS of young and middle-aged adults. The disease affects different parts of the brain and spinal cord and produces diverse symptoms such as shaky movements of the limbs, unsteady gait, speech pronounciation defects, spastic weakness, optic nerve inflammation causing blurred vision, and spinal cord inflammation leading to numbness or partial paralysis of legs and trunk. The demyelination is thought to be caused by autoreactive T cells that are reactive to myelin antigens (McFarlin and McFarland, 1982; Waksman, 1988). Effective prevention and cure and better treatment for this and other myelin-related diseases requires a better understanding of the processes involved in the formation and maintenance of normal myelin. A powerful approach toward acquiring this better understanding has been the analysis of dysmyelinating mutants in various animal species. Correlation of the observed phenotypes of these mutants with their genotypes has provided several clues regarding the function and significance of the myelin proteins in normal myelination processes.

This chapter will review and assess the information derived from several naturally occurring dysmyelinating mutants. It will also evaluate the usefulness of hybrid double mutants achieved by interbreeding and of transgenic techniques in elucidating the functional roles of the myelin proteins.

X-LINKED MUTATIONS

Many dysmyelinating animal mutants pass on their defect in an X-linked recessive pattern. These include the jimpy *(jp)*, jp^{msd} (msd—myelin synthesis deficient), and rumpshaker *(rsh)* mouse mutants, the myelin deficient *(md)* rat, Pelizaeus–Merzbacher disease (PMD) in humans, the shaking pup mutation in dogs, and the Type AIII hypomyelinogenesis congenita mutation in pigs. All the mutants in this category show varying degrees of hypomyelination in their CNS, while their PNS remains unaffected. The phenotype exhibited by these mutants is pleiotropic with most myelin components being greatly reduced relative to normal controls. It was therefore difficult to point to the gene for any one component as the likely candidate for the primary defect. However, when the human PLP gene was found to be situated on the X chromosome (Willard and Riordan, 1985), attempts to locate the primary defects in these X-linked dysmyelinating mutants became focused on PLP. The hypothesis that PLP gene defects were responsible for the X-linked hypomyelination phenotype was supported by the correlation between the absence of PLP from normal PNS myelin and the observation that PNS myelin was unaffected in these X-linked mutants. Furthormore, the site of the *jp* mutation had previously been assigned by linkage studies to the same region of the mouse X-chromosome to which the PLP gene on the human X-chromosome was localized (Willard and Riordan, 1985).

The primary defects in all the dysmyelinating mutants mentioned above, with the exception of the *rsh* mouse and Type AIII hypomyelinogenesis congenita, have now been identified as point mutations in the PLP gene. All of these, with the exception of the *jp* mouse, are single amino acid substitutions due to single base changes. These single amino acid substitutions are scattered throughout the protein, and in one case there is a conservative alanine to valine substitution. It might seem surprising that such a substitution could lead to the breakdown of the entire process of myelination until we look at it in the light of the extraordinary evolutionary conservation of PLP. The PLP amino acid sequence is invariant between mouse, rat, and human (Macklin et al., 1987a; Milner et al., 1985; Kronquist et al., 1987). Bovine and canine PLP differ from human PLP by two and one amino acids, respectively (Lees et al., 1983; Hudson and Nadon, 1990). All three are conservative substitutions. No innocuous polymorphisms have ever been observed within a species. Such stringent conservation of sequence between species must mean that there is very little latitude for any part of the protein to mutate and still retain its function. It therefore ceases to be surprising that single amino acid substitutions in various parts of the protein should lead to loss of myelination.

These X-linked mutants share the characteristics of hypomyelination and low levels of expression of PLP and other myelin components, but they differ significantly in some features, including severity of phenotype. This section attempts to correlate these phenotypic variations with the different changes in the PLP gene in each mutant.

Jimpy (jp) *Mouse*

The most severely affected and also the most extensively studied of the X-linked dysmyelinating mutants is the *jp* mouse. It was first described by Phillips in 1954. *Jp* mice develop tremors in their hind limbs at about 10 days of age. The hind limbs are weak and in some cases completely paralyzed by about 20 days. By the fourth week of life they start to have generalized tonic seizures that usually last about a minute. These seizures are followed by a brief period of total inactivity, and then by resumption of prior behavior. Most affected animals die by about 30 days after birth, usually following a tonic seizure (Sidman et al., 1964).

Post mortem examination of *jp* brains shows almost a complete lack of myelin throughout the CNS (Sidman et al., 1964). Of the small amount of myelin that is present, most is uncompacted. PNS myelin, on the other hand, appears normal (Sidman et al., 1964). The brains of these mutants contain markedly reduced levels of myelin components. Of the myelin-associated lipids, the greatest deficits are in the content of sulfatides and cerebrosides (Sidman et al., 1964). The protein components of myelin are also greatly reduced. Most myelin-associated enzyme activities are very low. The level of 2′,3′-cyclid nucleotide 3′-phosphodiesterase (CNPase) in 20-day-old *jp* mice is about 10% of normal controls. The myelin structural proteins are also greatly reduced. Myelin basic protein is present at less than 2% of control levels at 20 days, and PLP is undetectable in Western blots. Myelin-associated glycoprotein (MAG)is reduced to about 5% of control levels (Yanagisawa and Quarles, 1986).

The reduction in levels of myelin components is probably due to the observed paucity of mature oligodendrocytes in the *jp* CNS. *Jp* brains have high numbers of oligodendrocyte precursor cells and high proliferation rates, but low numbers of mature oligodendrocytes (Skoff, 1982; Privat et al., 1982), indicating either that *jp* oligodendrocytes die at a frequency higher than normal or that there is some kind of a block in their maturation. It has been shown that at the time of oligodendrocyte proliferation and the onset of myelination, the number of dying glial cells was about 10% of the total number of glial cells in *jp* brains, compared with only about 0.5% to 2.7% in normal brains. The dying glial cells were identified ultrastructurally as oligodendrocytes (Knapp et al., 1986).

Oligodendrocytes taken from *jp* brains preserve their phenotype in culture (Sandru et al., 1980; Bartlett et al., 1988, 1985). They have shortened life spans and show poor myelination of cocultured *jp* or normal axons. However, if these *jp* oligodendrocytes are grown in medium conditioned by normal astrocytes, they survive for longer periods of time and myelinate cocultured axons (Bartlett et al., 1988). This partial rescue of the phenotype of *jp* oligodendrocytes shows that the defect is sensitive to environmental factors. It has been suggested that oligodendrocyte death in *jp* brains is secondary to axon degeneration. However, this is unlikely since *jp* axons can be myelinated in culture by normal oligodendrocytes (Billings-Gagliardi et al., 1983).

In females heterozygous for the *jp* mutation, there is an early deficit in the number of oligodendrocytes, which appears to be compensated for by an increased oligodendrocyte proliferation rate during myelination (Rosenfeld and Friedrich, 1986; Bartlett and Skoff, 1989). The rate of thymidine incorporation by oligodendrocytes in 14-day heterozygote brains is about four times that in normal brains. Adult heterozygotes have normal rates of thymidine incorporation (Bartlett and Skoff, 1989). This correlates well with the observed myelin deficit in the CNS of heterozygotes at various ages. At 2 weeks the spinal cord has a 31% myelin deficit, at 1 month a 12% deficit, and by 5 months, the amount of myelin is equal to normal controls (Bartlett and Skoff, 1986).

Astrocyte abnormalities in *jp* brains have also been reported. The number of glial fibrillary acidic protein (GFAP)–stained astrocytes is abnormally high in brains from 12-day-old *jp* mice (Ulrich et al., 1983). Astrocyte abnormalities have been noted not only during myelination but also at earlier stages when no mature oligodendrocytes are present. Astrocytic processes form an abnormally complex network and appear in unusual places, dividing axons into very small groups. As myelination progresses, these processes seem to insinuate themselves between axons and the rare myelin sheaths or within the myelin layers. These observations suggested that the primary defect in *jp* mice may involve disturbed cell–cell interactions in the CNS (Omlin and Anders, 1983).

Hypotheses regarding the primary defect in *jp* mice had been diverse due to the complex phenotype presented by these mutants. However, the mapping of the human PLP gene to the X chromosome (Willard and Riordan, 1985) allowed these speculations to become more focused. As previously mentioned, the *jp* defect had already been localized by linkage studies to the same region of the mouse X chromosome to which the PLP gene was mapped on the human X chromosome. Northern blots showed that *jp* PLP mRNA is shorter than normal PLP mRNA by about 100 bp (Fahim and Riordan, 1986; Gardinier et al.,1986), and has an abundance of only about 5% of control levels in 20-day-old animals (Gardinier et al., 1986). Sequencing of PLP cDNA revealed that the *jp* PLP cDNA has a 74-bp deletion beginning in the codon for amino acid 207 in the normal protein (Nave et al., 1986). This deletion causes a frame shift in the latter part of the open reading frame, creating a stop codon 28 nucleotides upstream of the normal stop codon. The altered open reading frame specifies a 242 amino acid protein that is 34 amino acids shorter than the normal PLP and has an altered carboxyl terminus (Nave et al., 1986).

No major deletion or rearrangements were detected in the *jp* PLP gene. In fact, a synthetic oligonucleotide corresponding to the deletion in the *jp* cDNA hybridized normally with *jp* genomic DNA (Nave et al., 1986). It was therefore apparent that the 74 nucleotides missing from the cDNA were present in the genomic DNA and that the genomic defect was probably in a splice site, leading to incorrect splicing of the 74 nucleotides. Sequencing of the intron/exon junctions of the PLP gene revealed an A to G transition in a position that is 100% conserved in the acceptor splice site preceding exon V (Macklin et al., 1987b). This mutation apparently leads to incorrect processing of the PLP RNA such that exon V is omitted from the mature mRNA.

The mutant PLP produced as a result of this aberrant splicing is detectable by immunostaining with anti-PLP antibodies. Most of the staining is restricted to the oligodendrocyte cell bodies. Positively stained myelin fibers are rare. The distribution of the staining within the oligodendrocyte is abnormal. In normal oligodendrocytes, the Golgi apparatus is strongly stained by anti-PLP antibodies, with moderate staining of the rough endoplasmic reticulum (RER). In contrast, the RER of *jp* oligodendrocytes is heavily labeled, and the staining of the Golgi apparatus is absent. It is therefore probable that synthesis of the mutant PLP molecules takes place normally in the RER, but further processing that requires their transport into the Golgi apparatus is blocked due to the abnormal sequence of the protein (Roussel et al., 1987).

It is possible that the abnormal processing of the mutant PLP is responsible for the premature death of *jp* oligodendrocytes. Trapping of the mutant protein in the RER could have toxic effects on the cell. Alternatively, it has been suggested that perhaps PLP is needed for the differentiation or maturation of oligodendrocytes, and in the absence of normal PLP the oligodendrocytes fail to reach maturation and die prematurely. The problem with either of these theories is that oligodendrocyte abnormalities and increased death rates are observed in *jp* brains before expression of PLP is detectable in normal brains (Skoff and Knapp, 1990). It is possible, however, that a small amount of PLP or its alternatively spliced isoform, DM-20, undetectable by the methods used, has an essential function in the normal maturation of oligodendrocytes that is unrelated to its structural role in mature myelin where it is present in abundance. Speculations regarding a nonstructural role for PLP or DM-20 are supported by studies showing that PLP apparently has ionophoric activity when incorporated into lipid vesicles (Lin and Lees, 1982; de Cozar et al., 1987). It is worth noting that DM-20, while present in much smaller amounts than complete PLP in mature brains, it actually detectable earlier and is present in larger amounts than PLP in the premyelination stage (Gardinier and Macklin, 1988). It may therefore by DM-20 rather than PLP that may be required for oligodendrocyte maturation.

Although most *jp* oligodendrocytes have shortened life spans and fail to myelinate, a small number constituting about 5% of the population appear to be different. These have normal life spans and lay down multilayered, compacted myelin sheaths (Skoff and Knapp, 1990). These myelin sheaths are not stained with antibodies to PLP, and yet appear structurally to be almost normal. The only structural abnormality detected is a condensation of the intraperiod line (Duncan et al., 1989). This defect is consistent with the currently accepted model of the topology of PLP in the membrane, according to which a major portion of the PLP molecule lies in the extracellular space (Hudson et al., 1989a). This suggests that the role of PLP may be to maintain proper spacing between membrane layers by interactions between the extracellular domains of the PLP molecule in adjacent layers that come into abnormally close apposition to each other in the absence of PLP. It is obvious, however, that this is a very simplistic view of the role of PLP, since it does not address the question of why, while some oligodendrocytes are able to survive and myelinate, the majority fail to myelinate and die prematurely.

Myelin Deficient (md) *Rat*

Of all the other known X-linked dysmyelinating mutants, the *md* rat, a mutant of the Wistar rat strain (Csiza and de Lahunta, 1979), is the one that most closely resembles the *jp* mouse phenotypically. *Md* rats develop visible head tremors at around 12 to 15 days of age. By 17 to 21 days they start to go into generalized tonic seizures at the slightest disturbance (Csiza and de Lahunta, 1979). The *jp* mouse and *md* rat mutations are comparable in severity. Both animals are short-lived, with death usually occurring within 30 days after birth. Both have a severe paucity of myelin in their CNS.

In addition to a paucity of myelin, several axon, astrocyte, and oligodendrocyte abnormalities have been observed in the *md* rat CNS. The mean axon area in the optic nerve of *md* rats at 19 days of age and at 43 to 46 days is similar to that in normal animals. However, the percentage of axonal area occupied by mitochondria is increased in *md* rats at both ages. The mitochondrial density in the axons is increased, as is the mean area occupied by each axonal mitochondria (Dentiger et al., 1985).

Astrocyte abnormalities are similar to those in *jp* mice, with astrocytic processes intruding within bundles of axons, between axons and myelin sheaths, and within the layers of the rare multilayered myelin sheaths (Rosenbluth, 1987). When normal fetal spinal cord fragments are transplanted into *md* rat spinal cord, patches of apparently normal myelin develop with no intrusion by host astrocytes, indicating that the astrocyte abnormalities observed are secondary to a defect in some other cell type (Rosenbluth et al., 1989). Some of the multilayered myelin sheaths that are present in *md* rat brains appear to have abnormalities in the paranodal region, which might be responsible for the intrusion of astrocytic processes from either end of the internode into the internodal periaxonal space, or between myelin layers (Rosenbluth, 1987).

Oligodendrocyte abnormalities are observed in the optic nerve as early as 3 days after birth, preceding the normal onset of myelination. Oligodendrocytes appear vacuolated and immature, and have distensions in the RER. In older animals the number of oligodendrocytes is reduced, but the problem of oligodendrocyte death is less severe than in *jp* mice (Dentiger et al., 1985).

Animals heterozygous for the *md* gene show myelin mosaicism in the CNS with patches of normal myelin interspersed by unmyelinated areas. The oligodendrocytes found associated with the abnormal patches have the distended RER characteristic of *md* oligodendrocytes (Duncan et al., 1987a). This is in contrast to *jp* heterozygotes, in which the affected oligodendrocytes die prematurely, and the function of myelination is taken over by oligodendrocytes expressing the normal PLP allele.

Like in *jp* mice, oligodendrocyte-specific proteins are greatly reduced in the brains of hemizygous *md* rats, with PLP being the most severely reduced. Myelin basic protein, MAG, and CNPase are present at 1.1%, 1.8%, and 11% of control levels, respectively, in 25-day-old *md* rat brains, whereas PLP is undetectable (Yanagisawa et al., 1986). These findings indicate that the primary defect in *md* rats is in the oligodendrocytes, probably in the PLP gene, and that the axonal and astrocytic abnormalities observed are secondary effects.

The few patches of multilayered compact myelin found in *md* rat CNS have an abnormally fused intraperiod line similar to that found in *jp* mice (Duncan et al., 1987b). This is a further indication that PLP in *md* rats is defective or absent. The primary defect in *md* rats is in fact a single base mutation in the PLP gene leading to a threonine to proline substitution at residue 74 of the mature protein (Simons and Riordan, 1990). This residue is predicted to be in an α-helical transmembrane segment of the protein (Hudson et al., 1989a). Keeping in mind the rigid structure of proline and the rareness of its occurrence in α-helical domains of proteins, it is likely that this mutation substantially changes the interaction of the protein with the membrane.

The myelin found in *md* rat CNS does not stain with antibodies against PLP but does stain with MBP antibodies (Zeller et al., 1989). It is therefore apparent that the mutant PLP does not reach its destination in the myelin membrane. The distensions in the RER of *md* oligodendrocytes are probably a result of the mutant PLP being trapped in the RER due to abnormal interaction with the membrane as a consequence of the threonine to proline substitution in the transmembrane segment.

Jimpy^{msd} (jp^{msd}) *mouse*

The myelin synthesis–deficient mouse known as *jp*^{msd}, which is allelic to *jp,* was first described in 1970 by Meier and MacPike. Like *jp*, it is an X-linked dysmyelinating disorder that leads to hind leg tremors and tonic seizures characteristic of the other mutants already described. Death of affected hemizygous males occurs within 30 days after birth even though the reduction in the amount of myelin is less severe than in *jp* mice. The *jp*^{msd} spinal cord has about twice as much myelin as *jp* (Billings-Gagliardi et al., 1980). *Jp*^{msd} mice also have slightly higher levels of PLP mRNA and protein than *jp*. The abnormal PLP protein in *jp* brains is not detectable in Western blots capable of detecting as little as 0.5% of control levels (Yanagisawa and Quarles, 1986), whereas both PLP and DM-20 are detectable in Western blots from *jp*^{msd} brains (Gardinier and Macklin, 1988). Since *jp* and *jp*^{msd} mutations arose in different mouse strains, it was thought possible that the difference in severity between the phenotypes of the two mutants may be merely due to differences in genetic background. However, the difference in amount of myelin persists when the two alleles are bred onto the same genetic background (Wolf et al., 1983), and is therefore probably a consequence of the difference in the nature of the two mutations.

In contrast to the gross alteration and truncation of the C-terminus of PLP in *jp* mice, the mutation in *jp*^{msd} mice is a single conservative amino acid substitution of a valine for an alanine due to a C to T transition in exon 6 of the PLP gene (Gencic and Hudson, 1990). Despite the extraordinary evolutionary conservation of PLP, the fact that a conservative substitution such as this can so completely disrupt the process of myelination is surprising. The mutated residue is situated in an extracellular segment predicted to be α-helical and thought to play a role in interactions with the adjacent membrane layer (Hudson et al., 1989a). While valine has a lower α-helix–forming propensity than alanine, it is not likely to cause disruption of an α-helix, especially when found in a segment with other strong α-helix formers, as it is in this case. It is therefore difficult to envision a mechanism by which this alanine to valine substitution could disrupt the structural role of PLP. Why then is myelination disrupted by this substitution? Could it be that PLP does in fact serve another function in addition to its structural role as suggested earlier?

The paucity of mature oligodendrocytes and increased oligodendrocyte death described for *jp* mice is also observed in jp^{msd} mice (Billings-Gagliardi et al., 1980). Although it is easy to envision how the grossly abnormal C-terminal in *jp* PLP or the disrupted transmembrane α-helix in *md* rat PLP could poison protein transport and targeting systems in the cell and thereby cause its death, such is not the case with jp^{msd} PLP. However, it is possible that the mutated residue is essential for some other function of PLP, such as being a factor in oligodendrocyte differentiation or maturation. This hypothesis is supported by quantitation of PLP and DM-20 proteins in jp^{msd} brains. The DM-20 to PLP ratio remains high throughout the life span of jp^{msd} mice and resembles that found in normal mice at premyelination ages (Gardinier and Macklin, 1988). This suggests that jp^{msd} oligodendrocytes are trapped in an immature stage. This supports the possibility that DM-20 or PLP may normally function as an oligodendrocyte maturation factor, since both PLP and DM-20 are altered in jp^{msd} mice.

Shaking Pups

The canine shaking pup is an X-linked hypomyelinating mutant that arose in a line of English Springer Spaniels (Griffiths et al., 1981b). The gross phenotype of the shaking pup is very similar to that already described for the other dysmyelinating mutants. Shaking pups are smaller than their normal littermates. They exhibit tremors beginning at 10 to 12 days after birth. They are, however, longer lived than the mouse and rat mutants described and can live for prolonged periods.

The longer life span may be attributable to the greater amount of myelin found in the shaking pup than in the other mutants. A larger proportion of axons are myelinated than in the mouse and rat mutants but the sheaths are reduced in thickness as well as in the length of the internodes relative to control animals (Griffiths et al., 1981a). There appears to be no correlation between axon diameter and the thickness of the myelin sheath (Duncan et al., 1983). The total amount of myelin is greatly reduced relative to controls. The yield of myelin fractions from the brain stems of 4-week-old and 16-week-old mutant pups is only about 2.4% of that from age-matched controls (Inuzuka et al., 1986).

There is no noticeable increase in oligodendrocyte death over normal control animals. Oligodendrocyte numbers, however, are greatly reduced. Many of the oligodendrocytes present have distensions in the RER, similar to those seen in *md* rats (Duncan et al., 1983).

Myelin-specific proteins are present at low levels in the CNS. Levels of MBP, CNP, and MAG are 2.6%, 10%, and 15% of controls respectively in 4-week-old and 16-week-old mutant pups (Inuzuka et al., 1986). Proteolipid protein is more severely reduced to less than 1% of control levels (Yanagisawa et al., 1987). The ratio of DM-20 to PLP is increased with DM-20 being present in a slightly higher amount than PLP in 4-week-old mutants (Yanagisawa et al., 1987). This is in contrast to normal animals in which PLP is much more abundant than DM-20. This inversion of the DM-20 to PLP ratio, resembling that in young premyelinating normal animals, is also found in jp^{msd} mice as mentioned earlier, and is probably indicative of immaturity of the oligodendrocytes.

Females heterozygous for the shaking pup mutation show myelin mosaicism in the optic nerve and spinal cord, similar to that found in *md* rats. Oligodendrocytes associated with the abnormal patches of myelin exhibit the distended RER trait, thought to be due to the trapping of mutant protein in the RER. Female heterozygote dogs have a marked tremor that develops around the onset of myelination but later disappears with age (Duncan et al., 1987a). This is probably due to an early deficit of oligodendrocytes expressing the normal allele, which is compensated for at a later age.

The primary defect in shaking pups is a single amino acid substitution in PLP and DM-20 at residue 36 of the mature protein, which is at the end of the first transmembrane segment. A proline is substituted for the histidine normally present at this position (Nadon et al., 1988). Interestingly, the *md* rat mutation also involves the introduction of a proline into a transmembrane segment. The probable effect of the introduction of a proline residue into a transmembrane segment has already been mentioned. It is likely that the PLPs made as a result of either of these mutations have grossly abnormal membrane interactions. It is interesting to note that the two mutations involving proline are the only two mutations studied that are associated with distensions of the RER. It therefore seems likely that the prolines in the first transmembrane segment of the mutant PLPs make a protein that is synthesized normally in the RER but cannot be translocated into the Golgi. However, immunostaining of tissue sections shows that in shaking pups, at least some of the mutant PLP synthesized does find its way to the myelin membrane (Duncan, 1990). In *jp* mice, which make PLP with a grossly abnormal C-terminus, transport of the mutant PLP into the Golgi also appears to the blocked. However, no distensions of the RER are observed.

Rumpshaker (rsh) *mouse*

The *rsh* mouse is a recently identified dysmyelinating mutant described by Mitchell et al in 1990. The *rsh* gene is closely linked to the *jp* locus and is probably the same gene. However, the severity differs greatly from *jp* mice. *Rsh* mice have a normal life span and breeding ability. They have a much greater amount of myelin in the CNS. The myelin is of reduced thickness but most of it appears to be structurally normal in other respects. Some of the sheaths present have an abnormal periodicity, and in some cases the major dense lines and intraperiod lines are indistinguishable (Mitchell et al., 1990).

Oligodendrocyte numbers are increased relative to controls. The oligodendrocytes appear to be immature as in *jp* mice, but unlike *jp* mice, oligodendrocyte death is infrequent. The amount of MBP is not markedly reduced, and the myelin sheaths stain strongly with MBP antibodies. The overall amount of MAG is reduced, but wherever myelin is present, MAG staining appears to be normal. In contrast to MBP and MAG, the amount of PLP is more markedly reduced but the deficit is much less severe than in *jp* mice or any of the other X-linked dysmyelinating mutants described. Proteolipid protein RNA is also reduced, being present at 40% to 60% of control levels at peak times. Most myelin sheaths do not stain with antibodies specific to PLP that do not recognize DM-20. Antibodies that recognize both PLP and DM-20, however, do react with the myelin sheaths, although with lower intensity than normal controls (Griffiths et al., 1990). This may reflect an elevated DM-20 to PLP ratio in *rsh* mice.

The primary defect in *rsh* mice is as yet unknown. It is likely to involve the PLP gene but the presence of myelin sheaths with normal periodicity indicates that the defect may not be in the coding region of the gene. Reduction in the proportion of PLP RNA that is polyadenylated in the *rsh* mouse suggests that the defect may be in RNA processing at the polyadenylation stage. Alternatively, the defect may be at some other stage of processing, affecting the DM-20 to PLP ratio in the sheath (Mitchell et al., 1990). Regardless, the identification of the defect will undoubtedly contribute to our understanding of the role of PLP/DM-20 in the myelin membrane.

Type AIII Hypomyelinogenesis Congenita

Type AIII *Hypomyelinogenesis Congenita* is an X-linked dysmyelinating disorder in Landrace pigs. The condition is characterized by a widespread muscular tremor in the newborn pig and unsteady gait at an early age. The myelin deficiency is less severe than in the mouse and rat mutants described and resembles the shaking pup. As in the shaking pup, a larger proportion of the axons are myelinated, but the majority of the sheaths are thin (Blakemore, 1974). Also, like in the shaking pup, the myelin found in the mutant pig does stain with PLP antibodies (Duncan, 1990). The primary defect in Type AIII pigs has not been determined.

Pelizaeus–Merzbacher disease

A group of human dysmyelinating disorders affecting the CNS have been lumped under the name of Pelizaeus–Merzbacher disease (PMD). The disease has been classified into several subtypes based on differences in inheritance patterns and on phenotypic heterogeneity. Two subtypes of PMD are consistent with a mutation in the PLP gene. These are the type I classical type and the type II connatal type. Both are X-linked disorders causing dysmyelination of the CNS, while the PNS remains unaffected (Zeman et al., 1964; Koeppen et al., 1987). They also phenotypically resemble the X-linked dysmyelinating animal mutants already described (Koeppen et al., 1988). Symptoms include nystagmus, speech problems, gait abnormality, and tremors (Zeman et al., 1964; Koeppen et al., 1987). Death usually occurs between 10 and 30 years of age, depending on severity of phenotype and rate of progression of the disease.

In the connatal form of the disease, some symptoms are observable at birth. The progression of the disease is faster than in the classical form, and death can occur before 10 years of age. The classical form of the disease has a later onset and slower progression, and patients may live up to 30 years. All cases of PMD do not fall unequivocally into one category or the other because there appears to be a spectrum of levels of severity (Hudson and Nadon, 1990; Raine, 1984b; Norton and Cammer, 1984).

In both classical and connatal PMD, post-mortem examination reveals severe paucity of myelin throughout the CNS. Oligodendrocyte numbers are severely reduced, as are the levels of the major myelin proteins. Antibodies to MBP and MAG reveal islands of myelin-like structure. These structures do not react with PLP antibodies (Koeppen et al., 1988; Koeppen et al., 1987).

The variable phenotypes are consistent with the possibility of different mutations in a gene on the X chromosome such as PLP. When DNA from a number of unrelated patients was first scanned for the possibility of gross rearrangements in the PLP gene, only one was found to exhibit such a change (Fahim and Riordan, 1986).

The primary defect in PMD has now been identified in at least three different lineages, one classified as connatal type and two cases of the classical form of the disease. All three are single amino acid substitutions in the PLP gene. The connatal type is a proline to leucine change at residue 14 of the protein, in the first transmembrane segment (Trofatter et al., 1989). The mutations in the two classical cases are a tryptophan to arginine substitution at position 162 in one case (Hudson et al., 1989b) and a proline to serine change at residue 215 of the protein in the other case (Gencic et al., 1989). Both these substitutions are in the extracellular domain of the protein (Hudson et al., 1989a).

The identification of these mutations provides a helpful clue regarding the reason for the variations in severity in different cases of PMD. The more severe, connatal case of the disease has a mutation in a transmembrane segment, which could have a more profound effect on the structural integrity of PLP than either of the extracellular mutations in the classical cases. Different point mutations in the PLP gene could generate a wide spectrum of degrees of myelin deficiency in the CNS and other phenotypic variations.

AUTOSOMAL MUTATIONS

Several autosomal mutations that lead to dysmyelination have been described in mice. These mutations have arisen at several different loci. Some, such as shiverer (*shi*) and myelin deficient (*mld*), are due to defects in the MBP gene. The twitcher mouse, which has a mutation in the galactosyl ceramidase I gene, is discussed in another chapter in this book. The quaking mouse probably has a defect in the oligodendrocyte cell cycle. Trembler (*Tr*) mice have an unknown mutation, probably in a protein that is unique to the PNS.

Myelin Basic Protein Mutations

Myelin basic protein (MBP) is a major structural protein of myelin found in both the CNS and the PNS. Myelin basic protein is abundant in CNS myelin, making up about 40% of the total protein. It is present in lower levels in the PNS. Myelin basic protein is localized at the cytoplasmic face of myelin membranes and is a component of the major dense line. It presumably plays a role in the compaction of the myelin sheath. In the mouse, the MBP gene is located on chromosome 18. Five MBP isoforms ranging in size from 14 to 21.5 kDa have been identified. They arise through alternative splicing from the same gene and are differentially regulated through development.

Two animal models of MBP mutations causing dysmyelination are known to exist. These are the shiverer (*shi*) and myelin-deficient (*mld*) mutations, both of which occur in mice. Both mutations lead to dysmyelination of the CNS. The PNS myelin appears to be functionally normal.

Extensive work has been done to characterize these two mutants. Transgenic techniques and interbreeding with PLP mutants to produce double mutants have yielded additional information regarding the role of MBP in myelination.

SHIVERER *(SHI)* MOUSE

The *shi* mouse has an autosomal recessive mutation that causes severe paucity of myelin in the CNS (Biddle et al., 1973). Peripheral nervous system myelin appears to be unaffected except for certain subtle abnormalities. Homozygous *shi/shi* mice develop generalized action tremors by the end of the second post natal week. These tremors get more severe with age. Tonic seizures start to occur by about 30 days of age, increasing in frequency and duration at later stages. The lifespan of a *shi* mouse is 50 to 100 days (Cheroff, 1981).

The levels of several proteins are abnormal in the *shi* CNS. The specific activities of certain enzymes are elevated. There is almost a threefold increase in the specific activity of 5′ nucleotidase in partially purified myelin from the spinal cords of *shi* mutants. The specific activity of CNPase and carbonic anhydrase are slightly elevated (Zimmerman and Cammer, 1983). Myelin associated glycoprotein levels are significantly reduced in immunoblots from brain homogenates. Immunocytochemistry reveals periaxonal staining with MAG antisera. The intensity of staining around individual axons, however, is reduced. The total number of immunostained axons are also reduced. The cytoplasm and perikarya of certain oligodendrocyte-like cells also immunostain with MAG antisera. This suggests that there may be a block in protein transport in *shi* oligodendrocytes since this kind of staining is not observed in oligodendrocytes in normal brains (Sheedlo and Siegel, 1986). Proteolipid protein and MBP levels are very low in *shi* brains (Cammer and Zimmerman, 1983). Myelin basic protein is less than 1% of normal. Myelin basic protein RNA is undetectable on Northern blots capable of detecting less than 1% of control levels (Roach et al., 1983).

The little myelin that exists in the *shi* CNS is uncompacted and lacks a major dense line. The myelin sheaths do not stain with antibodies against MBP (Dupouey et al., 1979; Privat et al., 1979). The absence of MBP from the myelin correlates well with the lack of a major dense line, since MBP is localized at the cytoplasmic face of normal myelin lamellae. Peripheral nervous system myelin, however, has a normal periodicity despite the absence of MBP (Kirschner and Ganser, 1980; Mikoshiba et al., 1981). This might be a reflection of the lower abundance of MBP in the PNS than the CNS in normal animals (Lees, 1984). In the PNS, the function of MBP may normally be supplemented by P_2, which is a basic protein unique to the PNS or P_0, which is also unique to the PNS. In the absence of MBP, as in the case of *shi* mice, P_2 or P_0 may sufficiently compensate for the absence of MBP to maintain the normal structure of the myelin. This abnormality in *shi* myelin has been exploited in order to study the behavior of normal myelinating oligodendrocytes. Transplantation of fragments of olfactory bulb from newborn normal mice into newborn *shi* brains led to myelination of very extensive areas in the test brain by the transplanted oligodendrocytes. Myelin laid down by transplanted oligodendrocytes was distinguishable from myelin from host oligodendrocytes due to the presence of MBP in myelin from the transplanted cells. The migration of the implanted oligodendrocytes over long distances before they synthesize myelin indicates that oligodendrocytes in normal development might behave in a similar manner (Lachapelle et al., 1983).

The primary defect in *shi* mice is a large deletion in the MBP gene. The deletion encompasses exons II through VII of the gene, which normally consists of 7 exons (Roach et al., 1985). That this mutation is in fact responsible for the *shi* phenotype has been demonstrated by correction of the phenotype by an MBP transgene (Readhead et al., 1987).

A normal MBP transgene was introduced into the germ-line of *shi* mice. *Shi* mice homozygous for the transgene had MBP mRNA and protein levels that were about 25% and 20% of normal endogenous levels, respectively. These mice exhibited a marked reduction in the tremor phenotype, had normal life spans, and produced compacted myelin with a normal major dense line. The number of axons myelinated was comparable to normal animals. *Shi* mice heterozygous for the transgene express about 8.5% of the normal level of MBP. These mice have slightly longer life spans, but do not show a significant alleviation of the tremors. Myelination in the optic nerves of these mice occurs only around a few large axons. These myelin sheaths are thin, but do have prominent major dense lines (Readhead et al., 1987). These results indicate that the minimum amount of MBP needed to synthesize close to a normal amount of myelin is somewhere between 8.5% and 20% of normal endogenous levels. This means that normal animals have a large excess of MBP.

It has been shown that a transgene capable of expressing only the smallest of the MBP isoforms can rescue the *shi* phenotype, indicating that all the isoforms of MBP may not be essential for the formation of normal myelin. The myelinated axons in the rescued mice were not uniformly distributed, and the sheaths were thinner than in normal mice. These abnormalities, however, may be due to an insufficient amount of MBP rather than the absence of the other isoforms (Kimura et al., 1989).

Interbreeding of *shi* mice with *jp* mice to produce double mutants of *shi/shi*jp/Y* genotype has yielded surprising results. The two mutations do not have an additive effect, but on the contrary, seem to suppress each other's effects partially. The double mutants have increased numbers of myelin sheaths relative to *jp*, but less than in *shi*. They have thicker sheaths and more prominent major dense lines than *shi*. The double mutants express more MBP than *shi* alone and more PLP than *jp* alone (Kerner and Carlson, 1986; Billings-Gagliardi et al., 1986). These results may mean that the roles of both MBP and PLP are more complex than simply maintaining the structural integrity of the myelin sheath, and that each protein has some additional function in the control of oligodendrocyte development and myelin formation (Billings-Gagliardi and Wolf, 1990).

MYELIN DEFICIENT *(MLD)* MOUSE

The *mld* mutation in mice is allelic to *shi*. It is an autosomal recessive mutation that causes dysmyelination in the CNS (Doolittle and Schweikart, 1977). *Mld* mice show the same tremors and seizures as *shi* mice. Life spans are longer than *shi* mice, with some *mld* mice living as long as 9 months (Shen et al., 1985). The dysmyelination, however, is less severe than in *shi* mice. Even when the two mutations are bred onto the same genetic background, some myelin sheaths in the *mld* CNS are MBP positive and have a major dense line in 21-day-old animals (Shen et al., 1985). This is in contrast to *shi* mice in which no major dense lines are observed except in older animals.

Myelin in the PNS of *mld* mice does not appear to be severely affected. The periodicity of the myelin is slightly altered, but the sheaths appear to be otherwise normal. The amount of myelin is slightly reduced. The level of MBP is very low in young animals but increases to almost normal values in the PNS myelin by 135 days of age (Ginalski-Winkelmann et al., 1983–84).

The CNS is more severely affected. Myelin sheaths are rare and uncompacted and have very little MBP in 30-day-old animals. Myelin basic protein levels in the myelin, however, continue to increase between 30 and 135 days, accompanied by the appearance of the major dense line and improved compaction of the myelin (Mattieu et al., 1984). This is in contrast to *shi* mice, in which no MBP is detected in the myelin sheaths, myelin remains uncompacted, and major dense lines are rare and usually interrupted.

The mutation in *mld* mice consists of a duplication of the MBP gene accompanied by an inversion of a large segment containing exons III to VII in the upstream copy of the gene (Akovitz et al., 1987; Popko et al., 1988; Okano et al., 1988). The downstream copy of the gene is intact. The transcription rates of these genes determined *in vitro* are comparable to the normal MBP gene (Okano et al., 1988). Both promoters in the *mld* MBP allele are active, the upstream abnormal copy of the gene being slightly more active than the downstream normal copy (Tosic et al., 1989; Tosic et al., 1990). Transcription of these two tandem genes produces significant amounts of normal MBP RNA and also of antisense MBP RNA from the upstream gene. Most of the normal RNA, however, is restricted to the nucleus and very little of it is found in the cytoplasm. None of the antisense RNA is detected in the cytoplasm. The low level of the MBP RNA in *mld* mice, therefore, is not due to low transcription rates, but probably due to formation of sense–antisense hybrids that are rapidly degraded (Tosic et al., 1990).

Transgenic mice with normal endogenous MBP genes and antisense MBP transgenes mimic the *shi* or *mld* phenotype, with tremors and shortened life spans. The severity of the phenotype depends on the level of expression of the transgene. The levels of endogenous MBP transcripts are reduced in these transgenic animals (Katsuki et al., 1988). This transgenic model provides support for the hypothesis that the antisense transcripts from the upstream MBP gene in *mld* mice are responsible for the reduction in MBP levels in *mld* mice.

Other Autosomal Mutations

TREMBLER *(TR)* MOUSE

Trembler (*Tr*) is an autosomal dominant mutation in mice reported by Falconer in 1951. It produces convulsions in young animals and a coarse action tremor and moderate quadriparesis in older animals. The CNS appears to be normal, but the PNS is severely hypomyelinated (Falconer, 1951). The abnormal myelin in the PNS appears to undergo cycles of breakdown followed by partial regeneration. Myelin breakdown products are visible (Ayers and Anderson, 1973). Homozygous and heterozygous *tr* mice are behaviorally indistinguishabale, but the degree of hypomyelination in the PNS varies (Henry and Sidman, 1988).

In heterozygous animals, an average of approximately 30% of axons large enough to be myelinated are myelinated. In homozygous animals, however, there are virtually no myelinated fibers. Surprisingly, these homozygous animals are long-lived and do not exhibit an increase in behavioral abnormalities relative to heterozygotes except a noticeable worsening of the quadriparies in some animals over 1 year old (Henry and Sidman, 1988).

Schwann cell behavior is abnormal in the mice. Schwann cell proliferation rates are increased to as much as 10 times that found in control animals (Perkins et al., 1981). The Schwann cells lay down multiple layers of excess basal laminae, forming an "onion bulb" structure. The myelin that is formed is uncompacted. The Schwann cells differentiate to the stage where they target specific axons and ensheath them, but appear to be unable to progress beyond this stage. Compact myelin lamellae are therefore not formed (Ayers and Anderson, 1973).

In addition to hypomyelination and Schwann cell abnormalities, axon diameters appear to be reduced in *tr* mice (Love, 1976). In addition, some abnormalities in slow axonal transport have been reported. Since axon diameter is partially dependent on the movement of components of the cytoskeleton by slow axonal transport, the reduction of axon diameters may be due to reduction in transport rates of certain cytoskeletal components (de Waegh and Brady, 1990). Nerve-graft transplantation experiments have indicated that these axonal abnormalities may be secondary effects due to association with abnormal Schwann cells. In transplantation experiments normal host axons ensheathed by transplanted *tr* Schwann cells are hypomyelinated. Conversely, transplanted *tr* axons ensheathed by normal host

Schwann cells have normal myelin. The primary defect in *tr* mice therefore appears to be in the Schwann cells (Aguayo et al., 1977; Pollard and McLeod, 1980).

Various abnormalities exist in the amounts of Schwann cell synthesized proteins in the *tr* CNS. Myelin basic protein levels peak at 12 days of age instead of at 16 days as in normal mice. The amount of MBP RNA at the peak is reduced to about 25% of normal levels, and is barely detectable at 8 or at 40 days of age (Garbay et al., 1989). The protein is present at less than 4% of normal at 12 days, and is undetectable in the adult. The amount of P_0 protein appears to be normal in 12-day-old *tr* mice but fails to increase sharply at the time of active myelination. In the adult *tr*, levels of P_0 have fallen to less than 20% of controls (Garbay et al., 1988). The amount of P_2 protein is greatly reduced in the adult. The amount of MAG is similar in *tr* and normal animals at young ages and is slightly higher than normal in *tr* adults. The apparent molecular weight of MAG is higher than normal in the *tr* PNS but not in the CNS (Inuzuka et al., 1985).

Various disturbances in lipid metabolism have been observed in *tr* Schwann cells. Total lipid content is severely reduced in the PNS (Yao and Bourre, 1985). Accumulation of most polar lipids is abnormal in the sciatic nerves of *tr* mice from the earliest stages of development. Cerebrosides are the most severely affected throughout development. In normal sciatic nerves, polar lipids accumulate rapidly between 9 and 18 days of age, after which the phospholipids plateau while glycolipids continue to accumulate slowly. In contrast, with the exception of cardiolipin, all lipids stop accumulating in the *tr* sciatic nerve after 18 days (Heape et al., 1986). Among the neutral lipid abnormalities is a much lower rate of cholesterol accumulation in the PNS than in normal PNS. Triacylglycerols appear normal at early ages, but their quantities decline significantly after 18 days of age (Juguelin et al., 1986).

The primary defect in *tr* mice is not known. The lipid abnormalities observed in the *tr* PNS suggest that the primary defect may involve some aspect of Schwann cell lipid metabolism. The myelin breakdown products observed in *tr* PNS suggest that the myelin laid down is abnormal and therefore degraded and then regenerated in a cyclical fashion (Yao and Bourre, 1985).

Quaking *(Qu)* Mouse

Qu is an autosomal recessive mutation on chromosome 17 in mice. Axial tremors in *qu* mice first become visible at about 12 days of age. Tonic seizures occur throughout the animal's life span. The life span is normal (Hogan and Greenfield, 1984).

Qu mice have severe dysmyelination in the CNS. There is very little myelin in the entire CNS. The PNS is also affected, but to a lesser degree. There is no obvious myelin deficit observable by light microscopy. Ultrastructural study, however, has shown that some peripheral nerves have unmyelinated segments. There is also marked reduction in the number of myelin lamellae relative to axon diameter, and the myelin that is present is poorly compacted (Samorajski et al., 1970; Wisniewski and Morell, 1971; Watanabe and Bingle, 1972). Older *qu* animals exhibit a breakdown of Schwann cell–axon junctions in the PNS.

Qu oligodendrocytes are normally located, retain their capacity for proliferation, and do not appear to be reduced in numbers, but appear morphologically to be immature (Zhabotinskii et al., 1983). The expression of MBP and PLP resembles that of immature oligodendrocytes. The DM-20 to PLP ratio is elevated (Agrawal et al., 1987). The ratio of the smaller MBPs to the larger MBPs increases with age in the CNS and PNS of both normal and *qu* mice, but the ratio in adult *qu* mice in both the CNS (Greenfield et al., 1971) and the PNS (Weiss et al., 1983) is approximately the same as that found in very young normal animals, reflecting immaturity of *qu* oligodendrocytes and Schwann cells. In addition to the ratios of the PLPs and MBPs being altered, the total amounts of PLP and MBP are also reduced (Sorg et al., 1986). In both the CNS and PNS, the active period of MBP deposition is delayed in *qu* mice relative to normal controls. MBP mRNA in *qu* brains is low at early stages but appears to be present in normal amounts by 21 days of age (Roth et al., 1985; Jacque et al., 1983). The protein is also present at low levels for the first 3 weeks and later increases to almost normal levels in whole brain homogenates. But the amount of MBP incorporated into myelin continues to be low, indicating that the newly synthesized MBP is not incorporated into myelin (Sorg et al., 1986; Campagnoni et al., 1987). Proteolipid protein mRNA and protein levels are also reduced (Campagnoni et al., 1987; Konat et al., 1988). 2′, 3′-cyclic nucleotide 3′-phosphodiesterase mRNA is markedly reduced at all ages (Kurihara et al., 1989).

Developmental regulation of the two alternatively spliced isoforms of MAG, p67MAG and p72MAG, appears to be lacking in *qu* mouse brains. In normal mouse brains, p67MAG is abundant in young animals with p72MAG being the predominant form in adults (Lai et al., 1987). In *qu* mice, very little p72MAG is formed at any age (Fujita et al., 1988; Fujita et al., 1990), accompanied by overexpression of p67MAG, which has a slightly higher than normal molecular weight (Frail and Braun, 1985; Fujita et al., 1990), indicating the possibility of abnormal glycosylation (Fujita et al., 1990). Total amount of MAG is reduced to about 16% of control levels in the CNS, but is present in amounts equal to or slightly exceeding normal levels in the PNS (Inuzuka et al., 1987).

In normal animals MAG is located in the periaxonal membrane of compact myelin. It is therefore thought to play a role in the interaction between axons and myelin membranes (Brady and Quarles, 1988). Myelin associated glycoprotein is also found in uncompacted myelin in the PNS of young animals. The process of myelin compaction in the PNS appears to involve the removal of MAG from the uncompacted membranes, accompanied by removal of cytoplasm from between the myelin layers. The periaxonal layer of myelin retains its MAG and also its cytoplasm, forming the periaxonal cytoplasmic collar (Trapp, 1988). It has been hypothesized that MAG is involved in the maintenance of a normal periaxonal space and the formation of the cytoplasmic collar around axons in the PNS (Trapp, 1988). This is supported by observations of remyelinating Schwann cells in *qu* mice. There is a positive correlation between the presence of MAG in the periaxonal membrane of *qu* Schwann cells and a normal periaxonal space of 12 to 14 nm, and also the existence of the periaxonal cytoplasmic collar (Trapp et al., 1984). With *qu* myelinating Schwann cells lacking MAG in the periaxonal membrane, the periaxonal space is enlarged and the periaxonal cytoplasmic collar is missing. The cytoplasmic side of the periaxonal membrane is fused with the cytoplasmic side of the next membrane, forming a major dense line instead of a cystoplasmic collar. Some *qu* Schwann cells, which do have MAG in the periaxonal membrane, have a normal periaxonal space and cytoplasmic collar (Trapp et al., 1984). This correlation between the presence of MAG and structural integrity of the periaxonal region confirms the involvement of MAG in the interaction between axons and myelinating cells.

In *qu* animals, uncompacted membranes that fail to convert to compact myelin contain MAG. These membranes retain cytoplasm between their cytoplasmic faces and maintain a 12 to 14-nm gap between their extracellular faces (Trapp et al., 1984). This gap is similar in size to a normal periaxonal space. This indicates that removal of MAG from membranes is required to allow extrusion of cytoplasm from between myelin membranes, and that the large immunoglobulin-like extracellular domain of MAG forces a gap of 12 to 14 nm between membrane layers. It is therefore likely that failure of compaction of *qu* myelin is due to failure of removal of MAG from the membranes. The primary defect, however, is not in MAG since the *qu* gene and the MAG gene are on different chromosomes (D'Eustachio et al., 1988; Hogan and Greenfield, 1984). The mutation may involve a gene product needed for the removal of MAG.

DISCUSSION

The variety of animal models of X-linked dysmyelination has led to some understanding of the role of PLP in myelin. Lack of PLP appears to have no effect on PNS myelin. This is as expected since PLP is not a component of PNS myelin. However, it has been reported that PLP is in fact expressed in Schwann cells in the PNS, although it is not incorporated into the PNS myelin (Puckett et al., 1987). The role of this PLP in Schwann cells is unknown, but its presence raises the possibility that animals without functional PLP may exhibit some changes in PNS myelin that are as yet undetected. Alternatively, perhaps the function, if any, of the PLP in the PNS is not disrupted by the mutations that render it nonfunctional with respect to its structural role in the CNS myelin.

Central nervous system myelin is severely affected by mutations in PLP. Strict conservation of the sequence of PLP appears to be essential for normal formation of CNS myelin. Even conservative substitutions in the PLP sequence (Jp^{msd} mouse) can cause severe disruption of CNS myelination.

Mutations that affect transmembrane segments of PLP (*md* rat and shaking pup), and the *jp* mouse mutation, which grossly alters the protein, appear to block the intracellular transport of the protein. Distensions of the RER in the case of *md* rat and shaking pup indicate that the abnormal protein is trapped in the RER. In the shaking pup, however, some PLP is detected in myelin sheaths. Therefore, some of the PLP does reach its destination in the oligodendrocyte plasma membrane.

The extreme paucity of oligodendrocytes in all the PLP mutants (with the exception of the *rsh* mouse, which may have a regulatory rather than structural defect in the PLP gene) has led to speculations regarding a second role for PLP in addition to the accepted structural one. It has been suggested that PLP and/or DM-20 is essential for the maturation and survival of oligodendrocytes. DM-20 is a good candidate for this role since it is expressed earlier than PLP and is present in amounts higher than PLP at the premyelination stage.

The MBP mutants (*shi* and *mld*) also have a severe myelin deficiency in the CNS, but are longer lived than the PLP mutants, and do not have the paucity of oligodendrocytes that occurs in PLP mutants. This indicates that the decrease in oligodendrocytes observed in PLP mutants is not a consequence of the myelin deficiency but is in fact a direct consequence of the absence of normal PLP.

The lack of a major dense line in the MBP-negative myelin in the CNS of the MBP mutants is consistent with the location of MBP on the cytoplasmic side of myelin membrane and its structural role in maintaining interactions between apposed cytoplasmic faces of myelin lamellae. The normal structure of PNS myelin, including normal periodicity, despite the lack of MBP in *shi* mice, is indicative of the ability of other basic proteins in the PNS to take over the strucural role of MBP.

The characteristics of MBP/PLP double mutants are difficult to interpret. The partial reciprocal suppression of phenotype indicates that the roles of MBP and PLP, and their effects on the regulation of myelination, are far more complex than previously suspected.

Study of the *qu* mutant has provided strong correlation between the presence of MAG and the formation of normal interactions between axons and the periaxonal membrane. Myelin associated glycoprotein appears to be required to establish interaction between axons and oligodendrocytes or Schwann cells, and to maintain a normal periaxonal gap. Its removal from uncompacted membranes is required to allow extrusion of cytoplasm and compaction of the myelin.

The study of these dysmyelinating mutants has led to a better understanding of the roles of the major myelin proteins. However, the complex process of myelination is still far from clearly understood. Much remains to be learned regarding the regulation of the onset of myelination, the targeting of axons, the formation of multiple layers of membrane, and the compaction of the lamellae.

REFERENCES

Agrawal H, Agrawal D, Yoshimura T, Benjamins J (1987): In vitro acylation of myelin PLP and DM-20 in the quaking mouse brain. *Neurochem Res* 12:783–786

Aguayo A, Attiwell M, Trecarten J, Perkins S, Bray G (1977): Abnormal myelination in transplanted trembler mouse Schwann cells. *Nature (Lond.)* 265:73–75

Akowitz A, Barbarese E, Scheld K, Carron J (1987): Structure and expression of myelin basic protein gene sequences in the mld mutant mouse: Reiteration and rearrangement of the MBP gene. *Genetics* 116:447–464

Ayers M, Anderson R (1973): Onion bulb neuropathy in the trembler mouse: a model of hypertrophic interstitial neuropathy (Dejerine-Sottas) in man. *Acta Neuropathol* 24:54–70

Bartlett W, Knapp P, Skoff R (1985): Failure of oligodendrocytes from jimpy brain to differentiate in vitro. *Soc Neurosci Abstr* 11:1067

Bartlett W, Knapp P, Skoff R (1988): Glial conditioned medium enables jimpy oligodendrocytes to express properties of normal oligodendrocytes: production of myelin antigens and membranes. *Glia* 1:253–259

Bartlett W, Skoff R (1986): Expression of the jimpy gene in the spinal cords of heterozygous female mice. I. An early myelin deficit followed by compensation. *J Neurosci* 6:2802–2812

Bartlett W, Skoff R (1989): Expression of the jimpy gene in the spinal cords of heterozygous female mice. 2. Oligodendroglial and endothelial cell hyperplasia. *Brain Res Dev Brain Res* 47:1–11

Biddle F, March E, Miller J (1973): Research news. *Mouse News Lett* 48:24–26

Billings-Gagliardi S, Adcock L, Lamperti E, Schwing-Stanhope G, Wolf M (1983): Myelination of jp, jpmsd, and qk axons by normal glia in vitro: ultrastructural and autoradiographic evidence. *Brain Res* 268:255–266

Billings-Gagliardi S, Adcock L, Wolf M (1980): Hypomyelinated mutant mice: Description of jpmsd and comparison with jp and gk on their present genetic backgrounds. *Brain Res* 194:325–338

Billings-Gagliardi S, Wolf M (1990): Implications of double mutant phenotypes. *Ann NY Acad Sci* 605:215–229

Billings-Gagliardi S, Wolf M, Kirschner D, Kerner A (1986): Shiverer and jimpy double mutant mice. II. Morphological evidence supports reciprocal intergenic suppression. *Brain Res* 374:54–62

Blakemore W (1974): Ultrastructural observations on the spinal cord of a Landrace pig with congenital tremor type AIII. *Res Vet Sci* 17:174–178

Brady R, Quarles R (1988): Developmental and pathophysiological aspects of the myelin-associated glycoprotein. *Cell Mol Neurobiol* 8:139–148

Cammer W, Zimmerman T Jr (1983): Distribution of myelin-associated enzymes and myelin proteins into membrane fractions from the brains of adult shiverer and control (+/+) mice. *Brain Res* 265:73–80

Campagnoni A, Sorg B, Roth H, Kronquist K, Newman S, Kitamura K, Campagnoni C, Drandall B (1987): Expression of myelin protein genes in the developing brain. *J Physiol Paris* 82:229–238

Cheroff G (1981): Shiverer: an autosomal recessive mutant mouse with myelin deficiency. *J Hered* 72:128–130

Csiza C, de Lahunta A (1979): Myelin-deficiency (md), a neurologic mutant in the Wistar rat. *Am J Pathol* 95:215–224

D'Eustachio P, Colman D, Salzer J (1988): Chromosomal location of the mouse gene that encodes the myelin-associated glycoproteins. *J Neurochem* 50:589–593

de Cozar M, Lucas M, Monreal J (1987): Ionophoric properties of the proteolipid apoprotein from bovine brain myelin. *Biochem Int* 14:833–841

de Waegh S, Brady S (1990): Altered slow axonal transport and regeneration in a myelin deficient mutant mouse: the trembler as an in vivo model for Schwann cell axon interactions. *J Neurosci* 10:1855–1865

Dentinger M, Barron K, Csiza C (1985): Glial and axonal development in optic nerve of myelin deficient rat mutant. *Brain Res* 344:255–266

Doolittle D, Schweikart K (1977): Myelin deficient, a new neurological mutant in the mouse. *J Hered* 68:331–332

Duncan I (1990): Dissection of the phenotype and genotype of the X-linked myelin mutants. *Ann NY Acad Sci* 605:110–121

Duncan I, Griffiths I, Munz M (1983): "Shaking pup:" A disorder of central myelination in the Spaniel dog. III. Quantitative aspects of glia and myelin in the spinal cord and optic nerve. *Neuropathol Appl Neurobiol* 9:355–368

Duncan I, Hammang J, Goda S, Quarles R (1989): Myelination in the jimpy mouse in the absence of proteolipid protein. *Glia* 2:148–154

Duncan I, Hammang J, Jackson K (1987a): Myelin mosaicism in female heterozygotes of the canine shaking pup and myelin-deficient rat mutants. *Brain Res* 402:168–172

Duncan I, Hammang J, Trapp B (1987b): Abnormal compact myelin in the myelin-deficient rat: absence of proteolipid protein correlates with a defect in the intraperiod line. *Proc Natl Acad Sci USA* 84:6287–6291

Dupouey P, Jacque C, Bourre J, Cresselin F, Privat A, Baumann N (1979): Immunocytochemical studies of myelin basic protein in shiverer mouse devoid of major dense line of myelin. *Neurosci Lett* 12:113–118

Fahim S, Riordan J (1986): Lipophilin (PLP) gene in X-linked myelin disorders. *J Neurosci Res* 16:303–310

Falconer D (1951): Two new mutants, "trembler" and "reeler" with neurological actions in the house mouse (Mus musculus L). *J Genet* 50:192–201

Frail D, Braun P (1985): Abnormal expression of the myelin-associated glycoprotein in the central nervous system of dysmyelinating mutant mice. *J Neurochem* 45:1071–1075

Fujita N, Sato S, Ishiguro H, Inuzuka T, Baba H, Kurihara T, Takahashi Y, Miyatake T (1990): The large isoform of myelin-associated glycoprotein is scarcely expressed in the quaking mouse brain. *J Neurochem* 55:1056–1059

Fujita N, Sato S, Kurihara T, Inuzuka T, Takahashi Y, Miyatake T (1988): Developmentally regulated alternative splicing of brain myelin-associated glycoprotein mRNA is lacking in the quaking mouse. *FEBS Lett* 232:323–327

Garbay B, Domec C, Fournier M, Bonnet J (1989): Developmental expression of the Po glycoprotein and basic protein mRNAs in normal and trembler mutant mice. *J Neurochem* 53:907–911

Garbay B, Fournier M, Sallafranque M, Muller S, Boiron F, Heape A, Cassagne C, Bonnet J (1988): Po, MBP, histone and DNA levels in sciatic nerve. Postnatal accumulation studies in normal and trembler mice. *Neurochem Pathol* 8:91–107

Gardinier M, Macklin W (1988): Myelin proteolipid protein gene expression in jimpy and jimpy (msd) mice. *J Neurochem* 51:360–369

Gardinier M, Macklin W, Diniak A, Deininger P (1986): Characterization of myelin proteolipid mRNAs in normal and jimpy mice. *Mol Cell Biol* 6:3755–3762

Gencic S, Abuelo D, Ambler M, Hudson L (1989): Pelizeaus-Merzbacher disease: an X-linked neurologic disorder of myelin metabolism with a novel mutation in the gene encoding proteolipid protein. *Am J Hum Genet* 45:435–442

Gencic S, Hudson L (1990):Conservative amino acid substitution in the myelin proteolipid protein of jimpy^{msd} mice. *J Neurosci* 10:117–124

Ginalski-Winkelmann H, Omlin F, Matthieu J (1983-84): Myelin basic protein deficit in the PNS of mld mutant mice recovers during development. *Dev Neurosci* 6:118–126

Greenfield S, Norton W, Morell P (1971): Quaking mouse: isolation and characterization of myelin protein. *J Neurochem* 18:2119–2128

Griffiths I, Duncan I, McCulloch M (1981a): "Shaking pups": a disorder of central myelination in the spaniel dog. II. Ultrastructural observations on the white matter of the cervical spinal cord. *J Neurocytol* 10:847–858

Griffiths I, Duncan I, McCulloch M, Harvey M (1981b): Shaking pups: a disorder of central myelination in the Spaniel dog. I. Clinical, genetic and light microscopical observations. *J Neurol Sci* 50:423–433

Griffiths I, Scott I, McCulloch M, Berrie J, McPhilemy K, Cattanach B (1990): Rumpshaker mouse: a new ex-linked mutation affecting myelination: evidence for a defect in PLP expression. *J Neurocytol* 19:273–283

Heape A, Juguelin H, Fabre M, Boiron F, Cassagne C (1986): A quantitative developmental study of the peripheral nerve lipid composition during myelinogenesis in normal and trembler mice. *Brain Res* 390:181–189

Henry E, Sidman R (1988): Long lives for homozygous trembler mutant mice despite virtual absence of peripheral nerve myelin. *Science* 241:344–346

Hogan E, Greenfield S (1984): Animal models of genetic disorders of myelin.In: *Myelin,* Morrell P, ed. New York: Plenum Press, pp 489–534

Hudson L, Friedrich V, Behar T, Dubois-Dalcq M, Lazzarini R (1989a): The initial events in myelin synthesis: orientation of proteolipid protein in the plasma membrane of cultured oligodendrocytes. *J Cell Biol* 109:717–727

Hudson L, Nadon N (1990): Amino acid substitutions in proteolipid protein that cause dysmyelination. In: *Myelin—A Treatise,* Martenson R, ed. Caldwell, NJ: Telford Press pp 259–310

Hudson L, Puckett C, Bernedt J, Chan J, Gencic S (1989b): Mutation of the proteolipid protein gene PLP in a human X chromosome-linked myelin disorder. *Proc Natl Acad Sci USA* 86:8128–8131

Inuzuka T, Duncan I, Quarles R (1986): Myelin proteins in the CNS of "shaking pups." *Brain Res* 392:43–50

Inuzuka T, Johnson D, Quarles R (1987): Myelin-associated glycoprotein in the central and peripheral nervous system of quaking mice. *J Neurochem* 49:597–602

Inuzuka T, Quarles R, Heath J, Trapp B (1985): Myelin-associated glycoprotein and other proteins in trembler mice. *J Neurochem* 44:793–797

Jacque C, Delassale A, Raoul M, Baumann N (1983): Myelin basic protein deposition in the optic and sciatic nerves of dysmyelinating mutants quaking, jimpy, trembler, mld and shiverer during development. *J Neurochem* 41:1335–1340

Juguelin H, Heape A, Boiron F, Cassagne C (1986): A quantitative developmental study of neutral lipids during myelinogenesis in the peripheral nervous system of normal and trembler mice. *Brain Res* 390:249–252

Katsuki M, Sato M, Kimura M, Yokoyama M, Kobayashi K, Nomura T (1988): Conversion of normal behavior to shiverer by myelin basic protein antisense cDNA in transgenic mice. *Science* 241:593–595

Kerner A, Carson J (1986): Shiverer and jimpy double mutant mice. I. Biochemical evidence for reciprocal intergenic suppression. *Brain Res* 374:45–53

Kimura M, Sato S, Akatsuka A, Nozawa-Kimura S, Takahashi R, Yokoyama M, Nomura T, Katsuki M (1989): Restoration of myelin formation by a single type of myelin basic protein in transgenic shiverer mice. *Proc Natl Acad Sci USA* 86:5661–5665

Kirschner D, Ganser A (1980): Compact myelin exists in the absence of basic protein in the shiverer mutant mouse. *Nature (Lond.) 283:207–210*

Kirschner D, Ganser A (1984): Diffraction studies of molecular organization and membrane interactions in myelin. In: *Myelin,* Morrell P, ed. New York: Plenum Press, pp 51–95

Knapp P, Skoff R, Redstone D (1986): Oligodendroglial cell death in jimpy mice. An explanation for the myelin deficit. *J Neurosci* 6:2813–2822

Koeppen A, Barron K, Csiza C, Greenfield E (1988): Comparative immunocytochemistry of Pelizaeus-Merzbacher disease, the jimpy mouse, and the myelin-deficient rat. *J Neurol Sci* 84:315–327

Koeppen A, Ronca N, Greenfield E, Hans M (1987): Defective biosynthesis of proteolipid protein in Pelizaeus-Merzbacher disease. *Ann Neurol* 21:159–170

Konat G, Trojanowska M, Gantt G, Hogan E (1988): Expression of myelin protein genes in quaking mouse brain. *J Neurosci Res* 20:19–22

Kronquist K, Crandall B, Macklin W, Campagnoni A (1987): Expression of myelin proteins in the developing human spinal cord: cloning and sequencing of human proteolipid protein cDNA. *J Neurosci Res* 18:395–401

Kurihara T, Takahashi Y, Fujita N, Sato S, Miyatake T (1989): Developmental expression of 2′, 3′-cyclic-nucleotide 3′-phosphodiesterase mRNA in brains of normal and quaking mice. *Brain Res Mol Brain Res* 5:247–250

Lachapelle F, Gumpel M, Baulac M, Jacque C, Duc P, Baumann N (1983): Transplantation of CNS fragments into the brain of shiverer mutant mice: extensive myelination by implanted oligodendrocytes. I. Immunohistochemical Studies. *Dev Neurosci* 8:325–334

Lai C, Brow M, Nave K, Noronha A, Quarles R, Bloom F, Milner R, Sutcliffe J (1987): Two forms of IB236/myelin-associated glycoprotein, a cell adhesion molecule for postnatal neural development, are produced by alternative splicing. *Proc Natl Acad Sci USA* 84:4337–4341

Lees B (1984): Proteins of myelin. In: *Myelin,* Morrell P, ed. New York: Plenum Press, pp 197–223

Lees M, Chao B, Lin L-F, Samiullah M, Laursen R (1983): Amino acid sequence of bovine white matter proteolipid. *Arch Biochem Biophys* 226:643–656

Lin L, Lees M (1982): Interactions of dicyclohexlcarbodiimide with myelin proteolipid. *Proc Natl Acad Sci USA* 79:941–945

Love P (1976): Hereditary hypertrophic neuropathy in the trembler mouse. Part 2. Histopathological studies: light microscopy. *J Neurol Sci* 30:343–368

Macklin W, Campagnoni C, Deininger P, Gardinier M (1987a): Structure and expression of the mouse myelin proteolipid protein gene. *J Neurosci Res* 18:383–394

Macklin W, Gardinier M, King K, Kampfi K (1987b): An AG GG transition at a splice site in the myelin proteolipid protein gene in jimpy mice results in the removal of an exon. *FEBS Lett* 223:417–421

Matthieu J, Omlin F, Ginalski-Winkelmann H, Cooper B (1984): Myelination in the CNS of mld mutant mice: Comparison between composition and structure. *Brain Res* 315:149–158

McFarlin D, McFarland H (1982): Multiple Sclerosis (first of two parts) *N Engl J Med* 307:1183–1188

Meier H, MacPike A (1970): A neurological mutation (msd) of the mouse causing a deficiency of myelin synthesis. *Exp Brain Res* 10:512–525

Mikoshiba K, Kohasaka S, Takamateuk T (1981): Neurochemical and morphological studies on the myelin of peripheral nervous system from shiverer mutant mice: absence of basic proteins common to the central nervous system. *Brain Res* 204:455–460

Milner R, Lai C, Nave K-A, Lenoir D, Ogata J, Sutcliffe J (1985): Nucleotide sequences of two mRNAs for rat brain myelin proteolipid protein. *Cell* 42:931–939

Mitchell L, Griffiths I, Brophy P (1990): Rumpshaker Mouse: a probable mutation of the PLP gene. *Ann NY Acad Sci* 605:394–397

Nadon N, Duncan I, Hudson L (1988): Molecular analysis of the shaking pup mutation. *J Neurosci* 14:829A

Nave K, Lai C, Bloom F, Milner R (1986): Jimpy mutant mouse: a 74-base deletion in the mRNA for myelin proteolipid protein and evidence for a primary defect in RNA splicing. *Proc Nat Acad Sci USA* 83:9264–9268

Norton W, Cammer W (1984): Chemical pathology of diseases involving myelin. In: *Myelin,* Morrell P, ed. New York: Plenum Press, pp 369–403

Okano H, Tamura T, Miura M, Hoyama A, Ikenaka K, Oshimura M, Mikoshiba K (1988): Gene organization and transcription of duplicated MBP genes of myelin deficient (shi mld) mutant mouse. *EMBO J* 7:77–83

Omlin F, Anders J (1983): Abnormal cell relationships in Jimpy mice: electron microscopic and immunocytochemical findings. *J Neurocytol* 12:767–784

Perkins C, Aguayo A, Bray G (1981): Schwann cell multiplication in trembler mice. *Neuropathol Appl Neurobiol* 7:115–126

Phillips R (1954): Jimpy, a new totally sex-linked gene in the house mouse. *Z Vererbungskjre* 86:322–326

Pollard J, McLeod J (1980): Nerve grafts in the trembler mouse: an electrophysiological and histological study. *J Neurol Sci* 46:373–383

Popko B, Puckett C, Hood L (1988): A novel mutation in myelin deficient mice results in unstable myelin basic protein gene transcripts. *Neuron* 1:221–225

Privat A, Jacque C, Bourre J, Dupouey P, Baumann N (1979): Absence of major dense line in the myelin of the mutant mouse shiverer. *Neurosci Lett* 12:107–112

Privat A, Valat J, Lachapelle F, Baumann N, Fulcrand J (1982): Radioautographic evidence for the protracted proliferation of glial cells in the central nervous system of jimpy mice. *Dev Brain Res* 2:411–416

Puckett C, Hudson L, Ono K, Friedrich J, Benecke M, Dubois-Dalcq M, Lazzarini R (1987): Myelin specific proteolipid protein is expressed in myelinating Schwann cells but not incorporated into myelin sheaths. *J Neurosci Res* 18:555–558

Raine C (1984a): Morphology of myelin and myelination. In: *Myelin,* Morrell P, ed. New York: Plenum Press, pp 1–50

Raine C (1984b): The neuropathology of myelin diseases. In: *Myelin,* Morrell P, ed. New York: Plenum Press, pp 259–310

Readhead C, Popko B, Takahashi N, Shine H, Saavedra R, Sidman R, Hood L (1987): Expression of a myelin basic gene in transgenic shiverer mice: correction of the dysmyelinating phenotype. *Cell* 48:703–712

Roach A, Boylank H, Prusiner S, Hood L (1983): Characterization of clones cDNA representing rat myelin basic protein: absence of expression in brain of shiverer mutant mice. *Cell* 34:799–806

Roach A, Takahashi N, Pravtcheva D, Ruddle F, Hood L (1985): Chromosomal mapping of mouse myelin basic protein gene and structure and transcription of the partially deleted gene in shiverer mutant mice. *Cell* 42:149–155

Rosenbluth J (1987): Abnormal axoglial junctions in the myelin-deficient rat mutant. *J Neurocytol* 16:497–509

Rosenbluth J, Hasegawa M, Schiff R (1989): Myelin formation in myelin-deficient rat spinal cord following transplantation of normal fetal spinal cord. *Neurosci Lett* 97:35–40

Rosenfeld J, Friedrich V Jr (1986): Oligodendrocyte production and myelin recovery in heterozygous jimpy mice: an autoradiographic study. *Int J Dev Neurosci* 4:179–187

Roth H, Hunkeler M, Campagnoni A (1985): Expression of myelin basic protein genes in several dysmyelinating mouse mutants during early postnatal brain development. *J Neurochem* 45:572–580

Roussel G, Neskovic N, Trifilieff E, Artault J, Nussbaum J (1987): Arrest of proteolipid transport through the Golgi apparatus in Jimpy brain. *J Neurocytol* 16:195–204

Samorajski T, Freide R, Reiner P (1970): Hypomyelination in the quaking mouse: A model for the analysis of disturbed myelin formation. *J Neuropathol Exp Neurol* 29:507–523

Sandru L, Siegrist H, Wiesmann U, Herchkowitz L (1980): Development of oligodendrocytes in jimpy brain cultures. In: *Neurological Mutations Affecting Myelination,* Baumann N, ed. Elsevier / North-Holland, New York and Amsterdam, pp 469–474

Sheedlo H, Siegel G (1986): Myelin associated glycoprotein (MAG) in the CNS of adult shiverer (Shi/Shi) mouse. *J Neurosci Res* 16:629–641

Shen X, Billings-Gagliardi S, Sidman R, Wolf M (1985): Myelin deficient (shi mld) mutant allele: morphological comparison with shiverer (shi) allele on a B6C3 mouse stock. *Brain Res* 360:235–247

Sidman R, Dickie M, Appel S (1964): Mutant mice (quaking and jimpy) with deficient myelination in the central nervous system. *Science* 144:309–310

Simons R, Riordan J (1990): The myelin-deficient (*md*) rat has a single base substitution in exon 3 of the myelin proteolipid protein gene. *J Neurochem* 54:1079–1081

Skoff R (1982): Increased proliferation of oligodendrocytes in the hypomyelinated mutant mouse jimpy. *Brain Res* 248:19–31

Skoff R, Knapp P (1990): Expression of the jimpy phenotype in relation to proteolipid protein appearance. *Ann NY Acad Sci* 605:122–134

Sorg B, Agrawal D, Agrawal H, Campagnoni A (1986): Expression of myelin proteolipid protein and basic protein in normal and dysmyelinating mutant mice. *J Neurochem* 46:379–387

Tosic M, de Rivaz J, Roch J, Roach A, Matthieu J (1989): Expression of the downstream copy of the myelin basic protein gene in myelin deficient mutant mice. *J Neurochem* 52(suppl 1):S46

Tosic M, Roach A, de Rivaz J, Dolivo M, Matthieu J (1990): Post-transcriptional events are responsible for low expression of myelin basic protein in myelin deficient mice: role of natural antisense RNA. *EMBO J* 9:401–406

Trapp B (1988): Distribution of the myelin-associated glycoprotein and Po protein during myelin compaction in quaking mouse peripheral nerve. *J Cell Biol* 107:675–685

Trapp B, Quarles R, Suzuki K (1984): Immunocytochemical studies of quaking mice support a role for the myelin-associated glycoprotein in forming and maintaining the periaxonal space and periaxonal cytoplasmic collar of myelinating Schwann cells. *J Cell Biol* 99:594–606

Trofatter J, Dlouhy S, De Myer W, Conneally P, Hodes M (1989): Pelizaeus–Merzbacher disease: tight linkage to proteolipid protein gene exon variant. *Proc Natl Acad Sci USA* 86:9427–9430

Ulrich J, Matthieu J, Herschkowitz N, Kohler R, Heitz P (1983): Immunocytochemical investigations of murine leukodystrophies. A study of the mutants 'jimpy' (jp) and 'myelin deficient' (mld). *Brain Res* 268:267–274

Waksman B (1988): Autoimmunity in demyelinating diseases. *Ann NY Acad Sci* 540:13–24

Watanabe I, Bingle G (1972): Dysmyelination in "quaking" mouse: electron microscopic study. *J Neuropathol Exp Neurol* 31:352–369

Weiss M, Greenfield S, Brostoff S, Hogan E (1983): Protein composition of PNS myelin: developmental comparison of control and quaking mice. *J Neurochem* 41:448–453

Willard H, Riordan J (1985): Assignment of the gene for myelin proteolipid protein to the X-chromosome in man and mouse: implications for X-linked inherited disorders of myelin. *Science* 230:940–942

Wisniewski H, Morell P (1971): Quaking mouse: ultrastructural evidence for arrest of myelinogenesis. *Brain Res* 29:63–73

Wolf M, Kardon G, Adcock L, Billings-Gagliardi S (1983): Hypomyelinated mutant mice. V. Relationship between jp and jpmsd re-examined on identical genetic backgrounds. *Brain Res* 271:121–129

Yanagisawa K, Duncan I, Hammang J, Quarles R (1986): Myelin-deficient rat: analysis of myelin proteins. *J Neurochem* 47:1901–1907

Yanagisawa K, Moller J, Duncan I, Quarles R (1987): Disproportional expression of proteolipid protein and DM-20 in the X-linked, dysmyelinating shaking pup mutant. *J Neurochem* 49:1912–1917

Yanagisawa K, Quarles R (1986): Jimpy mice: quantitation of myelin-associated glycoprotein and other proteins. *J Neurochem* 47:322–325

Yao J, Bourre J (1985): Metabolic alterations of endoneurical lipids in developing trembler nerve. *Brain Res* 325:21–27

Zeller N, Dubois-Dalcq M, Lazzarini R (1989): Myelin protein expression in the myelin-deficient rat brain and cultured oligodendrocytes. *J Mol Neurosci* 1:139–149

Zeman W, DeMyer W, Falls H (1964): Pelizeaus–Merzbacher disease. A study in nosology. *J Neuropathol Exp Neurol* 23:334–354

Zhabotinskii I, Chumasov E, Shefer V (1983): Hereditary demyelination in mutant quaking mice (breeding and light-microscopy). *Arch Pathol* 45:48–53

Zimmerman T, Cammer W (1983): Elevated specific activity of 5′ nucleotidase in a spinal cord myelin fraction from shiverer mice. Comparison with other myelin-associated enzymes and myelin proteins. *Biochem Biophys Acta* 760:377–380

Genetic Mutations Affecting Murine Cerebellar Structure and Function

DAN GOLDOWITZ AND LEONARD M. EISENMAN

Single gene mutations that affect central nervous system (CNS) structure and, in turn, behavior offer the clearest example of the involvement of genes in the development and function of the brain.

A relatively large number of genetic loci have been identified that affect the structural integrity of the mouse cerebellum (Sidman et al., 1965; Lyon and Searle, 1989). This, no doubt, is due to the ease of noticing a mutant phenotype that includes abnormalities of motor activity, the primary functional realm of the cerebellum. These mutations in the mouse provide the most comprehensive series of mutant loci (alleles) affecting a single mammalian CNS structure—allowing an unusual opportunity to dissect the genetic underpinnings of structure and function (Caviness and Rakic, 1978; Sotelo, 1980). The cerebellum is an ideal structure to pursue the study of genetically defined models of nervous system morphogenesis and function because it is relatively easy to identify (1) new mutations by abnormalities in motor behavior, and (2) the concomitant mutant cellular phenotype due to the simple and ordered cytoarchitecture of the cerebellum.

This chapter will discuss this thesis in three respects:

1. a brief introduction of the cellular biology of cerebellar development in the context of some of the neurological mutations found in mice
2. the use of cerebellar mutants to dissect the components of cerebellar circuitry that are related to normal and abnormal motor control
3. the use of mutant mice to explore the involvement of the cerebellum in associative behaviors.

CEREBELLAR DEVELOPMENT AND NEUROLOGICAL MUTANT MICE

Brief Introduction to the Normal Cellular and Developmental Biology of the Cerebellum

Several excellent monographs have appeared on the development of the cerebellum, and this brief introduction is borrowed from their pages (Larramendi, 1969; Ramón y Cajal, 1960; Miale and Sidman, 1961; Fujita, 1969; Altman and Bayer, 1985a,b,c).

Cell Generation and Migration

As with the rest of the brain, the cerebellum begins as a proliferative neuroepithelium (see Fig. 4.1). This neuroepithelium is located at the rhombic lip, a region of the hindbrain that sits above the fourth ventricle just posterior to the inferior colliculus. This ventricular region gives rise to the large neurons of the cerebellum, that is, the Purkinje, nuclear, and Golgi cells (but see Hausmann et al., 1985). Results from recent work using chick–quail transplantation chimeras have found that the Purkinje, basket, Golgi, stellate, and possibly granule cells that populate the anterior portion of the cerebellum all have a common origin in the mesencephalic vesicle (Martinez and Alvarado-Mallart, 1989; Hallonet et al., 1990). However, the common origin of Purkinje and molecular layer neurons found in bird chimeras is not in agreement with conclusions drawn from ^{3}H-thymidine birthdating studies in the rodent (e.g., Miale and Sidman, 1961; Altman and Bayer, 1985c).

The nuclear and Purkinje cells migrate outward to populate the cerebellum. The nuclear cells, which are produced first, occupy a mass in the superficial part of the cerebellar anlage, the nuclear transitory zone. The Purkinje cells, which begin to be produced shortly before the end of nuclear cell production, form the deeper cortical transitory zone. These events start to occur at day 12 of gestation (E12) with the day of impregnation designated as E0 (the normal gestational period of mice, *Mus musculus domesticus,* is 19 days). In later embryogenesis, Purkinje cells migrate to form a several cell–thick outer cortical plate that transforms during early postnatal development into a cellular monolayer, the Purkinje cell layer (see Fig. 4.1).

The granule cells (the most numerous cell in the CNS) of the cerebellum come from a secondary proliferative zone, the external granular layer (EGL), a feature that is unique to the cerebellum (Fig. 4.1). In the rodent it is believed that the other cerebellar microneurons also originate from this neuroepithelium, although recent work with chick–quail chimeras casts doubt on the generality of this premise. In rodents, the EGL arises from the posterolateral aspect of the cerebellar anlage, the germinal trigone. Cells from the germinal trigone begin to migrate over the cerebellar surface at E13. By E17, the entire cerebellar surface is covered by the proliferating neuroblasts of the EGL. The EGL persists for about 3 weeks of postnatal life, giving rise to postmitotic populations of neurons that migrate to their appropriate addresses: for the stellate and basket cells this is the molecular layer, whereas for the granule cell it is the internal granular layer (IGL) situated deep to the Purkinje cells. The Bergmann glia fibers, with radial processes extending from the Purkinje cell layer to the pial surface, are believed to provide the guides for granule cell migration to the internal granular layer (Rakic, 1971).

Purkinje Cell Development and Synaptogenesis

The largest part of differentiation takes place postnatally, as the synaptogenic period is intimately related to dendritic maturation. This is not to ignore the presence of early synaptic contacts in the late prenatal rodent cerebellum, but the

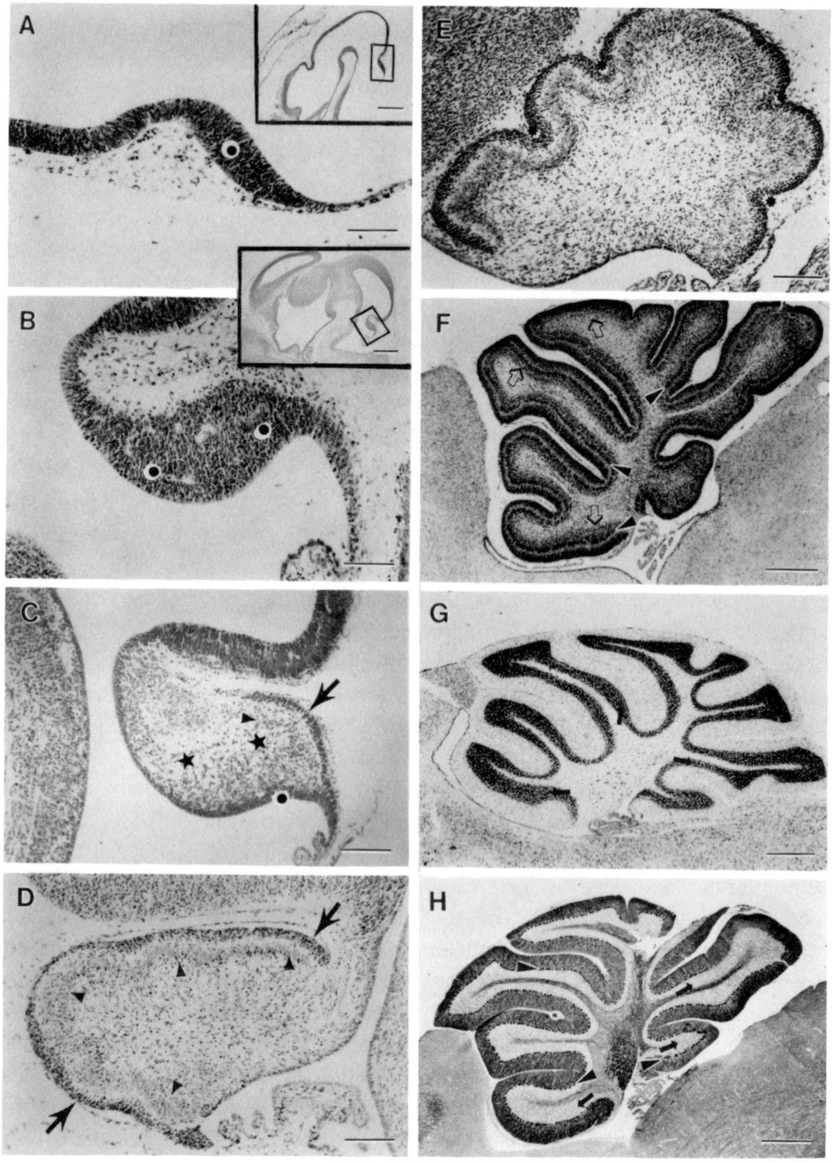

FIGURE 4.1. Photomicrographs of mouse cerebellar development. The cerebellum is shown at embryonic ages E11 **(A)**, E13–14 **(B)**, E15 **(C)**, E17 **(D)** and postnatal ages P0 **(E)**, P11 **(F)**, and adult **(G,H)**. At early embryonic ages **(A,B)** the cerebellum is principally constituted by a proliferative neuroepithelial layer *(dots)*. Insets in the upper right aspect of **A** and **B** show a low magnification view of the developing brain and illustrate the region shown at higher magnification *(boxed region)*. By E15 **(C)** there are regions clearly occupied by the Purkinje cells (*stars* and *arrowhead*) and nuclear cells. The nuclear cells occupy a nuclear transitory region (to the *left of the arrowhead* and *above the stars*). The

presynaptic and postsynaptic partners were unable to be identified with certainty (West and Del Cerro, 1976). The Purkinje cell, as the only efferent cell of the cerebellar cortex, is most dramatically altered during the course of postnatal development. Ramón y Cajal elegantly described the morphogenesis of the Purkinje cell (Ramón y Cajal, 1952, 1960). At birth these large neurons are clustered in a multilayered array and each cell is multipolar in shape. Postnatal development of these cells, although occurring in a similar sequence in all regions of the cerebellum, does not occur simultaneously but rather is advanced in some regions (e.g., the depths of the primary fissure) and delayed in other regions (e.g., hemispheres and apices of the vermal folia) (Heckroth et al., 1990; Goodlett et al., 1990). Between birth and P5 the multilayered arrangement is transformed into a monolayer and the cells exhibit long dendrons extending in all directions, with some extending up through the thin molecular layer to the EGL. One day later, at P6, the Purkinje cell soma has become smoother but still exhibits somatic spines. In addition, a dendritic arbor appears that usually arises from a single dendritic trunk originating from the apical surface of the soma. Over the succeeding days there is an increase in the height and branching of the Purkinje cell dendrite and a concomitant increase in the thickness of the molecular layer.

The two major afferents to the Purkinje cell, the parallel and climbing fibers, begin synaptogenesis in the perinatal period. Axons of granule cells, the parallel fibers, grow and form contacts with the tertiary branchlets of the Purkinje cell after the postmitotic differentiation of the first granule cells (Altman, 1972). This process continues until all the granule cells leave the EGL and settle in the IGL. The other major afferent to the Purkinje cell, the climbing fiber, originates entirely from the inferior olivary complex (Desclin, 1974). Recent evidence indicates that this afferent is present within the cerebellum at a very early age (E15–16) and

---

stars mark the cortical transitory zone. In addition, the arrow marks the emerging EGL. The EGL *(arrows)* has circled the perimeter of the cerebellum by E17 **(D)** with the Purkinje cells *(arrowheads)* forming a plate of cells deep to the EGL (arrows). Folia (asterisks) become obvious features of the cerebellum at P0 **(E)** and the anterior-to-posterior and dorsal-to-ventral dimensions of the cerebellum increase. **(F)** The EGL cells start to migrate deep to the Purkinje cells and form an internal granular layer *(open arrows),* and the Purkinje cells form a monolayer accompanied by the outgrowth of dendrites *(arrowheads).* These processes continue for about 2 more weeks, resulting in the appearance of the adult cerebellum **(G,H).** The mature cerebellum has an IGL *(arrows)* that is densely packed with granule cells **(G),** the Purkinje cells *(arrowheads)* occupy a monolayer, and their dendrites ramify into the molecular layer **(H).**

All material was embedded in paraffin, cut at a thickness of about 6 μm, and, with the exception of **H,** stained with cresyl violet. The material in **F** and **H** was stained with a Purkinje cell–specific antibody. In all photomicrographs the anterior cerebellum is to the left and the posterior cerebellum and associated choroid is to the right with the exception of **D,** which is reversed. The calibration bars have the following equivalents: **A–C** = 80 μm, **D–E** = 125 μm, **F** = 300 μm, **G–H** = 500 μm. The calibration bar of the inset in **A** = 370 μm and **B** = 600 μm.

forms simple synaptic contacts with Purkinje cells that are maintained until P2 (Mason and Blazeski, 1989). Further development of the climbing fiber involves the maturation of contacts on the processes emanating from the Purkinje cell soma. Over the course of the next few days, as the primary dendrite of the Purkinje cell begins to form, there is a delay between the main spurt of dendritic development that begins at P6 and the transposition of the climbing fiber from soma to dendrite that begins at P8. Following this cap stage, the climbing fiber continues to grow and ascends the dendrites until it entwines the primary and secondary branches of the Purkinje cell dendritic tree. The early postnatal development of the climbing fiber–Purkinje cell relationship is characterized by a multiple innervation of each Purkinje cell by climbing fibers. During further development this relationship changes until each adult Purkinje cell is innervated by a single climbing fiber (although each climbing fiber may innervate a number of Purkinje cells) (Ramón y Cajal, 1952, 1960; Eccles et al., 1966).

Genetic Mutations and the Deciphering of Development

As described above, work on cerebellar development has identified a series of stages through which this structure progresses to arrive at the trilaminar cortex and subcortical white matter nuclei that are characteristic of the mature cerebellum. A goal for the developmental neurobiologist is to establish a concordance between our hypothesized "key events" in development with the real developmental milestones traversed by the organism.

Neurological mutations allow one to determine the genetic and epigenetic instructions that actually regulate development. To use these genetic mutations in such a fashion it is imperative to determine the nature of gene action (Gruneberg, 1963). One standard approach to this problem relies on determining the temporal and spatial definition of aberrant phenotypes. The cell(s) that display the earliest detectable mutant phenotype is (are) logically considered as the most likely primary target(s) of the mutant gene. Many of the guesses, thus, come from a careful study of development in the mutant and our biases of "movers" in development. Excellent examples of this approach are the studies of the weaver (Rakic and Sidman, 1973a,b; Hirano and Dembitzer, 1973; Sotelo, 1975; Smeyne and Goldowitz, 1989), reeler (Mariani et al., 1977; Goffinet, 1983), staggerer (Landis and Sidman, 1978; Yoon, 1972, 1976), lurcher (Swisher and Wilson, 1977; Caddy and Biscoe, 1979; Heckroth et al., 1990), and nervous (Sotelo and Triller, 1979), to name but a few.

These studies do well to define the temporal progression of the mutant phenotypes, but can at best only speculate as to the cellular target(s) of the mutant gene. Another experimental approach that permits more conclusive insight into the cellular site of gene action is the use of experimental mouse chimeras. To produce these experimental mice, 8-cell embryos from a mutant mouse mating and a wild-type mating are combined and transferred to the uterus of a pseudopregnant female (Tarkowski, 1961; Mintz, 1964; McLaren, 1976). The mice that arise from this manipulation are composed of cells from both genotypes. The outcome of the interactions between these cells of the two genotypes provides evidence as to the cellular target(s) of the mutant gene. Any of the abnormal phenotypes expressed in the mutant can be examined in this manner, and if not shown to be an intrinsic target of the gene, may be in the cascade of secondary effects that follow the primary site(s) of gene action. For example, experimental chimeras have been used to identify the Purkinje cell as a direct target of the *pcd* (Mullen, 1977a), staggerer (Herrup and Mullen, 1979b), lurcher (Wetts and Herrup, 1982a,b), and weaver (Smeyne and Goldowitz, 1990) mutations. Still other studies using chimeras have indicated that Purkinje cell deficits in the reeler and nervous mice do not seem to be a direct (intrinsic) effect of these mutant genes (Mullen, 1977a, 1982; Mikoshiba et al., 1985; Terashima et al., 1986).

A cell biological and developmental timeframe can be established by combining this information with the descriptive details of mutant development to achieve a fairly rigorous, synthetic approach to genes and cerebellar development (see the section that follows and Fig. 4.3).

Neurological Mutations in Mice that Affect Cerebellar Development

There are several highly informative publications and chapters that catalogue and describe the various mutations that affect the cerebellum (Sidman et al., 1965; Sidman, 1968, 1987; Lyon and Searle, 1989; Landis and Landis, 1978). What we set out to do here is approach these mutants from the vantage point of a cerebellum that is to be constructed, considering the various cellular elements and developmental events that shape this structure and the mutations that disturb its development.

Thus, the objective in this section is to present neurological mutant mice in a contiuum of time and space. This has two purposes: (1) to further our understanding of the role of genetic information in laying down neuronal architecture, and (2) to provide insights into how these mutations can be used as tools to study cerebellar development. The end result is the construction of a linear plot of the timing of cerebellar development and the believed time of action of each mutation, with information regarding the primary cellular defect(s) that characterize the mutant phenotypes (Fig. 4.2).

Cell Proliferation and Early Differentiation

Mutant mice illuminate events critical to the production and development of cells that populate and form the cerebellum. Based on cell counts made in adult animals, certain genetic mutations have been identified that affect the production and/or continued survival of Purkinje cells. In reeler (Heckroth et al., 1988), staggerer (Herrup and Mullen, 1979b), and weaver (Blatt and Eisenman, 1985), the populations of Purkinje cells are markedly reduced. It is felt that these losses of Purkinje cells occur during the embryonic or early postnatal period as there has been no evidence to indicate a later postnatal loss of Purkinje cells. In weaver (gene symbol, *wv*), where this question has been examined in a preliminary

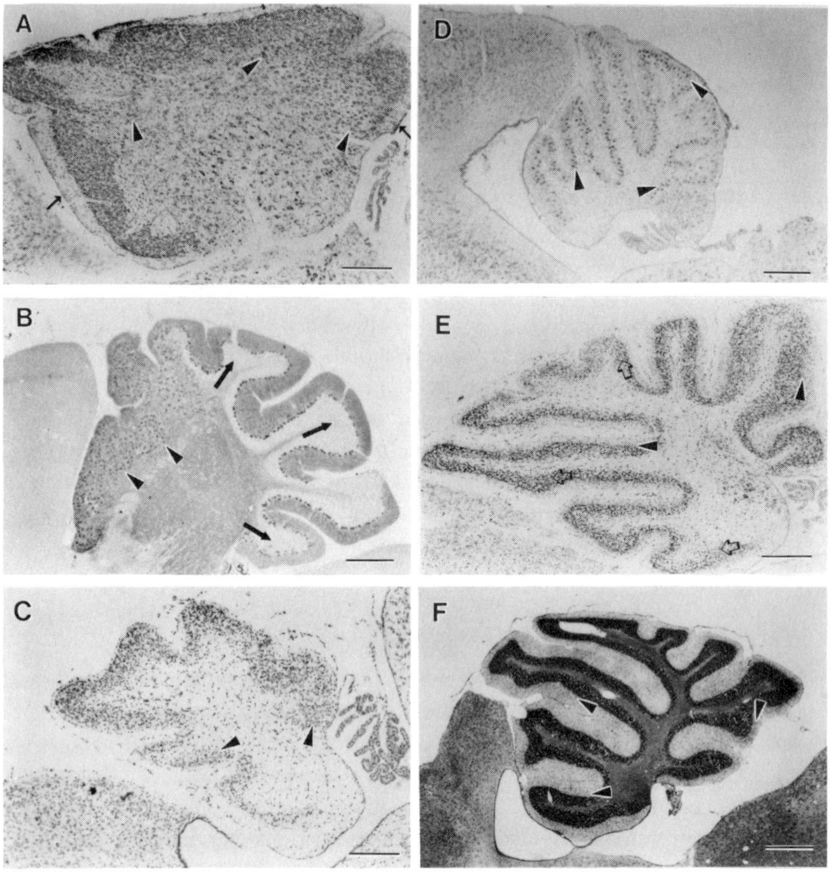

FIGURE 4.2. Examples of cerebellar morphology in various mutant mice: (A) reeler, **(B)** meander tail, **(C)** staggerer, **(D)** weaver, **(E)** lurcher, and **(F)** Purkinje cell degeneration. Each micrograph is variably enlarged to help illustrate the mutant phenotype. **A:** An off-midline sagittal section of a P15 homozygous reeler cerebellum. Note the thin EGL *(arrows)* and the ectopic masses of Purkinje cells *(arrowheads)*. **B:** An off-midline sagittal section of an adult meander tail cerebellum. This tissue has been immunocytochemically processed to illustrate the presence of large clusters of Purkinje cells *(arrowheads)* in the anterior lobules. The unstained granule cells that are present in normal numbers in the posterior lobules *(arrows)* are distinctly absent in the anterior lobules. **(C):** A parasagittal section of a postnatal day 35 staggerer cerebellum. Compare calibration bars to Fig. 4.1**G,H** or 4.2**F** to appreciate the dramatic decrease of cerebellar size. The granule cell population is almost totally depleted and there is a marked loss (far more than in weaver or reeler) of Purkinje cells *(arrowheads)*. **D:** A midsagittal section of an adult weaver cerebellum. The Purkinje cells *(arrowheads)* are highlighted due to the absence of granule cells. **E:** An off-midline sagittal section of a P23 lurcher cerebellum. The Purkinje cell layer *(arrowheads)* is missing most Purkinje cells and there is a decreased density of granule cells *(open arrows)*. **F:** A midline sagittal section of a 3-month-old Purkinje cell degeneration mutant. Purkinje cells are absent (although not apparent in this low power photomicrograph), and in their place are accumulations of glia cells *(arrowheads)*. At this age the density of granule cells appears relatively normal. The calibration bars allow a comparison of actual sizes (**A,C,E** = 200 μm; **B,D** = 300 μm; **F** = 500 μm). With the exception of **B,** all sections are stained with cresyl violet to demonstrate cellular details. In **B,** tissue is stained with a Purkinje cell–specific antibody.

fashion, the loss of Purkinje cells is apparent at postnatal day 4 (Rezai and Yoon, 1972). It is not known if this is due to a defective production of Purkinje cells or postmitotic event that negatively impacts cell survival (e.g., cell migration). As a contributor to the loss of cells, or independent of that phenomenon, mutations may cause a retardation of the developmental program. The best evidence for this comes from the staggerer mutant. In the cerebellum of this mutant several cell surface components that are markers of development have been shown to be locked in an immature state (Hatten and Messer, 1978; Trenkner, 1979; Wille and Trenkner, 1981). The link between developmental arrest and the expression of the staggerer gene is an intriguing but, as yet, unsolved problem.

Many of the mutations that have an obvious impact on the Purkinje cell [staggerer (gene symbol, *sg*), lurcher (gene symbol, *Lc*), and Purkinje cell degeneration (gene symbol, *pcd*)] are believed to have secondary effects in the death of granule cells, indicating the reliance of granule cells on interactions with Purkinje cells. This reliance is most clear when Purkinje cells are not available as a synaptic target for granule cells. It is also possible that the proliferation of granule cell precursors relies on trophic support from the underlying Purkinje cells (e.g., see interpretive studies concerning the abnormal EGL in the staggerer mouse by Messer and Hatch, 1984; Sonmez and Herrup, 1984).

The granule cell population may be a direct target for certain mutant genes. In meander tail and weaver there is a marked deficit in the granule cell population that, temporally, does not appear to be linked to deficits in other cells types.

In the autosomal recessive mutation meander tail (gene symbol, *mea*) the granule cells of the anterior lobe are decimated (Sidman, 1987; Ross et al., 1990), dying before egress to the internal granular layer. The Purkinje cells remain in a multilayered clump with stunted dendrites (Fig. 4.2). In meander tail there is rudimentary foliation in the affected area, somewhat reminiscent of the reeler mutant.

The best documented case of a mutation intrinsically affecting the granule cell is that of weaver (Rakic and Sidman, 1973a). In the vermal portion of the cerebellum there is a near total loss of granule cells (Fig. 4.2). Granule cell loss is attenuated in the hemispheres. The weaver mutation was originally considered to be a problem of granule cell migration associated with abnormal glia [based on findings by Rezai and Yoon (1972) on the delay in migration in heterozygous mice, and in the studies indicating an abnormality in weaver Bergmann glia by Rakic and Sidman (1973a,b)]. It now seems clear, however, that the granule cells of the cerebellar vermis are a direct target of the weaver gene (Goldowitz and Mullen, 1982; Hatten et al., 1986; Goldowitz, 1989). The glial defect appears to be secondary to the defective granule cells (Goldowitz, 1989). The congenital loss of cerebellar granule cells during human development is seen in the syndrome of granule cell hypoplasia (Harding, 1984).

The autosomal recessive mutation reeler (gene symbol, *rl*) has been a fascinating example of a genetic element that affects the development of the brain in a global manner (Caviness, 1977), and may be responsible for a generalized decrease in the production of cortical neurons (see Heckroth et al., 1988 for such a hypothesis, although the key experiments to test this point have yet to be carried

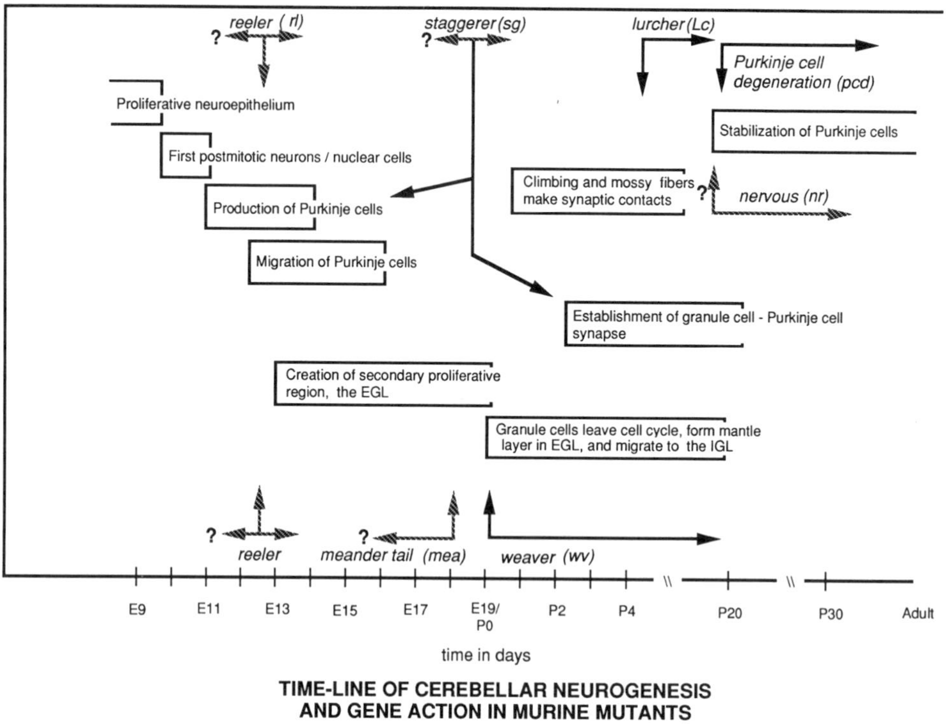

FIGURE 4.3. Time line of cerebellar neurogenesis and gene action in murine mutants. The focus of this time line is on the two dominant cell types of the cerebellum, the Purkinje and granule cells. The timing of Purkinje cell development and the genes that affect that development are shown in the upper part of the figure. Granule cell developmental epochs and genes that affect that development are highlighted in the bottom of the figure. The solid arrows indicate well documented findings, either toward the cell type (or cellular process) in the vertical direction or the time of action in the horizontal direction. The stippled arrows reflect questionable cellular sites or times *(question marks)* of gene action.

out). In reeler the EGL is much less extensive than in wild-type mice; there is far less surface area to cover, and the thickness of the EGL is no greater. Although this has not been documented in a quantitative manner, this zone appears to give rise to a much smaller than normal granule cell population. Abnormal cellular patterns are evident during embryogenesis (around E14; Goffinet, 1983). In addition to the apparent decreases in cell number noted above, the development of cerebellar foliation is rudimentary. Although there have been numerous hypotheses of how the reeler gene acts to create the mutant phenotype, there is no clear answer at present. The most viable suggestion appears to arise from work on the cerebral cortex that suggests an abnormality in radial glial–neuronal interactions is responsible for the reeler defect (Pinto-Lord et al., 1982; Gadisseux and Evrard, 1985). Indeed, the analysis of reeler ↔ wild-type chimeras indicates that a single affected cell type cannot explain the results in the chimeric cerebellum or cerebral cortex (Mullen, 1977b; Mikoshiba et al., 1985). The reeler phenotype is more likely caused by aberrant interactions between neurons or neurons and glia (e.g., see Goffinet, 1984; Terashima et al., 1986). The problem of how these defects may result in decreased numbers of cortical cells has not been addressed. In any case, human counterparts of the murine reeler syndrome have been described (Bordarier et al., 1986; Caviness, 1980).

Cell Maintenance

Several genes have been identified that are critical for the maintenance of Purkinje cells, that is, after birth and migration have successfully occurred. These are the dominant mutation lurcher *(Lc),* the recessive mutations Purkinje cell degeneration *(pcd),* nervous *(nr),* and the recently described mutant mouse tambaleante *(tbl)* (Wassef et al., 1987). In lurcher, cells begin to die about 1 week postnatally, and cell death continues for the next 2 months until virtually all of the Purkinje cells die (Caddy and Biscoe, 1979). In *pcd* and *nr,* Purkinje cell death starts during the weaning period and progresses for some time afterward (Mullen et al., 1976; Landis and Mullen, 1978; Landis, 1973). As noted earlier, the lurcher and *pcd* mutations are known to act intrinsic to the Purkinje cells, whereas the mode of action of the *nr* gene has not yet been determined (Mullen, 1982). What is particularly intriguing about the *pcd, nr,* and *tbl* mutants is that Purkinje cells die in reproducible (specific for each mutant) longitudinal patterns (Wassef et al., 1987). This is most striking in the nervous mouse, where discrete parasagittal bands of Purkinje cells survive for some time. These patterns of cell deletion just might offer a novel means to dissect the functional significance of the parasagittal banding of cells and afferents in the cerebellum.

In all of these mutants there is also a varying loss of granule cells. The death of granule cells seems to follow the initial death of Purkinje cells. The earlier the loss of Purkinje cells (as in lurcher), the greater the loss of granule cells or, conversely, as Purkinje cells are lost later in development the granule cell population is less affected. These findings fit well with the large experimental literature that has shown that granule cell survival is critically dependent on postnatal interactions with the Purkinje cell.

There are more mutations that affect the survival and dendritic development of the Purkinje cell; these mutants, however, have unclear temporal and cellular definition of gene action and/or include a more widespread (pleiotropic?) effect on the cerebellum. These mutants include leaner (an allele of tottering) (Herrup and Wilczynski, 1982), stumbler (Caddy et al., 1981; Caddy and Sidman, 1981), hyperspiny Purkinje cell (Guenet et al., 1983), and brindled (Yamano and Suzuki, 1985) as well as several others (Sidman, 1987).

Other Cells in the Cerebellum as Targets of Mutant Genes

Only a few studies have looked at other neuronal types in the cerebellar mutants to determine how mutant gene action affects their development (e.g., see Rakic and

Sidman, 1973b; Sotelo, 1975, Landis and Sidman, 1978). In these studies there has been no evidence for major disruptions of the Golgi, basket, or stellate cells, although there were developmental changes such as incomplete dendritic maturation. In terms of glia, there is no clear evidence to impute the Bergmann glia cell as a target for any of the known mutations, although abnormal adhesivity between glia and neurons has been hypothesized in the reeler mutant (see above).

Cell–Cell and Afferent–Target Interactions

Neurological mutations present a means, with the acuity and reproducibility of nature, to delete a target population and study the cellular and synaptic rearrangements that follow such a perturbation. Such an analysis presents an elegant, unique way to dissect the individual events that work to create the cerebellum (Rakic, 1976; Sotelo, 1980).

The granule cell's parallel fiber synapse onto the Purkinje cell is a key player in the blossoming of the Purkinje cell dendritic arbor and is important in the continued survival of the granule cell. In the staggerer mouse, there is an absence of tertiary spine formation, a specific absence of almost all parallel fiber–Purkinje cell contacts, and a massive degeneration of granule cells (Sotelo and Changeux, 1974; Landis and Sidman, 1978). There is good evidence from experimental chimeras that the Purkinje cell is the primary target of the mutant gene (Herrup and Mullen, 1979b), and that the granule cell is only secondarily affected (Herrup, 1983) [although see Yoon (1976) for another viewpoint]. Thus, the staggerer mutation provides an excellent model in which to study the role of the postsynaptic element in synapse formation.

Mutant mice also provide models to examine the reverse relationship; that is, what happens to the postsynaptic element in the absence of granule cells. In the weaver mutant, despite the absence of granule cells in the vermal portion of the cerebellar cortex, Purkinje cells develop membrane thickenings (Hirano and Dembitzer, 1973).

Due to the temporal variability of the different mutations, in terms of when the Purkinje cell is affected, it is possible to analyze the sequence of events that are important in establishing the relationship between the granule cell (and its parallel fiber) and the Purkinje cell. In the staggerer it appears that the presynaptic element forms but there is an absence of Purkinje cell tertiary spines, the postsynaptic site. As a consequence, synaptogenesis does not occur, and there is near total, retrograde granule cell death. Removal of the postsynaptic sites during the process of synaptogenesis, as occurs in the lurcher (Caddy and Biscoe, 1979), results in the partial loss of granule cells, whereas the loss of the postsynaptic partner well after the period of synaptogenesis, as occurs in the *pcd* and nervous mutants, leaves the presynaptic cell population relatively intact (see Fig. 4.2F).

These differential effects of the timing of target cell loss on the presynaptic population of granule cells strongly suggest that there is a critical period in development during which there occurs a stabilization of the presynaptic population, in the face of the loss of the postsynaptic target. This issue has been ingeniously probed using chimeras made between wild-type mice and mutants that have intrinsic defects in the Purkinje cells. The resulting mice have a varying number of surviving wild-type Purkinje cells, along with varying amounts of granule cells. In staggerer chimeras there was found to be a linear relationship between wild-type Purkinje cells and granule cells, nicely supporting a numerical matching hypothesis (Herrup and Sunter, 1987). This was not found to be the case, however, in lurcher↔wild-type chimeras, where there was an increased survival of granule cells (especially in chimeras that were composed largely of mutant cells) (Wetts and Herrup, 1983; Vogel et al., 1989). Important differences in these mutants are the time and mode of mutant gene action: in staggerer there is an early effect on Purkinje cells and no formation of granule cell–Purkinje cell contacts whereas in lurcher there is a later effect on Purkinje cells allowing an initial interaction between granule and Purkinje cells (Heckroth et al., 1990). Additional factors during synaptogenesis, such as competition between afferents, may also regulate the final population of presynaptic neurons (Vogel and Herrup, 1989).

Mutant mice have provided important tools to study the development of the climbing fiber–Purkinje cell synaptic relationship. In a number of mutations, in which there is an early reduction in the number of Purkinje cells, the immature state of multiple innervation of Purkinje cells by climbing fibers is retained in the adult mutant (Crepel et al., 1980; Mariani, 1982). In the lurcher mutant mouse, there is a most interesting defect in the relationship between the climbing fiber and the Purkinje cell (Heckroth et al., 1990). Climbing fibers in lurcher form contacts with the somatic spines of Purkinje cells, failing to relinquish these immature contacts and climb onto the dendritic tree. Since the Purkinje cell has been shown to be a direct target of the lurcher gene (Wetts and Herrup, 1982a), it is parsimonious to suggest that the Purkinje cell does not emit the proper cellular cue to initiate the more mature form of dendritic contact. A similar defect in the ascent of climbing fibers onto the Purkinje cell dendritic tree has been found in the hyperspiny Purkinje cell mutant mouse (Guenet et al., 1983). These mutations offer models to study critical events in the establishment of one of the key synaptic relationships in the cerebellum.

In other mutations that primarily affect granule cell development (e.g., weaver and meander tail), the positioning of Purkinje cells and their dendritic arbor is abnormal. These features of the Purkinje cell thus rely on extrinsic interactions for their full expression. Reeler is an excellent model system to study the relative effects of afferent interactions on Purkinje cell development. In this mutant there is a graded placement of Purkinje cells (from normally positioned to ectopically situated in the deep white matter) resulting in varying degrees of interaction with the granule cell population, and with the concomitant suffering of the dendritic tree. Sadler and Berry (1989) have elegantly shown that the initial phase of dendritic growth is normal in reeler (and weaver) Purkinje cells. However, in the absence of parallel fiber contacts, the later phases of Purkinje cell dendritic development (the proliferation of dendritic segments and their subsequent remodelling) do not occur.

CEREBELLUM, GENETIC MUTANTS, AND MOTOR CONTROL

Our appreciation of the role of the cerebellum in the overall behavior of mammals originate from studies in human patients with cerebellar lesions. These studies have clearly demonstrated the involvement of this structure in the control of motor activity, the regulation of muscle tone, and equilibrium (Holmes, 1939). The apparently simple input–output functions that subserve motor activity and the crystallin-like array of cerebellar cytoarchitecture encouraged the analogy between the cerebellum and a computative machine (Eccles et al., 1967). Although this model was well ahead of the explosion in computational neurosciences, we are not much closer to the relationship between cytoarchitectural units and motor control. Lesions in different areas of the cerebellum severely affect all of these functions. Indeed, it is the appearance of motor deficits that has enabled the identification of numerous cerebellar mutants. The intriguing potential exists to use these various mutants, that have more or less specific defects in cerebellar architecture, to analyze the differences in motor performance that result from different alterations in cerebellar morphology and organization.

Presumably, the different qualities of ataxia, as engendered in the thoughtful names given to the various mutant mice (e.g., weaver, nervous, stumbler, staggerer, reeler, and lurcher) relate to differences in motor syndromes (see Sidman et al., 1965) and to final cerebellar cell composition and circuitry in each of the mutant strains. Ataxia can be severe, as in leaner, or mild, as in nervous and *pcd*. There can be a fine tremor, as in weaver, a coarse tremor, as in reeler, or a dynamic tremor of the neck, as in lurcher. The trunk may be kept low to the ground, as in staggerer and reeler, or the trunk may be arched and demonstrate up-and-down movements, as in lurcher [or maybe also stumbler (Caddy et al., 1981)]. Hindquarters may be the most obviously affected part of the body, as in rolling mouse Nagoya (Oda, 1981), reeler, and staggerer, whereas in other mutants there is a hyperextension that involves all the limbs (weaver). Staggerer demonstrates a strange backing-up behavior (Sidman et al., 1965; Lalonde, 1987b), and each of the mutants have interesting responses when they are held upside-down by the tail [e.g., Lalonde, 1987a,b; but see Triarhou and Ghetti (1987) for an important caveat concerning such observations]. Inherent in the diversity of mutant phenotypes among the cerebellar mutants is the opportunity to identify structural defects that are specifically associated with motor abnormalities.

Unfortunately, this approach to teasing apart the relationship between structure and function has not been pursued with the vigor it deserves. There have been studies that have identified motor defects in various mutant mice (e.g., Lalonde, 1987a,b), but only one study, using the lurcher mutant, has exploited in a careful manner the analytic power that mutant mice offer in making correlations between structure and function (Fortier et al., 1987). The careful behavioral analysis of motor performance in lurcher by Fortier et al. (1987), studying the component parts of movement with video and electromyography (EMG) recordings, suggests that the rhythmic motor behaviors associated with scratching and swimming are little affected in lurcher, whereas the motor patterns associated with walking exhibit disorganization of the step cycle and irregular EMG patterns. Maintenance of equilibrium and forward progression are characteristic of treadmill walking, but these are not essential to performing scratching and swimming behaviors. These studies in the lurcher, a mouse that is basically a cerebellar decorticate animal, would suggest that normal terrestial locomotion is dependent on the structural integrity of the cerebellar cortex (Fortier et al., 1987). Further studies that analyze and compare the motor performance in other mutants with different cerebellar phenotypes would certainly help to dissect the functional domains of the cerebellum.

One can take advantage of the differential site of action of the various mutant genes, such as the anterior lobe of the cerebellum in meander tail and leaner, the vermis in weaver (Herrup and Trenkner, 1987), and parasagittal arrays of Purkinje cells in nervous (Wassef et al., 1987). Of course, it is critical to keep in mind the cascade of effects that the original genetic lesion has on the organization and structure of the CNS. It would be naive to consider the most obvious mutant phenotype as the sole deficit in a given mutant. An example of this shortsightedness would be to consider only the granule cell loss in weaver and to overlook the effect of this gene on Purkinje cells (Blatt and Eisenman, 1985; Smeyne and Goldowitz, 1990), and on the dopaminergic cells of the substantia nigra (Schmidt et al., 1982; Roffler-Tarlov and Graybiel, 1986). Given this important consideration, the various neurological mutant mice offer an untapped resource for studying the structural correlates of motor function.

CEREBELLUM, COGNITION, AND NEUROLOGICAL MUTANTS

Lesions of the cerebellum most obviously result in motor disturbances. The severe motor deficits that follow destruction of the cerebellum may easily mask more associative-type roles of the cerebellum. During the early era of brain stimulation work, studies of cerebellar stimulation fit well with the view that the cerebellum was involved in "elementary" motor responses (as reviewed in Berntson et al., 1973). It was not until the 1970s that these ideas were seriously challenged. Brain stimulation studies showed that cerebellar stimulation could lead to specific, integrated, and directed behaviors (e.g., grooming, eating, and predatory behaviors) (Berntson et al., 1973; Ball et al., 1974; Reis et al., 1973). Electrophysiological and anatomical data had accrued to indicate cerebellar influence and connectivity extending to the limbic system (see Snider and Maiti, 1976, for a review). A number of more recent studies strongly point to the cerebellum's involvement in reflex adaptation and motor learning. These studies used various behavioral endpoints such as gain change in the vestibulo-ocular reflex (VOR)

(Ito, 1982, 1984a,b), learning of a classical conditioned nictitating membrane response (McCormick et al., 1982; McCormick and Thompson, 1984; Yeo et al., 1985a,b), bradycardia (Supple and Leaton, 1990), and the acquisition of more complicated motor patterns that involve numerous muscle groups and joints (Gilbert and Thach, 1977; Black et al., 1990). In concert with these findings are correlative reports in the clinical literature of structural alterations in cerebella of autistic (Gaffney et al., 1987; Courchesne et al., 1987, 1988; Murakami et al., 1989; but see also Heh et al., 1989) and schizophrenic patients (Snider, 1982). Thus, studies in laboratory animals and humans over the last two decades have revealed a new domain of cerebellar function: learned motor behaviors (Berntson and Torello, 1982; Leiner et al., 1989).

Although these studies are fascinating and suggest that the cerebellum may be involved with cognitive aspects of behavior, in addition to performance, there is a good deal of controversy regarding these assertions (Welsh and Harvey, 1989; Kelly et al., 1990; Lisberger, 1988). Cogent issues raised in this controversy concern the difficulties in dissociating (1) performance variables when measuring motor learning or modifiability and (2) changes in behavior that result from alterations in tonic activities of brain stem structures due to cerebellar removal, rather than direct interruption of the pathways involved in the plastic changes. These important distinctions have been approached in some studies in the goldfish by using long survival times after cerebellectomy (Baker et al., 1988; Weiser et al., 1988).

In these studies, goldfish lose the ability to modify their VOR after cerebellectomy but, after a long delay of at least 1 month, these same fish can adapt their VOR. As indicated above, mutant mice present us with all gradations of cerebellar malformations, or "natural lesions." It would be most interesting to analyze the behavior of these animals in adaptation, or motor learning, paradigms. Since these alterations in cerebellar structure are present from early on, the nervous system of the adult mutant could be viewed as a relatively stabilized neural system; that is, that the direct and indirect sequelae of the mutation would already have been expressed, and the "lesioned" CNS stabilized. How would such animals perform under conditions that normally produce reflex adaptation or complex motor or associative learning?

Neurological mutant mice offer a valuable resource in detailing the role of the cerebellum in behavior. Cerebellar mutants present the researcher with fully reproducible, natural lesions that encompass various cerebellar abnormalities and associated syndromes. Only in the past few years have mutant mice been used in behavioral studies. Many of these studies have catalogued the behavioral deficits of several of the cerebellar mutant mice on specific tasks (e.g., spontaneous alternation and habituation) (Lalonde, 1986; Lalonde et al., 1986a,b, 1987, 1988). Another approach has been to examine learned behaviors (radial arm maze and active avoidance) of different mutant mice to try to dissect a component of cerebellar structure that is related to a specific aspect of behavioral performance (Goldowitz and Koch, 1986).

In many mutants the genetic "lesion" is rather specific and occurs early in development. As such, adult mutant mice with relatively specific cerebellar abnormalities could be used to determine the essential nature of the cerebellum in behavioral paradigms, such as involving motor learning or modifiability. As an example, murine mutants could be analyzed for their ability to adapt their VOR. This reflex results in an equal and opposite rotation (gain = 1) of the eyes in response to head rotation. The gain of this reflex can be modified by creating a long-term, retinal image slip. The circuitry for the VOR consists of a three-neuron link involving vestibular input via the vestibular nerve to the vestibular nuclei, and the latter's projection to the extraocular nuclei and their subsequent innervation of the extraocular muscles. Modifiability of the VOR is thought to involve a side loop through the cerebellar cortex (specifically, the flocculus) (Ito, 1982; Robinson, 1987). The essential nature of this side loop can be examined by careful testing of selected neurological mutants and determining their ability to modify the gain of their VOR. As an example of this approach, the lurcher mutant, an animal that loses all of its Purkinje cells and therefore is a cerebellar decorticate, could be tested for VOR modifiability. The prediction would be that this mutant would be unable to modify the gain of its VOR, if indeed such adaptation involves a side loop through the cerebellar cortex.

The use of experimental mouse chimeras made between mutant and wild-type mice is another potentially exciting avenue to pursue cerebellar involvement in associative functions. Several features of the chimeric situation recommend the use of chimeras. First, although the mutant mouse in question may have severe motor abnormalities, a mutant–normal chimera can have normal motor performance (e.g., *Lc*, Wetts and Herrup, 1982a; *pcd*, Mullen, 1977a; *sg*, Herrup and Mullen, 1979b). This, no doubt, is due to a complement of normal cells that contributes to the formation of the cerebellum. In the same context, many of the secondary cellular defects that follow from the primary cellular defect may be ameliorated in the chimeric situation and be removed from consideration as variables in the experimental design. Second, each chimera is composed of varying proportions of mutant and normal cells. Thus, in a series of chimeras, there would be a range of unique experimental animals ranging from mainly mutant to predominantly normal. In theory, it would be possible to determine if a linear relationship exists between the increasing percentages of normal cells (e.g., Purkinje cells in a lurcher–normal chimera) and performance measures on a behavioral task. In this manner, one may be able to make structural–functional correlations between neuronal populations and behaviors. One can imagine using such an approach with the VOR paradigm discussed above.

With the increasing evidence of a cerebellar involvement in associative behaviors (Inhoffetal, 1989; Bracke-Tolkmitt, et al., 1989), the murine neurological mutant might well be considered as a means to explore this cerebellar function more fully. It is our feeling that the important and fascinating issues of the cerebellum's role, or lack thereof, in learned behaviors can be clarified from results of carefully designed experiments, combined with a careful selection of the mutants and behavioral paradigms.

Acknowledgments. The comments of Richard Mullen were helpful in the preparation of this manuscript. Our research is supported by NIH (NS 23475 to DG and NS 22093 to LME) and the State of Tennessee Neuroscience Center of Excellence (DG).

REFERENCES

Altman J (1972): Postnatal development of the cerebellar cortex in the rat. III. Maturation of the components of the granular layer. *J Comp Neurol* 145:465–514

Altman J, Bayer SA (1985a): Embryonic development of the rat cerebellum. I. Delineation of the cerebellar primordium and early cell movements. *J Comp Neurol* 231:1–26

Altman J, Bayer SA (1985b): Embryonic development of the rat cerebellum. II. Translocation and regional distribution of the deep neurons. *J Comp Neurol* 231:27–41

Altman J, Bayer SA (1985c): Embryonic development of the rat cerebellum III. Regional differences in the time of origin, migration, and settling of Purkinje cells. *J Comp Neurol* 231:42–65

Baker R, Weiser M, McElligott JG (1988): Adaptive gain control of the vestibulo-ocular reflex in goldfish. I. Hemicerebellectomy. *Soc Neurosci Abstr* 14:169

Ball GG, Micco DJ Jr, Berntson GG (1974): Cerebellar stimulation in the rat: complex stimulation-bound oral behaviors and self-stimulation. *Physiol Behav* 13:123–127

Berntson GG, Potolicchio SJ Jr, Miller NE (1973): Evidence for higher functions of the cerebellum: eating and grooming elicited by cerebellar stimulation in cats. *Proc Nat Acad Sci USA* 70:2497–2499

Berntson GG, Torello MW (1982): The paleocerebellum and the integration of behavioral function. *Physiol Psychol* 10:2–12

Black JE, Isaacs KR, Anderson BJ, Alcantara AA, Greenough WT (1990): Learning causes synaptogenesis, whereas motor activity causes angiogenesis, in cerebellar cortex of adult rats. *Proc Natl Acad Sci USA* 87:5568–5572

Blatt GJ, Eisenman LM (1985): A qualitative and quantitative light microscopic study of the inferior olivary complex of normal, reeler, and weaver mutant mice. *J Comp Neurol* 232:117–128

Bordarier C, Robain L, Rthore M-O, Dulac O, Dhellemes C (1986): Inverted neurons in agyria. *Hum Genet* 73:374–378

Bracke-Tolkmitt R, Linden A, Canavan AGM, Rockstroh B, Scholz E, Wessel K, Diener HC (1989): The cerebellum contributes to mental skills. *Behav Neurosci* 103:442–446

Caddy KWT, Biscoe TJ (1979): Structural and quantitative studies on the normal C3H and Lurcher mutant mouse. *Phil Trans R Soc London (Biol)* 287:167–201

Caddy KWT, Sidman RL (1981): Purkinje cells and granule cells in the cerebellum of the stumbler mutant mouse. *Dev Brain Res* 1:221–236

Caddy KWT, Sidman RL, Eicher E (1981): Stumbler, a new mutant mouse with cerebellar disease. *Brain Res* 208:251–255

Caviness VS Jr (1977): Reeler mutant mouse: a genetic experiment in developing mammalian cortex. In: *Society for Neuroscience Symposia Vol 2,* Cowan WM, Ferrandeli JA, eds. Bethesda: Society of Neuroscience, pp 27–46

Caviness VS Jr (1980): The developmental consequences of abnormal cell position in the *reeler* mouse. *TINS* 3:31–33

Caviness VS Jr, Rakic P (1978): Mechanisms of cortical development: a view from mutations in mice. *Annu Rev Neurosci* 1:297–326

Courchesne E, Hesselink JR, Jernigan TL, Yeung-Courchesne R (1987): Abnormal neuroanatomy in a nonretarded person with autism: unusual findings with magnetic resonance imaging. *Arch Neurol* 44:335–341

Courchesne E, Yeung-Courchesne R, Press GA, Hesselink JR, Jernigan TL (1988): Hypoplasia of cerebellar vermal lobules VI and VII in autism. *N Engl J Med* 318:1349–1354

Crepel F, Delhaye-Bouchard N, Guastavino J-M, Sampaio I (1980): Multiple innervation of cerebellar Purkinje cells by climbing fibers in the staggerer mutant mouse. *Nature* 283:483–484

Desclin JC (1974): Histological evidence supporting the inferior olive as the major source of cerebellar climbing fibers in the rat. *Brain Res* 77:365–384

Eccles JC, Ito M, Szentagothai J (1967): *The Cerebellum as a Neuronal Machine*. New York: Springer–Verlag

Eccles JC, Llinás R, Sasaki K (1966): The excitatory synaptic action of climbing fibers on the Purkinje cells of the cerebellum. *J Physiol (Lond.)* 182:268–296

Fortier PA, Smith AM, Rossignol S (1987): Locomotor deficits in the mutant mouse, Lurcher. *Exp Brain Res* 66:271–286

Fujita S (1969): Autoradiographic studies on histogenesis of the cerebellar cortex. In: *Neurobiology of Cerebellar Evolution and Development*, Llinás R, ed. Chicago: American Medical Association Education and Research Foundation, pp 743–747

Gadisseux JF, Evrard P (1985): Glial-neuronal relationship in the developing central nervous system. *Dev Neurosci* 7:12–32

Gaffney GR, Tsai L, Kuperman S, Minchin S (1987): Cerebellar structure in autism. *AJDC* 141:1330–1332

Gilbert PFC, Thach WR (1977): Purkinje cell activity during motor learning. *Brain Res* 128:309–328

Goffinet AM (1983): The embryonic development of the cerebellum in normal and reeler mutant mice. *Anat Embryol* 168:73–86

Goffinet AM (1984): Events governing the organization of post-migratory neurons: studies on brain development in normal and reeler mice. *Brain Res Rev* 7:261–296

Goldowitz D (1989): The weaver granuloprival phenotype is due to intrinsic action of the mutant locus in granule cells: evidence from homozygous weaver chimeras. *Neuron* 2:1565–1575

Goldowitz D, Koch J (1986): Performance of normal and neurological mutant mice on radial arm maze and active avoidance tasks. *Behav Neural Biol* 46:216–226

Goldowitz D, Mullen RJ (1982): Granule cell as a site of gene action in the weaver mouse cerebellum: evidence from heterozygous mutant chimeras. *J Neurosci* 2:1474–1485

Goodlett CR, Hamre KM, West JR (1990): Regional differences in the timing of dendritic outgrowth of Purkinje cells in the vermal cerebellum demonstrated by MAP2 immunocytochemistry. *Dev Brain Res* 53:131–134

Gruneberg H (1963): *The Pathology of Development: A Study of Inherited Skeletal Disorders in Animals*. New York: John Wiley & Sons

Guenet JL, Sotelo C, Mariani J (1983): Hyperspiny Purkinje cell, a new neurological mutation in the mouse. *J Hered* 74:105–108

Hallonet MER, Teillet MA, Le Douarin NM (1990): A new approach to the development of the cerebellum provided by the quail-chick marker system. *Development* 108:19–31

Harding AE (1984): *The Hereditary Ataxias and Related Disorders*. Edinburgh: Churchill Livingstone

Hatten ME, Liem RKH, Mason CA (1986): Weaver mouse cerebellar granule neurons fail to migrate on wild-type astroglial processes in vitro. *J Neurosci* 6:2676–2683

Hatten ME, Messer A (1978): Postnatal cerebellar cells from Staggerer mutant mice express embryonic cell surface characteristic. *Nature* 276:504–506

Hausmann B, Mangold U, Sievers J, Berry M (1985): Derivation of cerebellar Golgi neurons from the external granular layer: evidence from explanation of external granule cells in vivo. *J Comp Neurol* 232:511–522

Heckroth JA, Goldowitz D, Eisenman LM (1988): Purkinje cell reduction in the reeler mutant mouse: a quantitative immunohistochemical study. *J Comp Neurol* 279:546–555

Heckroth JA, Goldowitz D, Eisenman LM (1990): Olivocerebellar fiber maturation in normal and lurcher mutant mice: defective development in lurcher. *J Comp Neurol* 279:546–555

Heh CWC, Smith R, Wu J, Hazlett E, Russel A, Asarnow R, Tanguay P, Buchsbaum M (1989): Positron emission tomography of the cerebellum in autism. *Am J Psychiatry* 146:242–245

Herrup K (1983): Role of staggerer gene in determining cell number in cerebellar cortex. I. Granule cell death is an indirect consequence of staggerer gene action. *Dev Brain Res* 11:267–274

Herrup K, Mullen RJ (1979a): Regional variation and absence of large neurons in the cerebellum of the staggerer mouse. *Brain Res* 172:1–12

Herrup K, Mullen RJ (1979b): Staggerer chimeras: Intrinsic nature of Purkinje cell defects and implications for normal cerebellar development. *Brain Res* 178:443–457

Herrup K, Sunter K (1987): Granule cell death during cerebellar development of Staggerer mouse chimeras: a test of the numerical matching hypothesis. *J Neurosci* 7:829–836

Herrup K, Trenkner E (1987): Regional differences in cytoarchitecture of the weaver cerebellum suggest a new model for weaver gene action. *Neurosci* 23:871–885

Herrup K, Wilczynski SL (1982): Cerebellar cell degeneration in the leaner mutant mouse. *Neurosci* 7:2185–2196

Hirano A, Dembitzer HM (1973): Cerebellar alterations in the weaver mouse. *J Cell Biol* 56:478–486

Holmes G (1939): The cerebellum of man. *Brain* 62:1–30

Inhoff AW, Diener HC, Rafal RD, Ivry R (1989): The role of cerebellar structures in the execution of serial movements. *Brain* 112:565–581

Ito M (1982): Cerebellar control of the vestivulo-ocular reflex—around the flocculus hypothesis. *Annu Rev Neurosci* 5:275–296

Ito M (1984a): The modifiable neuronal network of the cerebellum. *Jpn J Physiol* 34:781–792

Ito M (1984b): *The Cerebellum and Neural Control*. New York: Raven Press

Kelly TM, Zuo CC, Bloedel JR (1990): Classical conditioning of the eyeblink reflex in the decerebrate-decerebellate rabbit. *Behav Brain Res* 38:7–18

Lalonde R (1986): Delayed spontaneous alternation in weaver mutant mice. *Brain Res* 398:178–180

Lalonde R (1987a): Motor abnormalities in weaver mutant mice. *Exp Brain Res* 65:479–481

Lalonde R (1987b): Motor abnormalities in staggerer mutant mice. *Exp Brain Res* 68:417–420

Lalonde R, Botez MI, Boivin D (1986a): Spontaneous alternation and habituation in a t-maze in nervous mutant mice. *Behav Neurosci* 100:350–352

Lalonde R, Lamarre Y, Smith AM, Botez MI (1986b): Spontaneous alternation and habituation in lurcher mutant mice. *Brain Res* 362:161–164

Lalonde R, Manseau M, Botez MI (1987): Spontaneous alternation and habituation in Purkinje cell degeneration mutant mice. *Brain Res* 411:187–189

Lalonde R, Manseau M, Botez MI (1988): Spontaneous alternation and exploration in weaver mutant mice. *Behav Brain Res* 31:111–114

Landis DMD, Landis SC (1978): Several mutations in mice that affect the cerebellum. *Adv Neurol* 21:85–105

Landis DMD, Sidman RL (1978): Electron microscopic analysis of postnatal histogenesis in the cerebellar cortex of staggerer mutant mice. *J Comp Neurol* 179:831–864

Landis SC (1973): Changes in neuronal mitochondrial shape in brains of nervous mutant mice. *J Hered* 64:193–196

Landis SC, Mullen RJ (1978): The development and degeneration of Purkinje cells in *pcd* mutant mice. *J Comp Neurol* 177:125–143

Larramendi LMH (1969): Analysis of synaptogenesis in the cerebellum of the mouse. In: *Neurobiology of Cerebellar Evolution and Development,* Llinás R, ed. Chicago: American Medical Association Education and Research Foundation, pp 803–843

Leiner HC, Leiner AL, Dow RS (1989): Reappraising the cerebellum: what does the hindbrain contribute to the forebrain? *Behav Neurosci* 103:998–1008

Lisberger SG (1988): The neural basis for learning of simple motor skills. *Science* 242:728–735

Lyon MF, Searle AG, eds. (1989) *Genetic Variants and Strains of the Laboratory Mouse,* 2nd ed. Oxford: Oxford University Press

Mariani J (1982): Extent of multiple innervation of Purkinje cells by climbing fibers in the olivocerebellar system of weaver, reeler, and staggerer mutant mice. *J Neurobiol* 13:119–126

Mariani J, Crepel F, Mikoshiba K, Changeux JP, Sotelo C (1977): Anatomical physiological and biochemical studies of the cerebellum from reeler mutant mice. *Phil Trans R Soc Lond B* 281:1–28

Martinez S, Alvarado-Mallart RM (1989): Rostral cerebellum originates from the caudal portion of the so-called "mesencephalic" vesicle: a study using chick/quail chimeras. *Eur J Neurosci* 1:549–560

Mason CA, Blazeski R (1989): Climbing fibers contact target Purkinje cells in embryonic mouse cerebellum. *Soc Neurosci Abstr* 15:959

McCormick DA, Clark GA, Lavond DG, Thompson RF (1982): Initial localization of the memory trace for a basic form of learning. *Proc Natl Acad Sci USA* 79:2731–2735

McCormick DA, Thompson RF (1984): Neuronal responses of the rabbit cerebellum during acquisition and performance of a classically conditioned nictitating membrane-eyelid response. *J Neurosci* 4:2811–2822

McLaren A (1976): *Mammalian Chimeras*. Cambridge: Cambridge University Press

Messer A, Hatch K (1984): Persistence of cerebellar thymidine kinase in Staggerer and Hypothyroid mutants. *J Neurogenet* 1:239–248

Miale I, Sidman RL (1961): An autoradiographic analysis of histogenesis in the mouse cerebellum. *Exp Neurol* 4:277–296

Mikoshiba K, Yokoyama M, Nishimura O, Katsuki M, Nomura T, Tsukada Y (1985): Mosaic expression of the Reeler and normal phenotypes in the cerebral cortex in Reeler-normal chimeras at a late embryonic stage. *Dev Growth Differ* 27:737–744

Mintz B (1964): Formation of genetically mosaic mouse embryos, and early development of lethal (t^{12}-t^{12})-normal mosaics. *J Exp Zool* 157:273–292

Mullen RJ (1977a): Site of pcd gene action and Purkinje cell mosaicism in cerebella of chimaeric mice. *Nature* 270:245–247

Mullen RJ (1977b): Genetic dissection of the CNS with mutant-normal mouse and rat chimeras. In: *Society for Neuroscience Symposium,* Vol II, Cowan WM, Ferendelli JA, eds. Bethesda: Society of Neuroscience, pp 47–65

Mullen RJ (1982): Analysis of CNS development with mutant mice and chimeras. In: *Genetic Approaches to Developmental Neurobiology,* Tsukada Y, ed. Tokyo: University of Tokyo Press

Mullen RJ, Eicher E, Sidman RL (1976): Purkinje cell degeneration, a new neurological mutation in the mouse. *Proc Natl Acad Sci USA* 73:208–212

Murakami JW, Courchesne E, Press GA, Yeung-Courchesne R, Hesselink JR (1989): Reduced cerebellar hemisphere size and its relationship to vermal hypoplasia in autism. *Arch Neurol* 46:689–694

Oda S (1981): A new allele of the tottering locus, rolling mouse Nagoya, on chromosome no. 8 in the mouse. *Jpn J Genet* 56:295–299

Pinto-Lord MC, Evrard P, Caviness VS Jr (1982): Obstructed neuronal migration along radial glial fibers in the neocortex of the Reeler mouse: a Golgi-EM analysis. *Dev Brain Res* 4:379–393

Rakic P (1971): Neuron-glia relationship during granule cell migration in developing cerebellar cortex. A Golgi and electronmicroscopic study in *macacus rhesus. J Comp Neurol* 141:283–312

Rakic P (1976): Synaptic specificity in the cerebellar cortex: study of anomalous circuits induced by single gene mutations in mice. *Cold Spring Harbor Symp Quant Biol* 60:333–346

Rakic P, Sidman RL (1973a): Sequence of developmental abnormalities leading to granule cell deficit in cerebellar cortex of weaver mutant mice. *J Comp Neurol* 152:103–132

Rakic P, Sidman RL (1973b): Organization of cerebellar cortex secondary to deficit of granule cells in weaver mutant mice. *J Comp Neurol* 152:133–162

Ramón y Cajal S (1952): *Histologie du Systeme Nerveux de l'Homme et des Vertebrates.* Madrid: Instituto Ramon y Cajal

Ramón y Cajal S (1960): *Studies on Vertebrate Neurogenesis,* Guth L, trans. Springfield, IL: Charles C Thomas

Reis DJ, Doba N, and Nathan MA (1973): Predatory attack, grooming, and consummatory behaviors evoked by electrical stimulation of cat cerebellar nuclei. *Science* 182:845–847

Rezai Z, Yoon CH (1972): Abnormal rate of granule cell migration in the cerebellum of "weaver" mutant mice. *Dev Biol* 29:17–26

Robinson DA (1987): The windfalls of technology in the oculomotor system. *Invest Ophthal Vis Sci* 28:1912–1925

Roffler-Tarlov S, Graybiel AM (1986): Expression of the weaver gene in dopamine-containing neural systems is dose-dependent and affects both striatal and nonstriatal regions. *J Neurosci* 6:3319–3330

Ross ME, Fletcher C, Mason CA, Hatten ME, Heintz N (1990): Meander tail reveals a discrete developmental unit in the mouse cerebellum. *Proc Natl Acad Sci USA* 87:4189–4192

Sadler M, Berry M (1989): Topological link-vertex analysis of the growth of Purkinje cell dendritic trees in normal, reeler, and weaver mice. *J Comp Neurol* 289:260–283

Schmidt MJ, Sawyer BD, Perry KW, Fuller RW, Foreman MM, Ghetti B (1982): Dopamine deficiency in the weaver mutant mouse. *J Neurosci* 2:376–380

Sidman RL (1968): Development of interneuronal connections in brains of mutant mice. In *Physiological and Biochemical Aspects of Nervous Integration,* Carlson FD, ed. Englewood Cliffs, NJ: Prentice-Hall. pp 163–193

Sidman RL (1987): Neuropathology, mutant studies. In: *Encyclopedia of Neuroscience,* Adelman G, ed. Boston: Boston Birkhäuser, pp 828–832

Sidman RL, Green MC, Appel S (1965): *Catalogue of the Neurological Mutants of Mice*. Cambridge: Harvard University Press

Smeyne RJ, Goldowitz D (1989): Development and death of external granular layer cells in the weaver mouse cerebellum: a quantitative study. *J Neurosci* 9:1608–1620

Smeyne RJ, Goldowitz D (1990): Purkinje cell loss is due to a direct action of the weaver gene in Purkinje cells: evidence from chimeric mice. *Dev Brain Res* 52:211–218

Snider SR (1982): Cerebellar pathology in schizophrenia—cause or consequence? *Neurosci Biobehav Rev* 6:47–53

Snider RS, Maiti A (1976): Cerebellar contributions to the Papez circuit. *J Neurosci Res* 2:133–146

Sonmez E, Herrup K (1984): Role of staggerer gene in determining cell number in cerebellar cortex. II. Granule cell death and persistence of the external granule cell layer in young mouse chimeras. *Dev Brain Res* 12:271–283

Sotelo C (1975): Anatomical, physiological and biochemical studies of the cerebellum from mutant mice. II. Morphological study of cerebellar cortical neurons and circuits in the weaver mouse. *Brain Res* 94:19–44

Sotelo C (1980): Mutant mice and the formation of cerebellar circuitry. *Trends Neurosci* Feb:33–35

Sotelo C, Changeux JP (1974): Transsynaptic degeneration 'en cascade' in the cerebellar cortex of Staggerer mutant mice. *Brain Res* 67:519–526

Sotelo C, Triller A (1979): Fate of presynaptic afferents to Purkinje cells in the adult nervous mutant mouse: a model to study presynaptic stabilization. *Brain Res* 175:11–36

Supple WF Jr, Leaton RN (1990): Cerebellar vermis: essential for classically conditioned bradycardia in the rat. *Brain Res* 509:17–23

Swisher DA, Wilson DB (1977): Cerebellar histogenesis in the Lurcher (Lc) mutant mouse. *J Comp Neurol* 173:205–217

Tarkowski AK (1961): Mouse chimaeras developed from fused eggs. *Nature (Lond.)* 190:857–860

Terashima T, Inoue K, Inoue Y, Yokoyama M, Mikoshiba K (1986): Observations on the cerebellum of norma-reeler mutant mouse chimera. *J Comp Neurol* 252:264–278

Trenkner E (1979): Postnatal cerebellar cells of Staggerer mutant mice express immature components on their surfaces. *Nature* 277:566–567

Triarhou LC, Ghetti B (1987): Neuroanatomical substrate of behavioural impairment in Weaver mutant mice. *Exp Brain Res* 68:434–436

Vogel MW, Herrup K (1989): Numerical matching in the mammalian CNS: lack of a competitive advantage of early over late-generated cerebellar granule cells. *J Comp Neurol* 283:118–128

Vogel MW, Sunter K, Herrup K (1989): Numerical matching between granule and Purkinje cells in lurcher chimeric mice: a hypothesis for the trophic rescue of granule cells from target-related cell death. *J Neurosci* 9:3454–3462

Wassef M, Sotelo C, Cholley B, Brehier A, Thomasset M (1987): Cerebellar mutations affecting the postnatal survival of Purkinje cells in the mouse disclose a longitudinal pattern of differentially sensitive cells. *Dev Biol* 124:379–389

Weiser M, McElligott JG, Baker R (1988): Adaptive gain control of the vestibulo-ocular reflex in the goldfish. II. Total cerebellectomy. *Soc Neurosci Abstr* 14:169

Welsh JP, Harvey JA (1989): Cerebellar lesions and the nictitating membrane reflex: performance deficits of the conditioned and unconditioned response. *J Neurosci* 9:299–311

West MJ, Del Cerro M (1976): Early formation of synapses in the molecular layer of the fetal rat cerebellum. *J Comp Neurol* 165:137–160

Wetts R, Herrup K (1982a): Interaction of granule, Purkinje, and inferior olivary neurons in lurcher chimeric mice. I. Qualitative studies. *J Embryol Exp Morphol* 68:87–98

Wetts R, Herrup K (1982b): Interaction of granule, Purkinje and inferior olivary neurons in lurcher chimeric mice. II. Granule cell death. *Brain Res* 250:358–362

Wetts R, Herrup K (1983): Direct correlation between Purkinje and granule cell number in the cerebella of lurcher chimeras and wild-type mice. *Dev Brain Res* 10:41–47

Wille W, Trenkner E (1981): Changes in particulate neuraminidase activity during normal and Staggerer mutant mouse development. *J Neurochem* 37:443–446

Yamano T, Suzuki K (1985): Abnormalities of Purkinje cell arborization in brindled mouse cerebellum. *J Neuropathol Exp Neurol* 44:85–96

Yeo CH, Hardiman MJ, Glickstein M (1985a): Classical conditioning of the nictitating membrane response of the rabbit. I. Lesions of the cerebellar nuclei. *Exp Brain Res* 60:87–98

Yeo CH, Hardiman MJ, Glickstein M (1985b): Classical conditioning of the nictitating membrane response of the rabbit. II. Lesion of the cerebellar cortex. *Exp Brain Res* 60:99–113

Yoon CH (1972): Developmental mechanism for changes in cerebellum of “staggerer” mouse, a neurological mutant of genetic origin. *Neurology* 22:743–754

Yoon CH (1976): Pleiotropic effect of the Staggerer gene. *Brain Res* 109:206–215

Genetically Determined Animal Models of Narcolepsy, A Disorder of REM Sleep

EMMANUEL MIGNOT, CHRISTIAN GUILLEMINAULT, WILLIAM C. DEMENT, AND F. CARL GRUMET

NARCOLEPSY: THE HUMAN SYNDROME

Narcolepsy is a neurological condition that affects 0.02% to 0.09% of Caucasians, North American blacks, and Japanese populations. It is seen in all races, although the percentage of affected subjects has not been investigated in all continents, and it involves both sexes. It was first described in humans by Gelineau (1880). Narcolepsy is a life-long disease that is characterized by abnormal sleep tendencies, including excessive daytime sleepiness, often disturbed nocturnal sleep, and pathological manifestations of rapid eye movement (REM) sleep. The REM sleep abnormalities include sleep-onset REM periods and the dissociated REM sleep inhibitory processes—cataplexy and sleep paralysis (pathological equivalents of REM sleep atonia). Excessive daytime sleepiness, cataplexy (abrupt loss of motor tone induced by emotions) and, less often, sleep paralysis and hypnagogic hallucinations are the major symptoms of the disease. The causes of human narcolepsy are currently unknown but may involve a defect of the specific mechanisms regulating REM sleep and wakefulness.

Since its initial description in 1880, narcolepsy has been reported to cluster in certain families (Krabbe and Magnussen, 1942; Bruhova and Roth, 1972; Yoss and Daly, 1960; Kessler et al., 1974; Honda et al., 1983a,b; Baraitser and Parkes, 1987). Still, the genetic mode of transmission for this disease is not yet clearly understood. Early experimenters suggested an autosomal dominant or polygenic mode of transmission for narcolepsy (Daly and Yoss, 1959; Bruhova and Roth, 1972; Kessler, 1976). Recently, as part of a Japanese program investigating the host factors of intractable disease, systematic human leukocyte antigen (HLA) typing was performed on narcoleptic patients. An interesting association between a genetic marker and narcolepsy was identified in a recent Japanese study (Honda et al., 1983b), showing that 100% of Japanese narcoleptic patients possessed the major histocompatibility complex (MHC) antigen HLA-DR2 (and secondarily, Dw2, DQwl, a haplotype now defined as DRw15, Dw2, DQw6 under "the new WHO nomenclature," 1990), compared to only 34% of controls. Subsequent studies (Langdon et al., 1986; Billiard and Seignalet, 1985; Poirier et al., 1986; Thomson, 1985; Neely et al., 1987) extended the HLA-DR2 association to

Caucasians and blacks as well, the association being less tight in black patients (Neely et al., 1987). The presence of the strong association between the HLA DR2 antigen and narcolepsy (100% in Japanese, at least 91% in North American Caucasians) raises the possibility that narcolepsy could be an autoimmune disorder, as many HLA-associated diseases have been found to be. Alternatively, a nonimmunologically related gene bearing the susceptibility for narcolepsy might be in linkage disequilibrium with the HLA locus.

Studies performed to date have, however, neither confirmed nor refuted this hypothesis. Parkes et al. (1986), Matsuki et al. (1988), Rubin et al. (1988), and Fredrikson et al. (1990) failed to demonstrate the existence of any immunological abnormalities classically found in other autoimmune disease (normal erythrocyte sedimentation ratio, negative C-reactive protein, normal complement levels, normal immunoglobulin concentration and lymphocyte subpopulations, absence of CSF oligoclonal bands) in narcoleptic patients. Similarly, numerous nonimmunological genes have been cloned within the MHC region but, to date, none seem to be directly related to sleep. The high frequency of association with DR2 has suggested that a narcolepsy gene, if not DR2 itself, is very tightly linked with the DR genes and should be located between DX and C4 within the MHC complex. The DRβ and DQβ polymorphic genes expressed in the HLA DRw15, Dw2, DQw6 haplotype have also been sequenced and do not differ from normal published sequences (Lock et al., 1988; Uryu et al., 1989).

The existence of an HLA-related susceptibility gene in human narcoleptic patients is not sufficient to explain the pathology. The gene frequency of the haplotype carried by a calculated mean of 95% of narcoleptic patients, DRw15, Dw2, DQw6, is 12.6% in the Caucasian population (Dupont, 1990) and 4% in the Japanese population (Sasazuki et al., 1977). We can then calculate the penetrance under a strict dominant model on the assumption that this haplotype is the disease susceptibility gene of 0.005 (Caucasian estimate) to 0.04 (Japanese estimate). These calculations suggest that DRw15, Dw2, DQw6 or the gene in linkage disequilibrium with this HLA haplotype is not sufficient to induce narcolepsy. Recent data confirm these assumptions. For example, among 334 unrelated narcoleptic patients investigated at Stanford University for familial patterns of this disease (Guilleminault et al., 1989), especially with respect to the disease-associated HLA-DR2 marker, 54% had no known family history of sleep disorders, 40% had at least one additional family member with an isolated daytime sleepiness complaint, and 6% had a positive family history of narcolepsy. The familial occurrence of narcolepsy was calculated to be 4.2%, but multicase families were rare; only two were found with three or more affected relatives. Family members in many instances share the same HLA-DR2 haplotype as the proband but never develop narcolepsy; however, the risk of disease for first degree relatives of patients is, in our population, 6 to 18 times greater than for unrelated individuals.

Although most patients were HLA-DR2 positive in this same study, our data projected that as many as 9% of unrelated North American Caucasian narcoleptics will be HLA-DR2 negative. Similarly, Neely et al. (1987) reported that as many as

30% of black narcoleptic patients might be DR2 negative, a result to consider with caution, since many patients included in this study did not present cataplexy. These data, taken together with the fact that 100% of the 206 Japanese patients typed to date by Matsuki et al. (1989) were DR2 positive, suggests a substantial ethnic influence on the HLA-DR2 association with narcolepsy. We also reported an HLA-DR2 narcoleptic family with a high penetrance susceptibility gene for narcolepsy–cataplexy showing no HLA haplotype linkage, the first HLA-DR2 negative narcoleptic family (Guilleminault et al., 1989). This pattern of transmission, also described by Singh et al. (1990) in one family, although rare, confirms the existence of non–HLA-linked susceptibility genes with high penetrance. Finally, there are currently (including ours) at least five pairs of monozygotic adult (>40 years) twins discordant for narcolepsy, all of whom are DR2 positive (Montplaisir and Poirier, 1987, 1988; Pollmacher et al., 1990; Guilleminault et al., 1989). The only well documented concordant pair of twins is DR2 negative (Douglas et al., 1989). [Schrader et al. (1980) reported a pair of twins who carry the Dw2 antigen and therefore are extremely likely to be DR2 positive, although HLA-DR typing was not performed.] Analysis of these and other data thus indicate that, although strongly associated with the disease, HLA DR2 Dw2 is neither sufficient nor necessary for the development of narcolepsy. The accumulated evidence supports a hypothesis that narcolepsy is multifactorial, including at least two genes, one related to DR2 and one not linked to the MHC, and strongly environmentally influenced (Guilleminault and Grumet, 1986; Guilleminault et al., 1989). This hypothesis is consistent with findings in the canine narcolepsy model where at least two types of narcolepsy expression are observed, one of which, in dobermans and labradors, is determined by a gene, *canarc-1,* not related to the MHC (Dean et al., 1989; Motoyama et al., 1990).

EXPERIMENTAL AND NATURAL MODELS OF NARCOLEPSY

Jouvet (1962), using brain stem transection experiments, has clearly established that REM sleep is primarily generated in the pontine brain stem. This specific sleep stage or some of its dissociated manifestations (atonia) can be initiated by the injection of cholinergic agents in the pontine reticular formation. It is therefore believed that some of the "executive" neurons that generate REM sleep are cholinergic cells originating from the dorsolateral pontine tegmentum and projecting to the pontine reticular formation. Furthermore, the selective lesions of these brain stem regions can also suppress REM sleep or some of its manifestations. All these experimental manipulations can produce abnormal manifestations of REM that mimic some of the symptoms of narcolepsy but never the totality of the syndrome. At this time, one can hypothesize that narcolepsy could result from a defect in one of these known systems or of a yet not well characterized mechanism regulating these pontine systems.

Knecht et al. (1973) and Mitler et al. (1974) have independently reported the existence of a canine disorder similar to human narcolepsy. The latter group subsequently tried to establish a colony of canines affected with this disorder. The first cases obtained were miniature poodles, and attempts to obtain a genetically affected descendance from these animals was unsuccessful. A few years later, however, several familial cases of narcolepsy in dobermans and labradors were discovered. These animals have been used to establish a colony of genetically transmitted narcoleptic canines, now numbering 80 (Mitler et al., 1980; Baker et al., 1982).

CLINICAL AND ELECTROPHYSIOLOGICAL ASPECTS OF CANINE NARCOLEPSY

Canine narcolepsy is characterized mainly by the existence of cataplexy, an easily observable symptom. Sleepiness, a more subjective symptom, has been documented in these animals using laboratory procedures and electroencephalographic (EEG) recordings.

Canine cataplexy

Cataplexy is defined as an abrupt loss of motor tone induced by emotions (in humans mainly laughter or anger [Guilleminault, 1976]). In narcoleptic canines, this symptom is usually induced by positive emotions, such as play, food, or sexual excitation, but it can also strike spontaneously. In the doberman, playful aggressivity is also an inducing factor, whereas cataplexy is rather inhibited by fear or anxiety. The loss of motor tone experienced during a cataplectic episode can strike the entire body or only a few muscles, thus defining a "complete" or a "partial" attack. Whether partial or complete, the cataplectic episodes can usually be reversed by a loud noise or direct body stimulation.

Partial attacks are usually characterized by a sudden weakness in the hind limbs or, occasionally, in the forelimbs. The muscle weakness strikes bilaterally and the animal will have to sit or lie down if the weakness also extends to its front legs. Often, the animal will try to fight against the muscle atonia, resulting in a bouncing movement between a standing and a sitting position. These partial attacks can resolve spontaneously or extend into a complete attack, the mean duration of the episode being generally less than 15 sec.

Complete attacks involve an abrupt flaccid paralysis of all the postural muscles, rendering the animal unable to stand or to sit. The dog is then completely still, lying in a sometimes awkward position (Fig. 5.1). The muscle atonia is complete, and an abolition of monosynaptic and polysynaptic spinal muscle reflexes, such as the T or the H reflexes, is observed. In a first phase, the eyes are open and the dog is able to follow an object with its eyes. Breathing is regular and deep, and the heart rate is variable. A tachycardia has been reported to precede cataplexy but the heart rate is not remarkable during the attack. When the attacks are prolonged, the eyes become glassy and unreactive. The attack will then often evolve into a typical

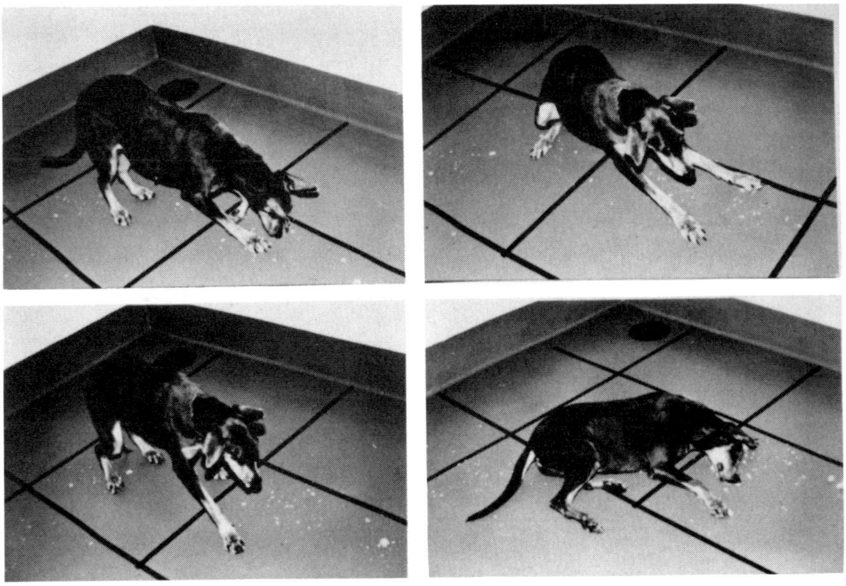

FIGURE 5.1. A narcoleptic doberman in the midst of a cataplectic attack. Notice that the attack was produced by excitation caused by the presentation of food. Photograph reproduced with permission from Nishino et al., *Shinkei kenkyu no Shuinko,* 1990;34:461–473.

REM sleep bout, with associated muscle atonia, rapid eye movements, muscle twitches and, sometimes, vocalizations.

The frequency of these episodes can vary substantially from one animal to another, even in the case of genetically related animals. Some canines show several hundred spontaneously occurring cataplectic episodes per day, whereas other individuals rarely show cataplexy, and only after extreme stimulation (Foutz et al., 1980; Kaitin et al., 1986).

Several tests have been established to quantify cataplexy in narcoleptic canines. These tests rest on the ability of food or play to induce cataplexy. In the Food Elicited Cataplexy Test (FECT), 6 to 12 pieces of wet food (1 cm) are placed 30 cm apart on the floor. The dog has been previously trained to eat each piece of food one after the other. A normal dog will usually eat all the pieces in 8 to 15 sec, whereas a narcoleptic canine, excited by the food, will have one or several cataplectic episodes lasting 2 sec to several minutes. The experimenter records the number of partial and complete cataplectic attacks and the total duration of the test (elapsed time). A second test, the Play Elicited Cataplexy Test (PECT), is also used sometimes to determine and quantify cataplexy. In this test, two animals are placed in the same room and allowed to play together for 10 min. Play induces multiple attacks in the animals and the number of complete and partial episodes occurring during the test is noted.

Electroencephalographic recordings with spectral analysis during canine cataplexy have been performed in various breeds and by several experimenters (Mitler and Dement, 1977; Kushida et al., 1985). The EEG profile is similar to wakefulness at the beginning of the episode (low voltage, mixed frequency) and then will evolve into a full-blown REM sleep episode with REM-like EEG and electromyographic (EMG) activity, associated hippocampal hypersynchronized theta activity (Kushida et al., 1985), pontogeniculocipital waves (PGO), and rapid eye movements (Mitler and Dement, 1977). Finally, it has been observed that canine cataplexy is often associated with a measurable increase in penile tumescence (J.M. Siegel, personal communication), a phenomenon often observed during REM sleep.

Sleep attacks

Sleep attacks are a major clinical feature of human narcolepsy. Some of the narcoleptic canines show typical drowsy attacks, sometimes induced by food. The animal will suddenly stop and wobble on its feet for 10 to 30 sec. The loss of motor tone is not the most obvious symptom in these attacks, and the animal appears more sleepy than cataplectic. On a few occasions, an animal having these attacks suddenly barked and bit the cage bars, a behavior that could be related to a hypnagogic hallucination. These attacks are relatively rare. Electroencephalographic recording during one of these "hallucinatory" behaviors was obtained in only one animal, due to the rarity of the event. Large bursts of slow waves were noted in association with the "sleepy" and abnormal ("hallucinatory") behavior but few conclusions can be derived from only one recording, except the elimination of an epileptic seizure.

Twenty-four hour EEG recordings of narcoleptic canines

The canidae ancestor of the dog was probably a nocturnal animal. The dog is therefore a nocturnal animal that humans have made diurnal. Sleep in these animals is naturally very fragmented, and only a weak circadian distribution of activity and sleep is observed in a light:dark cycle of 12hr:12hr (unpublished results using an actigraph). A comparison of sleep patterns in narcoleptic and control dogs has been performed by several authors in various breeds of narcoleptic and control canines (Mitler and Dement, 1977; Kaitin et al., 1986). These studies have shown that narcoleptic canines have more fragmented sleep than normal animals of the same breed. As has been reported in the nocturnal sleep recordings of human narcoleptics, the affected animals had more frequent sleep stage transitions and awakenings but no major circadian abnormalities.

Chronic sleepiness of the narcoleptic canines

Sleepiness is a subjective symptom, difficult to demonstrate in animals. In humans, objective measurements of sleepiness are performed using a Multiple

TABLE 5.1. Canine breeds affected with narcolepsy.

Breed	Number of cases independently reported[a]
Poodle	10[b]
Dachshund	6[b]
Beagle	3[b]
Doberman	3[c]
Irish setter	2
Cockapoo	2
St. Bernard	2
Labrador	1[c]
Wire-haired griffon	1
Cocker spaniels	1
Springer spaniel	1
Afghan	1
Airedale	1
Malamute	1
Welsh Corgi	1
Chihuahua-terrier mix	1

[a] Several cases were observed in a family, only one was recorded.
[b] Several affected individuals were bred but none of the offspring showed cataplexy (some of the individuals were observed until they were 3 years old).
[c] Several cases were observed in the same litter. Narcolepsy is transmitted in these breeds as an autosomal recessive trait.

Sleep Latency Test (MSLT). In this test, the patient is given the opportunity to take four or five 20-min naps every 2 hr during daytime, and an EEG recording of these naps is performed (APSS, 1986). A pathologically sleepy patient, such as a narcoleptic subject, will have a short sleep latency in each of these naps (mean less than 5 min) whereas the mean sleep latency of a normal volunteer will range from 8 to 20 min. The narcoleptic patients will also frequently show short REM sleep episodes during these 20-min daytime sleep episodes, a phenomenon unusual in normal volunteers.

This test cannot be applied as such to dogs. Lucas et al. (1979) developed a modified version of the MSLT to demonstrate sleepiness in narcoleptic canines using four poodles and two beagles, chronically implanted for EEG recordings (control and narcoleptics). The animals, previously acclimated to the recording chamber during 48 hr, were kept alert for 30 min and allowed to sleep for 30 min every hour during a 24-hr test period. Sleep was recorded continuously, and the obtained sleep latencies were much shorter for the narcoleptic canines than for the control dogs (0.8–3.8 min for the narcoleptics, 6.5–28.5 min for the controls).

Differences in symptomatology between races and natural evolution of the disorder in canines

Narcolepsy–cataplexy has been observed in Doberman pinschers, Labrador retrievers, miniature poodles, dachshunds, beagles, Saint Bernards, and miscellaneous other breeds (Table 5.1).It has also been noted in goats, donkeys, ponies, and miniature horses, and possibly exists in bulls.

The same symptoms are observed in all animals, but the severity clearly varies from one breed to another. The poodles and small breeds are usually more affected than dobermans and labradors (Baker et al., 1982). It is, however, difficult to determine if these differences are really breed-related. All poodles in the colony were given to Stanford after consultations with veterinarians, whereas all dobermans and labradors were born at Stanford University, the disorder being transmitted genetically.

Another difference between breeds is the reported age of onset (Baker and Dement, 1985). In poodles, dachshunds, and other breeds, the symptoms' onset seems to be later than in labradors and dobermans (7 weeks to 7 years), and the age at onset of symptoms can vary tremendously even within a breed (from 10 weeks to 6 years for the poodles, for instance). The clinical evolution is extremely variable in these animals, but, once established, with the exception of two animals, symptoms were continuously noted. Labradors and dobermans have a different long-term disease progression. These two breeds have been extensively investigated at the Stanford University Sleep Research Center: cataplexy may be clearly identified as early as 4 weeks of age, despite the fact that other dogs, sometimes belonging to the same litter, will show symptoms between the 3rd and 4th month only. The mean age of symptoms-onset has been calculated to be 8.6 ± 2.7 weeks of age for dobermans and 14.4 ± 5.6 weeks of age for labradors. Symptoms are very prominent during the first year of life, but a progressive attenuation of cataplexy and other symptoms occurs thereafter, and adult animals may not demonstrate any spontaneous symptoms. Pharmacological challenge tests are therefore necessary to identify the affected animals. The different clinical aspects of canine narcolepsy suggest that in dogs, as in man, there may be different forms of narcolepsy. Canine narcolepsy appears to present as a heterogeneous problem.

GENETIC TRANSMISSION OF NARCOLEPSY IN ANIMALS

As soon as several narcoleptic canines were donated to Stanford, Mitler et al. and then Baker tried to breed the narcoleptic dogs in order to establish the genetic transmission of the disorder. The first attempts involved four affected and unrelated poodles, and none of the offspring obtained in two crosses were affected. One of the offspring of the first generation was also backcrossed with his parent, and none of the resulting offspring were affected (Baker et al., 1982). The same experiment was also performed with a couple of beagles (including a backcross), and later dachshunds, with none of the offspring developing narcolepsy. We now believe, therefore, that narcolepsy in these breeds is a polygenic and/or environmentally influenced disorder.

In 1982, several dobermans and labradors, belonging to the same families and showing the symptoms of narcolepsy, were discovered, and constituted the beginning of our current narcoleptic canine colony. Since then, narcolepsy has been observed in 18 breeds, but familial cases were reported only for labradors and dobermans (Table 5.1). In addition, with the collaboration of the University of

California (Davis School of Veterinary Medicine), cases of narcolepsy in a pony and a miniature horse have been diagnosed (Foutz et al., 1980).

The genetic transmission of narcolepsy in dobermans and labradors is now well established (Foutz et al., 1980; Baker et al., 1982). Dobermans transmit the disease as an autosomal recessive trait with complete penetrance. More than 60 dogs born from two narcoleptic parents have been examined and all showed spontaneous cataplexy before the 4th month of age whereas more than 30 animals issues from three crosses of normal with affected dogs never showed any attacks (some have been observed up to 3 years of age). As a final test for recessiveness and for linkage study, backcrosses have been carried out between heterozygous animals and homozygous narcoleptic dobermans. To date, 26 backcross doberman animals (three litters, Fig. 5.2) have been born at Stanford University and examined for the appearance of symptoms during development. Of these animals, 15 showed cataplexy before 5 months of age, a result in agreement with the 2:2 distribution expected for the autosomal recessive transmission of the trait in an autosomal recessive model with full penetrance. Although fewer investigations have been performed on Labrador retrievers, the data also suggest an autosomal recessive transmission trait. Only three litters of animals (25 dogs) have been born from two narcoleptic labrador parents, and all offspring showed cataplexy. A backcross litter of 16 labradors was also born (Fig. 5.2), and the distribution of symptomatic versus asymptomatic offspring obtained is also consistent with an autosomal recessive model with full penetrance. It seems clear that the same locus is involved for the narcolepsy gene in both breeds, as pups born from affected labrador and doberman crosses are all affected (12 dogs from two different litters).

MAJOR HISTOCOMPATIBILITY COMPLEX CLASS II ANTIGENS IN CANINE NARCOLEPSY

The finding that in dobermans and labradors inheritance of narcolepsy syndrome is consistent with a single autosomal recessive gene, that we have called "*canarc-1*," led us to examine whether a canine MHC–narcolepsy association analogous to that observed in humans was present. Investigations of the presence of a specific disease-associated DLA-D (dog leukocyte class II antigen, analogous to HLA-D/DR) antigen were performed. Three distinct dog populations were examined: Doberman pinschers and Labrador retrievers with autosomal recessive inheritance, and small breed dogs in which the inheritance pattern is probably multigenic (Baker et al., 1982). Using homozygous typing cells (HTCs) obtained from dogs homozygous for a DLA-D antigen and capable of identifying D antigens in other dogs by their pattern of differential reactivity (Mixed Leucocyte Culture, MLC), it soon appeared that, unlike in man, few narcoleptic dogs shared any single D locus antigen. Also, the investigation revealed that narcoleptic dogs, unlike narcoleptic humans, do not possess a single, nearly unique disease-related DLA-D type. In canine narcolepsy then, expression of an almost unique MHC class II antigen does not appear to be a requirement for manifestation of the disease.

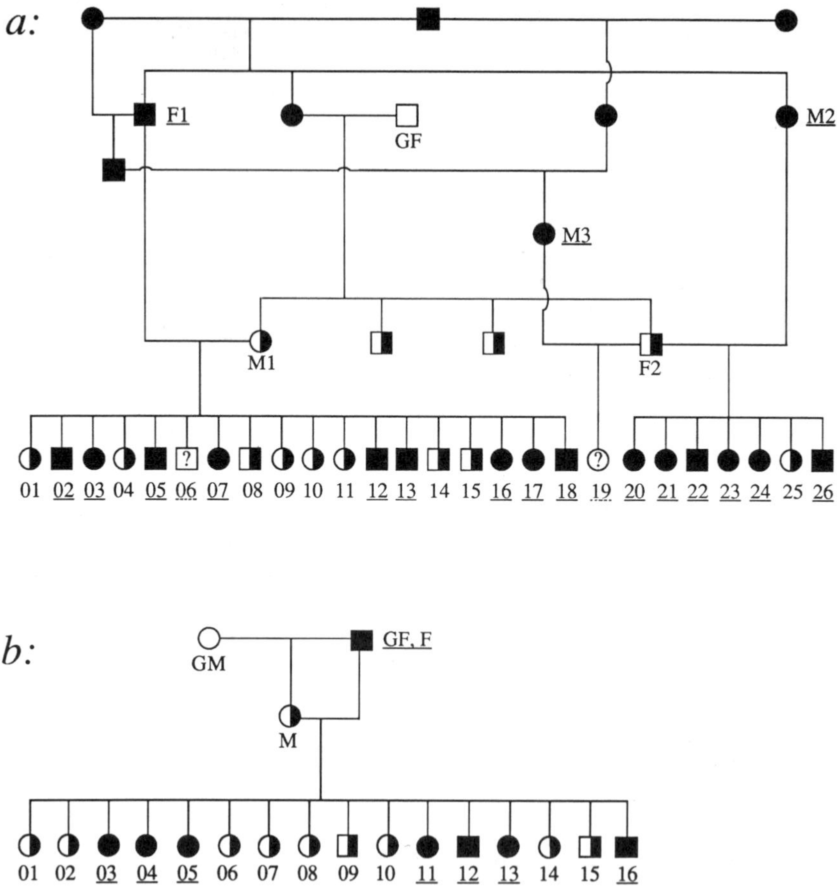

FIGURE 5.2. Genetic trees of the doberman (**A**) and labrador (**B**) backcrosses. Circle *(female)* and square *(male)* symbols show clinical and *canarc-1* status; filled, half filled, or empty represent respectively affected, carrier, and healthy control canines. GF, Grandfather; F, father; M, mother. Number were sequentially assigned to offspring and are underlined when narcoleptic. Individuals 06 and 19 were excluded from the study because of uncertain diagnosis.

We also examined the possibility of a genetic linkage between *canarc-1* and the DR locus. To do so, backcross litters of dobermans and labradors were generated (Fig. 5.2) to examine if the inheritance of MHC class II antigens was consistent with the inheritance of canine narcolepsy. Informative offspring were observed every 2 weeks for spontaneous and food-elicited cataplexy up to 5 months of age (underlined animals in Figs. 5.2 and 5.3). To confirm the diagnosis, backcross animals were also treated at 6 months of age with 0.1 mg/kg i.v. of physostigmine. This compound, a cholinomometic, is known to trigger cataplexy in narcoleptic (*canarc-1* homozygous) animals but is without effect in carrier (*canarc-1* het-

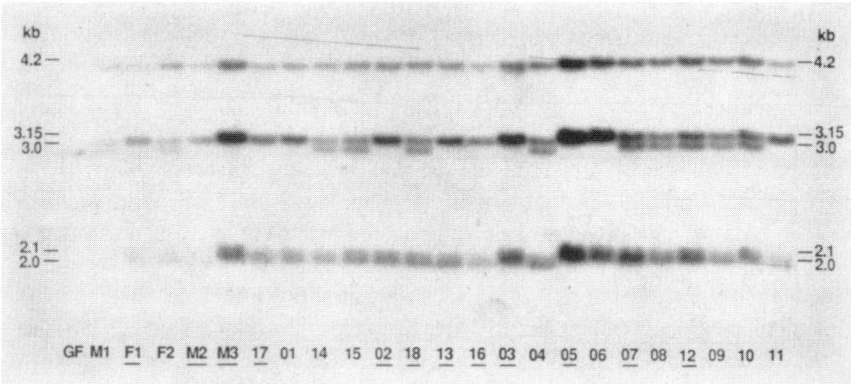

FIGURE 5.3. Southern hybridization of canine DNA (6 μg/lane, digested with AvaII) with the human HLA DRβ probe (2918.4, Bell et al., 1985) in part of the doberman family of Fig. 5.2. Methods are described in details in Mignot et al. (1991a). Note that the informative 3.0 kb band coming from the normal grandfather is segregating independently of the narcolepsy susceptibility in this crosses (double backcrosses with phase known). GF, grandfather, M1, *M2*, *M3*, mothers of the backcrosses; *F1*, F2, fathers of the backcrosses. Individual informative offspring are numbered as in Fig. 5.2. Animals showing spontaneous cataplectic attacks are underlined.

erozygous) animals. Using this additional stimulus, two previously asymptomatic animals showed cataplexy (individuals 06 and 19). These animals were excluded from the linkage study because their diagnosis assignments were unclear. The HLA DRβ clone #2918.4 was used as a probe (Bell et al., 1985), and an informative polymorphism was found with Pst1 in the Labrador family (4.9- and 12-kb bands) and with HinfI and AvaII (Fig. 5.2) in the Doberman family (2.0-kb and 3.0-kb, bands respectively). Twenty-six dobermans and 16 labradors informative for this marker were studied, and the results excluded the possibility of tight genetic linkage between *canarc-1* and the canine MHC (m = 40 informative offspring, k = 10 observed recombinants $Z < -2$ for $\theta < 4.8\%$) (Fig. 3). The autosomal recessive factor segregating in our canine population of dobermans and labradors, *canarc-1* is thus not related to the HLA system.

This result, however, does not exclude the participation of DLA in the pathophysiology of the disorder. A permissive genetic background (of the DLA system or of non–DLA-related genes) in our canine population could allow the canine non-MHC narcolepsy gene *canarc-1* to be expressed in a fully penetrant fashion. Similarly, it is also possible that the human equivalent of *canarc-1* is one of the factors underlying human narcolepsy but that additional genetic and/or environmental factors are needed to express the disorder in humans (see above). The recent discovery of human HLA DR2-negative narcoleptic cases, and the newly reported DR2-negative family, also provide an alternative to this model, demonstrating that the HLA predisposition is not a requirement for the expression of narcolepsy.

PHARMACOLOGICAL AND NEUROCHEMICAL INVESTIGATIONS IN CANINE NARCOLEPSY

Human and canine narcolepsy involve abnormal manifestations of REM sleep. Pharmacological manipulations of REM sleep have been performed for many years, and cholinergic and monoaminergic systems have been shown to be involved in the normal regulation of REM sleep. It was thus rational to perform pharmacological investigations of canine narcolepsy using agonists and antagonists of these systems. Our results to date indicate that drugs usually affect cataplexy in a way similar to that in which they modify REM sleep, that is, drugs that selectively increase/decrease REM sleep increase/decrease cataplexy. The use of the canine model, however, allows us to study a large number of compounds and is extremely specific. In addition, cataplexy is a behavior that occurs in the awake animal. It is therefore possible to observe the effects of drugs in the general behavior for the same dose necessary to elicit an effect on cataplexy (and presumably REM sleep). This makes it possible to determine whether the effects occur with or without any behavioral and/or peripheral side effects.

Cholinergic systems in the pontine brain stem have been known to be involved in the regulation of REM sleep for many years (for review see Siegel, 1989; Mignot et al., 1991b). Stimulation of these systems favors REM sleep, whereas pharmacological or mechanical lesions suppress REM sleep. In agreement with these results, cholinergic muscarinic stimulation using physostigmine or arecoline in narcoleptic animals elicits cataplexy, whereas cholinergic blockade using atropine improves the symptoms (Delashaw et al., 1979; Mignot et al., 1988a). This is a centrally mediated effect because methyl-atropine or neostigmine, which do not act centrally, have no effect on cataplexy. These investigations using cholinergic agents were associated with specific investigations of muscarinic receptors in various parts of the brain of narcoleptic dogs compared to normal dogs. Two different studies have indicated that muscarinic cholinergic receptors (subtype M2) present an upregulation in the pontine reticular formation (Boehme et al., 1984; Kilduff et al., 1986). This region is precisely the one where local injections of cholinergic agonists can elicit REM sleep or REM sleep atonic episodes (Siegel, 1989). This suggests that the abnormal REM sleep tendencies and abnormal muscle atonia demonstrated by narcoleptic dogs are specifically mediated by these hypersensitive cholinoceptive neurons.

Knowledge of the cholinergic involvement in canine cataplexy and REM sleep has not resulted in any new, improved therapy for narcoleptic patients. Human and canine narcolepsy are known to be clearly improved by amphetamine and amphetamine-like drugs (Guilleminault, 1976), drugs reported to enhance monoamine release globally (i.e., dopamine, norepinephrine, and serotonin) without affecting the cholinergic system. Also, many of the REM sleep-related manifestations of narcolepsy are improved by tricyclic antidepressant medications, drugs that act mainly on synaptic monoaminergic reuptake. These and other experimental data suggest that monoamines in addition to acethylcholine are involved in the control of REM sleep and narcolepsy. These global nonspecific

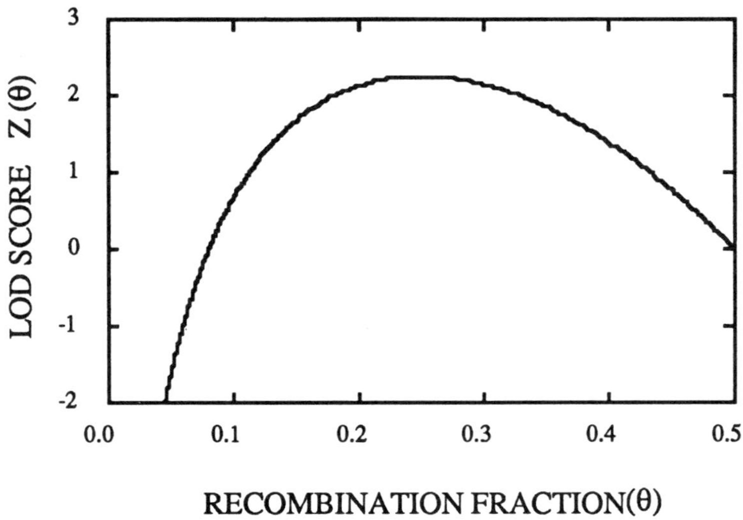

FIGURE 5.4. Absence of tight genetic linkage of *canarc-1* with DLA: LOD score values [Z(θ)] at various recombination fraction (θ). Results were obtained using the crosses reported in Fig. 5.2 (m = 40 informative offspring). LOD score equations were obtained from Morton (1955) and Ott (1985) and plotted using a SYSTAT program on MacIntosh II. Note that Z(θ) never reach +3, the threshold for significant linkage [Zmax(θ)=2.27 at θ=25%]. Linkage was excluded for the values of Z<-2, thus for θ<4.8%.

pharmacological effects on the monoaminergic systems were investigated in the narcoleptic canines using specific uptake inhibitors and receptor subtype-specific agonists and antagonists (Mignot et al., 1991b).

Serotonergic reuptake blockers (fluoxetine, zimelidine, indalpine, chlorimipramine, amitryptiline) and dopaminergic uptake blockers (amineptine, GBR12909) had a mild effect on canine cataplexy, whereas norepinephrine reuptake blocking agents (nixoxetine, tomoxetine, desipramine, protryptyline) efficiently blocked cataplectic attacks in dogs (Foutz et al., 1981; Renaud et al., 1991). The dopaminergic reuptake blocker, benzotropine, was also studied, and the compound slightly decreased canine cataplexy (Baker and Dement, 1985), a result difficult to interpret because of the anticholinergic properties of the compound. These results suggested a preferential involvement of norepinephrine in the control of cataplexy and in the therapeutic effects of the nonspecific monoaminergic compounds (Mignot et al., 1988b, 1991b). More specific investigations of selective alpha- and beta-adrenergic compounds on canine cataplexy were therefore performed. The effects of central and peripheral beta-1 and beta-2 adrenergic stimulation or blockade on cataplexy were examined using propanolol, albuterol, salbutamol, and isoproterenol. Our results indicate that beta-adrenergic receptors do not play a major role in the regulation of cataplexy.

We then examined the effects of alpha-1 and alpha-2 adrenergic compounds on canine cataplexy. Alpha-1 antagonists such as prazosin and phenoxybenzamine significantly worsen canine (Mignot et al., 1988a,b) and human (Guilleminault et al., 1989) cataplexy. Further studies allowed us to demonstrate the presence of at least two alpha-1 receptor subtypes in the canine central nervous system, called alpha-1a and alpha-1b (Mignot et al., 1989a,b). Drugs with a higher affinity for alpha-1b receptors have been found to induce or block cataplexy in dogs, suggesting an involvement of the putative alpha-1b receptors in the disorder. We now think that noradrenergic mechanisms are inhibitory of cataplexy (and REM sleep atonia) through the stimulation of selective alpha-1 adrenergic receptors subtypes. The alpha-2 adrenergic system was also investigated, and alpha-2 antagonist drugs have demonstrated their effectiveness in suppressing canine cataplexy (Nishino et al., 1990). Since the effects of alpha-2 antagonists are similar to those of alpha-1 agonists (i.e., an increase in alertness and a decrease in cataplexy is obtained), we think that the effects of alpha-2 antagonists are mediated through presynaptic alpha-2 receptors regulating noradrenergic activity (release and firing rate).

Narcoleptic canines have also been used to investigate neurochemical changes in normal and pathological brain tissues. Analysis of monoamine, metabolite, and receptor levels from narcoleptic and control dogs has shown a general increase in catecholamines in the narcoleptic dog brain (Mefford et al., 1983). Replication of this study confirmed that a significant increase of dopamine and DOPAC exist in the amygdala of narcoleptic dogs (Faull et al., 1986). Concentrations of dopamine (D2) and adrenergic (alpha-1) receptors were also measured and found to be elevated in the amygdala, but not in other brain structures, of the narcoleptic dog (Bowersox et al., 1987; Mignot et al., 1988a). These results suggest the critical involvement of the amygdala in the pathophysiology of narcolepsy. The role of this structure in emotional behaviors is well documented and could explain how cataplexy is precipitated by emotions. We also investigated the number of alpha-2 receptors in the locus coeruleus and found a significant increase in this structure (Fruhstorfer et al., 1989). Taken together with the results obtained with alpha-2 antagonists on canine cataplexy, this result could predispose the animal to sleepiness and cataplexy.

To better appreciate the interactions of noradrenergic and cholinergic systems, pharmacological co-administration experiments were performed using cholinergic and noradrenergic compounds. Results of these experiments suggest that noradrenergic effects need the integrity of cholinergic systems to be expressed. Pretreatment with atropine completely abolished the effects of prazosin (Mignot et al., 1988b), whereas physostigmine still increases cataplexy in the presence of nisoxetine (Foutz et al., 1981), a noradrenergic uptake blocker. This result suggests that catecholaminergic influence on cataplexy is mediated through a modulation of pontine cholinergic systems.

To integrate these various findings, we currently suggest that at least two systems are involved in the regulation of cataplexy: a pontine cholinergic system, starting from the pedunculopontine nucleus and projecting to the pontine reticular formation, and a noradrenergic system arising from the locus coeruleus and projecting to the pontine cholinergic neurons and the amygdala. Projections from the amygdala to the brain stem (locus coeruleus and/or pontine cholinergic neurons) would explain the role of emotions in triggering cataplexy. To test this model, we are currently injecting compounds into selected brain areas of the narcoleptic dog to examine their effects on cataplexy and sleep.

LINKAGE STUDIES IN CANINE NARCOLEPSY: COSEGREGATION OF *Canarc-1* WITH AN IMMUNOGLOBULIN μ-CHAIN SWITCH-LIKE GENE

The well established mode of transmission of canine narcolepsy offers a unique opportunity to identify and isolate a high penetrance gene responsible for the syndrome through reverse genetics *(canarc-1)*. The first step in the characterization of *canarc-1* is to find linkage markers. In order to do so, two approaches were taken, using either minisatellite probes or candidate genes.

Minisatellite DNA sequences (also called Variable Number of Tandem Repeats, or VNTRs) consist of a core sequence of 10 to 60 nucleotides repeated a few to several hundred times (Jeffreys et al., 1985a, 1985b, Nakamura et al., 1987, Wong et al., 1986, Georges et al., 1988, 1990). The number of repeats located at a given locus can vary tremendously from one individual to another, resulting in a high degree of polymorphism. These sequences are organized by small clusters that are spread over the genome, with a moderate preferential localization at proterminal chromosomal bands (Nakamura et al., 1988; Georges et al., 1990). The minisatellite probes can be used at high stringency to detect a single locus (VNTRs) (Nakamura et al., 1987, 1988) or at low stringency to detect a number of related sequences. In this case, each probe can detect 7 to 15 loci, each locus having up to 40 different alleles. The resulting Southern picture obtained consists of a large number of polymorphic fragments that are specific to any given individual ("DNA fingerprint," Fig. 5.5) and used for problems of human identification (Jeffreys et al., 1985a,b). All these characteristics make the minisatellite probes well suited for linkage studies in animal species for which no genetic map is available, such as the dog (Jeffreys and Morton, 1987, Lai et al., 1989). The probability of heterozygocity at each locus is high (95%) and most of the crosses studied are informative for the bands detected. Using this approach, Georges et al. (1990) calculated that by using five minisatellite probes in cattle, he was able to cover approximately 25% of the genome.

The candidate gene approach uses probes recognizing genes related to systems putatively involved in the pathophysiology of the disorder studied. If linkage is found, it is likely that *canarc-1* is related to the probe showing linkage. The candidate genes selected were related either to neurochemical systems implicated in canine narcolepsy (cholinergic and monoaminergic enzymes or receptors, see above) or to immune factors (because of the HLA association in humans). The difficulty of this approach is in detecting an informative polymorphism with the probe used.

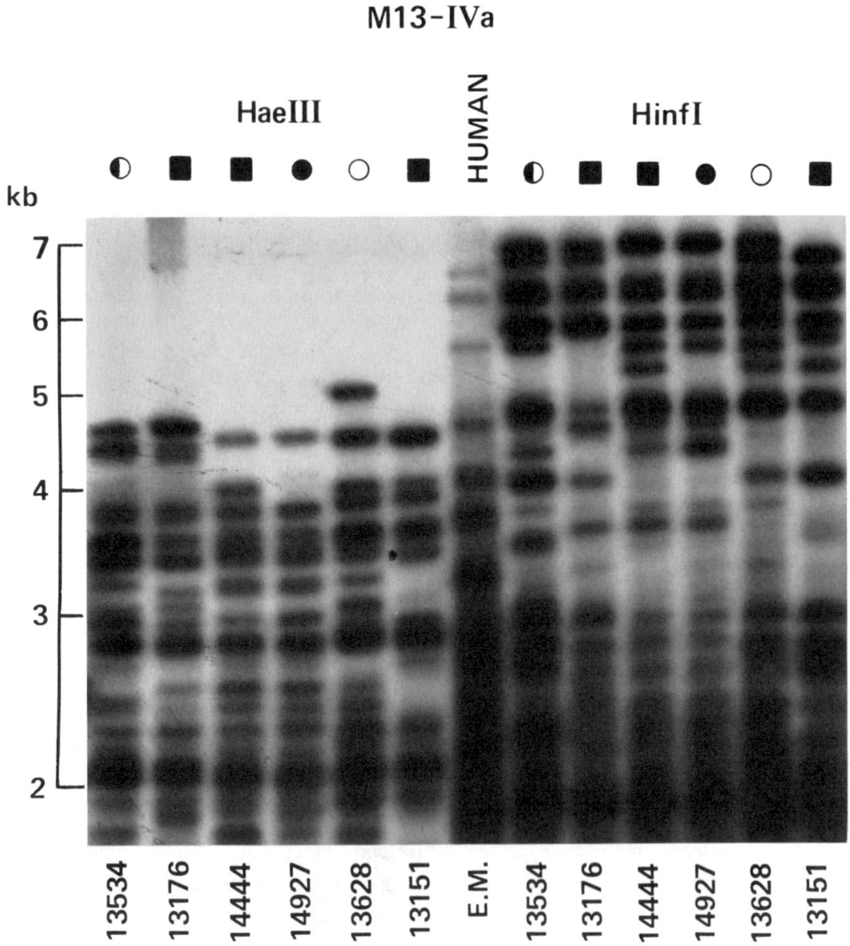

FIGURE 5.5. DNA fingerprints of various unrelated dobermans and labradors. DNA was extracted from blood lymphocytes, digested using Hae III *(left)* of Hinf I *(right)* and electrophoresed in 0.8% agarose gel for 48 hr at 40 V. DNA was then transferred onto a nitrocellulose membrane and Southern blots hybridized to ^{32}P-labeled bovine minisatelite probe (M13-IVa (a gift of Dr. Georges). Methods are described in more detail in Georges et al. (1988).

Using this double approach, we have recently identified a unique gene related to a human μ-switch immunoglobulin gene (Sμ) that is genetically linked with *canarc-1* (Mignot et al., 1991a). When applied on canine DNA, this probe detected several related genes at high stringency (at least three; two of them (strong hybridizing bands) being in linkage with each other but not with *canarc-1,* whereas the third locus (weak hybridization band) cosegregated with *canarc-1.* Twenty-four animals informative for the marker were examined and all animals

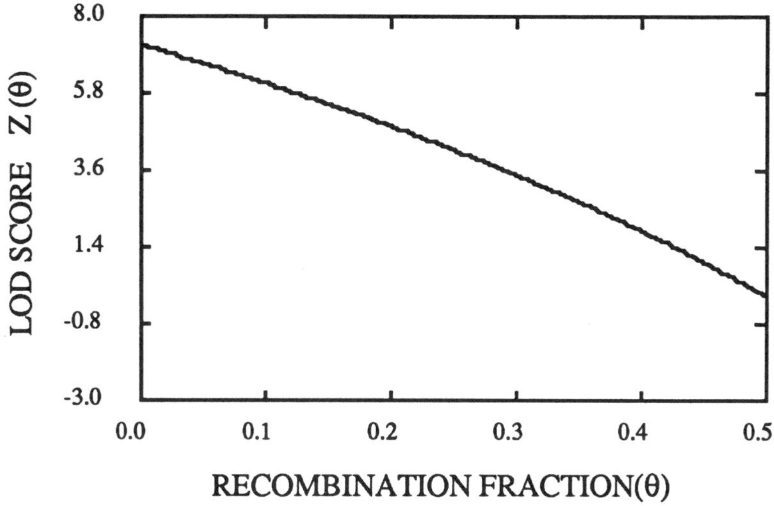

FIGURE 5.6. Genetic linkage of *canarc-1* with Sμ: LOD score values [Z(θ)] at various recombination fractions (θ). Results were obtained using the doberman crosses reported in Fig. 5.2. (m = 24 informative offspring). LOD score equations were obtained from Morton (1955) and Ott (1981) and plotted using a SYSTAT program on MacIntosh II. Note that linkage is statistically significant, Z(θ) being maximal at θ = 0% (no observed recombinant).

were concordant. The maximal LOD score is 7.2 at θ = 0% recombination, a highly significant result (Fig. 5.6).

Linkage of canine narcolepsy with an Sμ-like segment suggests several hypotheses. Immunological switch genes are involved in a complex somatic recombination process allowing B-cells to switch immunoglobulin class upon activation (Essert and Radbruch, 1990, Dupont, 1990). The band in linkage with *canarc-1* could either be a nonfunctional segment, or be involved in a somatic recombination process involving the immune system or another system. In humans and mice, most of the switch genes are clustered in only one region of a few hundred kb localized at the 3′ end of the immunoglobulin heavy chain locus (respectively 14q32 and 12) (Lai et al., 1989). The fact that our marker for *canarc-1* is not in linkage with the two other strong cross hybridizing loci detected with the probe suggests that *canarc-1* is not located within the canine immunoglobulin heavy chain cluster locus. A somewhat less likely explanation is that canine immunoglobulin switch genes are organized in unlinked clusters. Sμ-like genes located outside the main heavy chain complex have been isolated (Migone et al., 1983; Stockinger et al., 1986) and our marker could represent another example of this phenomenon.

The exact mechanism of action of *canarc-1* still remains to be determined. Although novel, nonimmunological functions could be proposed, the implication of immune genes in both humans and canines (i.e., HLA haplotype association in human and Sμ linkage in dogs) suggests an immunopathology in both narcoleptic canines and humans. We are now cloning the Sμ-like gene in linkage with *canarc-1* and will be using the clone as a probe in human narcoleptic patients to test the hypothesis of the involvement of *canarc-1* in the human pathology.

CONCLUSION

Pharmacological and biochemical studies of control and narcoleptic dogs will progressively enhance our therapeutic arsenal to treat human narcolepsy and our understanding of the neurochemical control of REM sleep. Also, the discoveries of a genetic marker for *canarc-1* will allow us to perform linkage analysis in humans, particularly in DR2-negative families. This could lead to the development of new diagnostic tools and ultimately to the identification of genes underlying the pathophysiology of narcolepsy.

Finally, we have emphasized that human and canine narcolepsy are clearly associated with an abnormal control of REM sleep, one of the three fundamental states of life. We hope that understanding narcolepsy may also shed light on the mechanisms and functions of the sleep process itself.

Acknowledgments. We thank Dr. Mike Lovett for his helpful guidance, Dr. Seiji Nishino for his comments, and Lewanne Sharp, Janis Arragoni, and Mike Gelb for their assistance. This work was supported by Grants NS 23724, NS 2710, and NS 15184 from the National Institute of Health to W.C. Dement.

REFERENCES

APSS Guidelines Committee, Carskadon MA, Chairman (1986): Association of Professional Sleep Societies Guidelines for the Multiple Sleep Latency Test (MSLT), a standard measure of sleepiness, *Sleep* 9:519–524

Baker TL, Foutz AS, McNerney V, Mitler MM, Dement WC (1982): Canine model of narcolepsy: genetic and developmental determinants *Exp Neurol* 75:729–762

Baker TL, Dement WC (1985): Canine narcolepsy-cataplexy syndrome: evidence for an inherited monoaminergic-cholinergic imbalance. In: *Brain Mechanisms of Sleep,* McGinty DJ, Drucker-Colin R, Morrison A, Parmengiani P eds. New York: Raven Press, pp 199–233

Baraitser M, Parkes JD (1987): Genetic study of the narcolepsy syndrome. *J Med Genet* 15:254–259

Bell JI, Estess P, ST John T, Saiki R, Watling DL, Erlich HA, McDevitt HO (1985): DNA sequence and characterisation of human class II major histocompatibility complex beta chains from the DR1 haplotype. *Proc Natl Acad Sci (USA)* 82:3405–3409

Billiard M, Seignalet J (1985): Extraordinary association between HLA-DF2 and narcolepsy. *Lancet 1:226–227*

Boehme R, Baker T, Meford I, Barchas J, Dement W, Ciaranello R (1984). Narcolepsy: Cholinergic receptor changes in an animal model. *Life Sci* 34, 1825–1828

Bowersox SS, Kilduff T, Faull K, Dement WC, Ciaranello RD (1987): Brain dopamine receptor levels elevated in canine narcolepsy. *Brain Res* 402:44–48

Bruhova S, Roth B (1972): Heredofamilial aspects of narcolepsy and hypersomnia. *Arch Suisse Neurol Neurochirurg Psychiatr* 110:45–54

Daly D, Yoss R (1959): A family with narcolepsy. *Proc Staff Meet Mayo Clinic* 34:313–320

Dean RR, Kilduff TS, Dement WC, Grumet FC (1989): Narcolepsy without unique MHC Class I antigen association: Studies in the canine model. *Hum Immunol* 25:27–35

Delashaw JB, Foutz AS, Guilleminault C, Dement WC (1979): Cholinergic mechanisms and cataplexy in dogs. *Exp Neurol* 66:745–757

Douglas AB, Harris L, Pazderka F (1989). Monozygotic twins concordant for the narcoleptic syndrome. *Neurol* 39:140–141

Dupont BO (1990) *Immunobiology of HLA,* Vols. 1 and 2. New York: Springer–Verlag.

Essert C, Radbruch A (1990): Immunoglobulin class switching: molecular and cellular analysis. *Annu Rev Immunol* 8:717–735

Faull KF, Zeller-DeAmicis LC, Radde L, Bowersox SS, Baker TL, Kilduff TS, Dement WC (1986): Biogenic amine concentrations in the brains of normal and narcoleptic canines: current status. *Sleep*. 9:107–110

Foutz AS, Delashaw JB, Guilleminault C, Dement WC (1981): Monaminergic mechanisms and experimental cataplexy. *Ann Neurol* 10:369–376

Foutz AS, Mitler MM, Dement WC (1980): Narcolepsy. *Vet Clin North Am Small Anim Practice* 10:65–80

Fredrikson S, Carlander B, Billiard M, Link H (1990): CSF Immune variable in patients with narcolepsy. *Acta Neurol Scand* 81,253–254

Fruhstorfer B, Mignot E, Bowersox S, Nishino S, Dement WC, Guilleminault C (1989): Canine narcolepsy is associated with an elevated number of alpha-2 receptors in the locus coeruleus. *Brain Res* 500:209–214

Gelineau JB (1880): De la narcolepsie. *Gaz des Hop (Paris)* 53:626–628, 54:635–637

Georges M, Lathrop M, Hilbert P, Marcotte A, Schwers A, Swillens S, Vassart G, Hanset R (1990): On the use of DNA fingerprints for linkage studies in bovine. *Genomics (USA)* 6:461–71

Georges M, Lequarré AS, Castelli M, Hanset R, Vassart G (1988): DNA fingerprinting in domestic animals using four different minisatelite probes. *Cytogenet Cell Genet* 47:127–131

Guilleminault C (1976): Cataplexy. In: *Narcolepsy,* Guilleminault C, Passouant P, Dement WC, eds. New York: Spectrum Publications Inc, pp 125–143

Guilleminault C, Grumet C (1986): HLA-DR2 and narcolepsy: not all narcoleptic cataplectic patients are DR2. *Hum Immunol* 17:1–2

Guilleminault C, Mignot WC, Grumet FC (1989): Familial patterns of narcolepsy. *Lancet* 335:1376–1379

Honda Y, Asaka A, Tanimura M, Furusho T (1983a): A genetic study of narcolepsy and excessive daytime sleepiness in 308 families with narcolepsy or hypersomnia proband. In: *Sleep/Wake Disorders; Natural History, Epidemiology and Long-Term Evolution,* Guilleminault C, Lugaresi E, eds. New York: Raven Press, pp 187–199

Honda Y, Asaka A, Tanaka Y, Juji T (1983b): Discrimination of narcolepsy by using genetic markers and HLA. *Sleep Res* 12:254

Jeffreys AJ, Morton DB (1987): DNA fingerprints of dogs and cats. *Anim Genet* 18:1–15

Jeffreys AJ, Wilson V, Thein SL (1985a): Hypervariable "minisatelite" regions in human DNA. *Nature* 314:67–73

Jeffreys AJ, Wilson V, Thein SL (1985b): Individual-specific fingerprints of human DNA. *Science* 316:76–79

Jouvet M (1962): Recherches sur les structures nerveuses et les mechanismes responsables des differentes phases du sommeil physiologique. *Arch Ital Biol* 100:125–206

Kaitin KI, Kilduff TS, Dement WC (1986): Sleep fragmentation in canine narcolepsy. *Sleep* 9:116–119

Kessler S (1976): Genetic factors in narcolepsy. In: *Narcolepsy,* Guilleminault C, Dement WC, Passouant P, eds. New York: Spectrum Publications Inc

Kessler S, Guilleminault C, Dement WC (1974): A family study of 50 REM narcoleptics. *Acta Neurol Scand* 50:503–512

Kilduff TS, Bowersox SS, Kaitin KI, Baker TL, Ciaranello RD, Dement WC (1986): Muscarinic cholinergic receptors and the canine model of narcolepsy. *Sleep* 9:102–107

Knecht CD, Oliver JE, Redding R, Selcer R, Johnson C (1973): Narcolepsy in a dog and a cat. *J Am Vet Med Assoc* 162:1092–1093

Krabbe E, Magnussen G (1942): Familial narcolepsy. *Acta Psychiatr Neurol* 17:149–173

Kushida CA, Baker TL, Dement WC (1985): Electroencephalographic correlates of cataplectic attacks in narcoleptic canines. *Electroencephalogr Clin Res* 61:61–70

Lai E, Wilson RK, Hood LE (1989): Physical maps of the mouse and human immunoglobulin-like loci. *Adv Immunol* 46:1–59

Langdon N, Welsh KI, Van Dam M, Vaugham RW, Parkes JD (1986): Genetic markers in narcolepsy. *Lancet* 2:1178–1180

Lock CB, So AKL, Welch KI, Parkes JD, Trowsdale J (1988). MHC class II sequences of an HLA-DR2 narcoleptic. *Immunogenetics* 27:449–455

Lucas EA, Foutz AS, Dement WC, Mitler MM (1979): Sleep cycle organization in narcoleptic and normal dogs. *Physiol Behav* 23:737–743

Matsuki K, Honda Y, Juji T (1989): HLA antigens in 206 Japanese Patients with Narcolepsy and 46 patients with essential hypersomnia. In: *Immunobiology of HLA,* Vol. II, New York: Springer–Verlag, pp 438–440

Matsuki K, Juji T, Honda Y (1988). Immunological features in Japan. In: *HLA and Narcolepsy,* Honda Y, Juji T, eds. New York: Springer–Verlag, pp 58–76

Mefford IW, Baker TL, Boehme R, Foutz A, Ciaranello R, Barchas J, Dement WC (1983): Narcolepsy: biogenic amine deficits in an animal model. *Science* 220:629–632

Mignot E, Bowersox S, Maddaluno J, Dement W, Ciaranello R (1989a). Evidence for multiple [^{3}H]-Prazosin binding sites in canine brain membranes. *Brain Res* 486: 56–66

Mignot E, Guilleminault C, Bowersox S, Frushorfer B, Nishino S, Maddaluno J, Ciaranello R, Dement WC (1989b): Central alpha-1 adrenoceptor subtypes in narcolepsy-cataplexy, a disorder of REM sleep. *Brain Res* 490:186–191

Mignot E, Guilleminault C, Bowersox S, Rappaport A, Dement WC (1988a): Effect of alpha-1 adrenoceptor blockade with prazosin in canine narcolepsy. *Brain Res* 444:184–188

Mignot E, Guilleminault C, Bowersox S, Rappaport A, Dement WC (1988b): Role of central alpha-1 adrenoceptors in canine narcolepsy. *J Clin Invest* 82:885–894

Mignot E, Nishino S, Valtier D, Renaud A, Guilleminault C, Dement WC (1991b): Pharmacology of canine narcolepsy. In: *Sleep 90,* Horne J, ed., pp 409-415.

Mignot E, Wang C, Rattazzi C, Gaiser C, Lovett M, Guilleminault C, Dement WC, Grumet C (1991a): Genetic Linkage of autosomal recessive canine narcolepsy with an immunoglobulin μ chain switch like segment. *Proc Natl Acad Sci (USA),* 88:3475-3478.

Migone N, Fefder J, Cann H, Van West B, Wang J, Takahashio N, Honjo T, Piazza A,

Cavalli-Sforza LL (1983): Multiple DNA polymorphisms associated with immunoglobulin μ-switch-like regions in man. *Proc Natl Acad Sci (USA)* 80:467–471
Mitler MM, Boysen BG, Campbell L, Dement WC (1974): Narcolepsy cataplexy in a female dog. *Exp Neurol* 45:332–340
Mitler MM, Dement WC (1977): Sleep studies on canine narcolepsy: pattern and cycle comparisons between affected and normal dogs. *Electroencephalogr Clin Res* 43:691–699
Mitler MM, Dement WC, Guilleminault C, Foutz AS (1980): Canine narcolepsy. In: *Animal Models of Neurological Disease*. Rose F, Behan P, eds. Kent, England: Pitman Medical Ltd.
Montplaisir J, Poirier G (1987): Narcolepsy in monozygotic twins. *Neurology* 37:1089–1092
Montplaisir J, Poirier G (1988): HLA in Narcolepsy in Canada. In: *HLA in Narcolepsy*, Honda C, Juji T, eds. Berlin: Springer–Verlag, pp 97–107
Morton NE (1955): Sequential tests for the detection of linkage. *Am J Hum Genet* 7:277–318
Motoyama M, Kilduff TS, Lee BSM, Dement WC, McDevitt HO (1990): Restriction fragments length polymorphisms in canine narcolepsy. *Immunogenetics* 29:124–126
Nakamura Y, Carlson M, Kapcho K, Kanamori M, White R (1988): New approach for isolation of VNTR markers. *Am J Hum Genet* 43:854–859
Nakamura Y, Leppert M, O'Connell P, Wolff R, Holm T, Culver M, Martin C, Fujimoto E, Hoff M, Kumlin E, White R (1987): Variable number of tandem repeat (VNTR) markers for human gene mapping. *Science* 235:1616–1622
Neely S, Rosenberg R, Spire JP, Antel J, Arnason BEW (1987): HLA antigens in narcolepsy. *Neurology* 37:1858–1860
Nishino S, Haak L, Sheperd H, Guilleminault C, Sakai T, Dement WC, Mignot E (1990): Effects of central alpha-2 adrenergic compounds on canine narcolepsy, a disorder of rapid eye movement sleep. *J Pharmacol Exp Ther* 253:1146–1152
Ott J (1985): *Analysis of human genetic linkage*. Baltimore John Hopkins University Press
Parkes JD, Lock C, Langdon N (1986): Narcolepsy and immunity. *Br Med J* 292:359–360
Poirier G, Montplaisir J, Decary F, Momege D, Lebrun A (1986): HLA antigens in narcolepsy and idiopathic hypersomnolence. *Sleep* 9:153–158
Pollmacher T, Schulz H, Geisler P, Kiss E, Albert ED, Swarzfischer F (1990): Monozygotic twins discordand for narcolepsy. *Sleep* 13:336–343
Renaud A, Nishino S, Arrigoni J, Guilleminault C, Mignot E, Dement WC (1991): Effects of selective monoamine uptake inhibitors and release enhancers on canine cataplexy. *Sleep Res*, 20:76.
Rubin RL, Hajdukovitch RM, Mitler MM (1988): HLA DR2 association with excessive somnolence in narcolepsy does not generalize to sleep apnea and is not accompanied by systemic autoimmune abnormalities. *Clin Immun Immunopathol* 49:149–158
Sasazuki T, McMichael A, Payne R, McDevitt HO (1977): HLA-D antigens in the Japanese population. *Tissue Antigen* 9:267–272
Schrader H, Gotlibsen OB, Tech EEG, Skomedal MD (1980): Multiple sclerosis and narcolepsy/cataplexy in a monozygotic twin. *Neurology* 30:105–108
Siegel JM (1989): Brainstem mechanisms generating REM sleep. In: *Principles and Practice of Sleep Medicine*, Kryger MH, Roth T, Dement WC eds. Philadelphia: WB Saunders
Singh SM, George CFP, Kryger MH, Jung JH (1990): Genetic heterogeneity in narcolepsy. *Lancet* 335:726–727
Stockinger H, Schmidtke J, Bostock C, Epplen JT (1986): Human DNA sequences isolated

with an immunoglobulin switch region probe: sequence, chromosomal localisation, and restriction fragment length polymorphism. *Hum Genet* 73:104–109

Thomson G (1985): The mode of inheritance of the HLA linked gene predisposing to narcolepsy. *Tissue Antigens* 26:201–203

Uryu N, Maeda M, Nagata Y, Matsuki K, Juji T, Honda Y, Kasai J, Ando A, Tsuji K, Inoko H (1989): No difference in the nucleotide sequence of the DQbeta beta1 domain between narcoleptic and healthy individuals with Dr2, Dw2. *Human Immunol* 24:175–181

WHO Nomenclature committee for factors of the HLA system 1989 (1990): Nomenclature for factors of the HLA system, 1989. *Immunogenetics* 31:131–140

Wong Z, Wilson V, Jeffreys A, Thein SL (1986): Cloning a selected fragment from a human DNA fingerprint: isolation of an extremely polymorphic minisatelite. *Nucl Acid Res* 14:4605–4617

Yoss RE, Daly DD (1960): Hereditary aspects of narcolepsy. *Trans Am Neurol Assoc* 85:239–240

Genetic Animal Models of Epilepsy

WOLFGANG LÖSCHER

By definition, human epilepsy, or epilepsies, are characterized by chronically recurring, spontaneous clinical seizures. Experimental research on epilepsy and antiepileptic drugs has mostly been done in mice and rats, in which seizures were induced by chemical or electrical means (Purpura et al., 1972; Löscher and Schmidt, 1988). Although these rodent models have proved useful for the development of new antiepileptic drugs, they are obviously not closely related to human epilepsy but represent models of seizure states rather than models of epilepsy. An ideal animal model for epilepsy should show the following characteristics: (1) the development of spontaneously occurring recurrent seizures, (2) a type of seizure similar in its clinical phenomenology to seizures occurring in human epilepsy, (3) the clinical seizures should be associated with epileptic-like activity in the electroencephalogram (EEG), (4) pharmacokinetics of antiepileptic drugs similar to those in man, thus allowing the maintenance of effective drug levels during chronic treatment, and (5) effective plasma concentrations of antiepileptic drugs similar to those required for control of human epilepsies. No model at present meets all these criteria. However, there are several more recent genetic animal models of epilepsy that resemble idiopathic epilepsy in humans more closely than any other experimental model. In these genetic animal models, seizures are either induced by specific sensory stimulation in genetically susceptible animals, such as baboons, gerbils, and DBA/2 mice, or the seizures occur spontaneously, for example, in epileptic dogs and rats. These animal models thus provide a means of investigating genetic, neurochemical, neuropathological, neurophysiological, and neuropharmacological aspects of epilepsy that no artificially produced seizure model can provide. However, despite the specific potential of genetic animal models of epilepsy for the development of new, highly specific antiepileptic drugs, most of them are only rarely used in preclinical evaluation of anticonvulsant drugs to date.

In this chapter, the different genetic animal models of epilepsy available to date will be described and data on the evaluation of anticonvulsant drugs in these models will be reviewed. Particular attention will be directed to the degree of specificity of the models for predicting antiepileptic activity against particular types of human epilepsy. Some interesting genetically predisposed animal species, for example, audiogenic seizure susceptible rabbits (Nellhaus, 1958), C57BL/10Bg mice with spontaneous seizures (Maxson et al., 1983), or AE mice with spontaneous seizures (Löscher et al., 1986) are not included in this chapter because anticonvulsant drugs have not been studied in detail in these models. For the developmental, genetic, and biochemical basis of the seizure susceptibility in genetic animal models, the interested reader is referred to numerous recent reviews that are quoted in the respective sections of this chapter.

The aim of this chapter is to illustrate the advantages and drawbacks of genetic animal models of epilepsy in order to encourage interested investigators to consider the potential of these models for the preclinical evaluation of new anticonvulsant drugs.

GENETIC ANIMAL MODELS OF EPILEPSY WITH SPONTANEOUS RECURRENT SEIZURES

By definition of epilepsy, animals with chronically recurring, spontaneous seizures represent ideal models for human epilepsy. However, it should be noted that none of the models that are described in the following meet all the criteria discussed in the introduction, especially regarding the pharmacokinetics of antiepileptic drugs. Various drugs are more rapidly eliminated in these species than in man, which should especially be considered in long-term studies on anticonvulsant drugs.

Epileptic Dogs

Among domesticated animals, the dog is affected with epilepsy far more than any other animal, making this a disease of considerable veterinary and comparative medical interest. Clinical and electroencephalographic observation strongly suggest that epilepsy in dogs closely approximates the disease in man (Croft, 1965; Holliday et al., 1970; Cunningham, 1971; Edmonds et al., 1979; Löscher et al., 1985). Accordingly, the usefulness of epileptic dogs as an animal model of human epilepsy has repeatedly been emphasized.

The most extensive clinical, electroencephalographic, genetical, pathophysiological, and neurochemical studies on inherited epilepsy in dogs have been performed in beagle dogs (Bielfelt et al., 1971; Edmonds et al., 1979). Actually, in colonies containing large numbers of beagle dogs it readily became apparent that a rather high proportion of the colony-reared dogs were afflicted with epilepsy and these dogs could thus be used for systematic studies of the disease (Bielfelt et al., 1971). However, for the evaluation of anticonvulsant drugs, beagle dogs are less suited because the elimination of drugs, at least of some of the common antiepileptics, is much more rapid in beagles than in other breeds of dogs (Frey et al., 1979). Although some experimental anticonvulsants have been evaluated in epileptic beagle dogs (Edmonds et al., 1979), basic studies on the anticonvulsant efficacy of antiepileptic drugs in these dogs are not yet available.

Evidence for genetic transmission of epilepsy in dogs has also been reported for a breed of laboratory dogs in Czechoslovakia (Martinek and Horak, 1970). However, seizures in these dogs were not of the spontaneous recurrent type but had to be induced by conditions that produce emotional excitement or fright. The only antiepileptic drug that has been evaluated in Horak's dogs is valproic acid, which was not effective in attenuating the seizures (Martinek and Arbeiter, 1980).

Besides beagle dogs and Horak's laboratory dogs, genetic and clinical epidemiological studies strongly suggest that genetically transmitted epilepsy occurs in numerous other breeds of dogs (Schweingruber et al., 1962; Croft, 1965; Van der Velden, 1968; Falco et al., 1974; Wallace, 1975; Löscher et al., 1985). For instance, analysis of the records of several hundred dogs of different breeds with epileptic seizures brought to the Department of Small Animal Diseases in Berlin indicated breed-related differences in the prevalence of epilepsy when the breeds of the epileptic dogs were compared to the distribution of breeds in the hospital population (Löscher et al., 1985). The highest prevalences were seen in miniature schnauzers (2.2%), bassets (2.2%), collies (1.7%), and cocker spaniels (1.4%). The total prevalence of dogs with epilepsy was 0.55%, which is similar to the value known for humans. In German shepherds (alsatians), for which inherited epilepsy has been described in a British study (Falco et al., 1974), prevalence was 0.5%, that is, not different from the total prevalence of epilepsy in dogs. Thus, the frequency with which veterinarians encounter epilepsy in German shepherds seems to relate primarily to the high frequency of this breed in the dog population. Further findings of our studies on epileptic dogs (Löscher et al., 1985) are summarized as follows:

1. Idiopathic epilepsy, as in man, is more common that symptomatic epilepsy in dogs.
2. The age of onset of idiopathic epilepsy is 4 years or younger in 60% of the animals. Seizures rarely appear before 6 months of age. In dogs with symptomatic epilepsy (due to brain tumors, head trauma, encephalitis, leptomeningitis, hydrocephalus, etc.), the age of onset is much higher in most cases.
3. Male dogs are about 2 times more often affected with idiopathic epilepsy than females.
4. Generalized tonic–clonic seizures of the grand mal type are by far the most common type of seizures in dogs but there are also other types of epileptic seizures, such as focal and myoclonic seizures. Grand mal seizures in dogs begin in a tonic phase with an apparent simultaneous loss of consciousness. The duration of this phase ranges from a few seconds to 2 or 3 min, and is followed by a clonic phase of similar duration. Urination, defecation, or salivation may occur during either the tonic or clonic phases. The clonic phase is replaced by a period of several minutes of running or walking movements. Subsequently, the dog may attempt to rise, may become responsive to handling, or may give other evidence of returning consciousness. When consciousness is regained, the dog may return to its normal attitude immediately or it may show postictal depression, acting fatigued, frightened, or dazed.

If this depression occurs, it usually lasts for only a few minutes after the seizure. In the EEG, the onset of unconsciousness and tonus is accompanied by a generalized discharge of spikes throughout the cerebral cortex. Generalized tonic–clonic seizures in dogs are not only of the primary generalized type but often occur as a secondary generalization of focal seizures.

The efficacy of chronic drug treatment with common antiepileptic drugs was studied at Berlin in some 80 epileptic dogs of different breeds with generalized tonic–clonic seizures (Löscher et al., 1985). The efficacy of drug therapy was evaluated in three categories: (1) complete control of seizures, (2) reduction of seizure frequency by 75% or more, and (3) no change in seizure frequency or a reduction of less than 75%.

Primidone, one of the major drugs for grand mal therapy in man (Schmidt, 1981), was given to 46 dogs at daily doses of 14 to 104 mg/kg for 6 to 60 months. During medication, effective plasma levels of phenobarbital, the main active metabolite of primidone, could be maintained. Complete control of generalized tonic–clonic seizures or reduction of seizure frequency by at least 75% was achieved in 40% of the dogs at average phenobarbital concentrations of 23 (range 5–49) μg/kg (Table 6.1). These data are in good agreement with values reported in trials of humans with generalized tonic–clonic seizures (Schmidt, 1981). Similar results to those with primidone were obtained following chronic treatment of 17 dogs with phenobarbital at daily doses of 2.5 to 13 mg/kg for 7 to 25 months (Table 6.1). Actually, our studies with primidone and phenobarbital indicated that phenobarbital is just as effective as primidone in dogs with grand mal epilepsy but exhibits fewer side effects, especially with regard to liver function (Schwartz-Porsche et al., 1985). It is of interest that following treatment with both primidone and phenobarbital, about 60% of the dogs did not show a significant reduction of seizure frequency (≥75%) according to the strict criteria used. Thus, the epileptic dog may be suitable for the investigation of drug resistance in epilepsy, which is a major difficulty in the management of the disease in man (Schmidt, 1981).

During chronic treatment with phenytoin, carbamazepine, and valproic acid for 2 to 8 months, no clear reduction of seizure frequency or severity was found (Table 6.1). This is explained by the short half-lives of these antiepileptic drugs in dogs (Löscher, 1978; Frey and Löscher, 1980, 1981), which did not allow the maintenance of effective plasma concentrations during chronic treatment, although high doses of the drugs were used (Table 6.1). On the other hand, pharmacokinetic studies have shown that primidone, phenobarbital, ethosuximide, and trimethadione have similar half-lives in dogs and humans (Frey and Schulz, 1970; El Sayed et al., 1978; Frey et al., 1979). Furthermore, owing to saturation kinetics or cumulation of active metabolites, maintenance of effective plasma concentrations in dogs is also possible with diazepam and clonazepam (Löscher and Frey, 1981; Al-Tahan et al., 1984) but, because of functional tolerance, the anticonvulsant effect is lost during chronic treatment of epileptic dogs (D. Schwartz-Porsche, personal communication).

TABLE 6.1 Anticonvulsant potency of common antiepileptic drugs in animals with epilepsy (spontaneous recurrent seizures).

	Rats with absence seizures	*Tottering* mice			Dogs with grand mal epilepsy		Human epilepsy: common clinical doses and therapeutic range[a]	
		Absence seizures		Focal seizures				
	ED (mg/kg i.p.)	ED (mg/kg i.p.)	EC (μg/ml)	ED (mg/kg p.o.)	ED (mg/kg/day p.o.)	EC (μg/ml)	ED (mg/kg/day p.o.)	EC (μg/ml)
Phenobarbital	2.5–10	25	17.5	n.d.	3–13	11–27	2–3	10–40
Primidone	n.d.	n.d.	n.d.	n.d.	14–71	5–49 (as phenobarbital)	10–20	10–40 (as phenobarbital)
Phenytoin	>160	>60	n.d.	n.d.	>56		5–6	3–23
Carbamazepine	>80	n.d.	n.d.	n.d.	>24		15–20	4–12
Ethosuximide	12.5–100	150	140	>500	n.d.	n.d.	15–20	30–120
Trimethadione	50–200	n.d.	n.d.	n.d.	n.d.	n.d.	30	600–800 (as dimethadione)
Valproic acid	50–200	n.d.	n.d.	>600	>190		15–20	40–100
Diazepam	0.5–2	1.4[b]		1.2 (ED 50)	n.d.	n.d.	0.2–0.4	0.2–0.5

[a] EDs and ECs of antiepileptic drugs in human epilepsy relate to those seizure types against which the respective drugs are particularly effective, i.e., partial (focal) seizures and generalized tonic–clonic (grand mal) seizures in the case of phenobarbital, primidone, phenytoin, and carbamazepine, and generalized minor (absence, myoclonic) seizures in the case of ethosuximide, trimethadione, valproic acid, and the benzodiazepines.

[b] Plasma levels of diazepam and its mean active metabolite desmethyldiazepam at the time of maximum effect (2 hr postinjection) were 0.003 and 0.05 μg/ml, respectively (Heller et al., 1983).

Data were taken from Schultz (1983), Heller et al. (1983), Loscher and Meldrum (1984), n.d., not determined.

In conclusion, epileptic dogs have several advantages as a model for human epilepsy: (1) dogs with generalized tonic–clonic seizures are the only grand mal model with spontaneous recurrent seizures. There are also other types of seizures common in dogs, which is an additional advantage of this animal model, (2) the prevalence of epilepsy in dogs is very similar to that in man, (3) maintenance of effective drug levels during chronic therapy is possible with several of the major antiepileptic drugs (barbiturates, primidone, benzodiazepines, trimethadione, ethosuximide) thus favoring studies on chronic drug treatment, (4) the efficacy of some of the common anti–grand mal antiepileptics (phenobarbital, primidone) is very similar in dogs and man, and (5) epileptic dogs are the only known model with spontaneous recurrent seizures in which mechanisms for drug resistance in epilepsy can be studied.

Unfortunately, there are some drawbacks of the epileptic dog model: (1) the naturally occurring seizures cannot be elicited at will by an investigator, which makes drug efficacy studies time-consuming, especially when the seizure frequency is low, (2) there are logistical difficulties in obtaining an adequately high number of animals for long-term studies, and (3) the pharmacokinetics of some major antiepileptic drugs differ markedly between dog and man, which makes the maintenance of therapeutic plasma concentrations of these drugs (valproic acid, carbamazepine, phenytoin) in dogs impossible.

Nevertheless, the epileptic dog model offers a number of interesting epidemiological, pharmacokinetic, and drug efficacy characteristics and may provide special advantages for the study of seizure development, age-related drug effects, and chronic drug treatment. Studies are under way now to use this model for the evaluation of new, potentially interesting anticonvulsant drugs. Furthermore, clinical and pharmacological characteristics of seizure types other than generalized tonic–clonic seizures, especially focal seizures, should be studied in more detail in dogs. Preliminary studies suggest that focal seizures are more difficult to treat than grand mal seizures in this species (Schwartz-Porsche et al., 1985), which is consistent with the clinical experience in human partial epilepsies (Schmidt, 1981).

Rats with Petit Mal (Absence) Epilepsy

Recently, rats with spontaneously occurring petit mal seizures have been described as a new model for absence (petit mal) seizures in human epilepsy (Vergnes et al., 1982; Micheletti et al., 1985a). The animals were selected from normal Wistar rats of a 20-year-old laboratory breeding colony in Strasbourg. About one third of the rats of this colony exhibited spontaneous spike-wave discharges in the EEG. The spontaneous bursts, 7 to 11/sec and 200 to 400 μV in amplitude, lasted from 0.5 to 40 sec and occurred hundreds of times a day. The paroxysmal discharges were accompanied by a sudden arrest in movement, twitching of the vibrissae, and single myoclonic jerks of face and neck muscles. These episodes usually occurred when the animal was in a state of quiet wakefulness but were sometimes seen also during movement, which was suddenly interrupted at the onset of the seizure activity and was resumed thereafter. The epileptic discharges were usually recorded in 6- to 12-month-old rats but persisted in the animals for their whole lifetime. The incidence of seizures could be increased markedly when epileptic rats were used for breeding, thus demonstrating the genetic transmission of the phenomenon (C. Marescaux, personal communication).

The efficacy of antiepileptic drugs in this model was evaluated in rats with chronically implanted cortical monopolar electrodes by comparing the duration of seizure discharges in 20-min periods before and after drug treatment (Micheletti et al., 1985a). As shown in Table 6.1, ethosuximide, trimethadione, valproic acid, and diazepam dose-dependently suppressed the seizure discharges in epileptic rats, while phenytoin and carbamazepine were inactive or even aggravated the seizure activity. Phenobarbital was effective at 2.5 to 10 mg/kg, but paradoxically exerted no anticonvulsant effect at 20 mg/kg. These data clearly demonstrate the validity of this model for human absence and/or myoclonic epilepsy: drugs that are effective against absence and myoclonic seizures in man (i.e., ethosuximide, valproic acid, and trimethadione) are highly active in rats, whereas the anti–grand mal drugs phenytoin and carbamazepine are inactive or even aggravating, which has also been reported in humans with absence seizures (Schmidt, 1981). Diazepam, which was very potent in the rats, is also active against absence seizures in humans but is not used for chronic therapy because of the development of tolerance (Schmidt, 1981). Phenobarbital, which showed a truncated dose-response in rats, is commonly used for treatment of grand mal and focal seizures in man, but some reports have also found it active against myoclonic seizures (cf. Micheletti et al., 1985a).

The epileptic rat model has also been used for evaluation of new anticonvulsant drugs, for instance those that act via activation of the inhibitory neurotransmitter gamma-aminobutyric acid (GABA) in the brain (Micheletti et al., 1985b). In contrast to results obtained in numerous other models of epilepsy (Löscher, 1985a,b,c), the GABA agonists muscimol, THIP (4, 5, 6, 7-tetrahydro-isoxazolo [5,4-c] pyridine-3-ol) and progabide, and gamma-vinyl GABA, an inhibitor of GABA degradation, were either not active or aggravated the seizure discharges in the epileptic rats. Similar results were obtained with baclofen. These results thus suggest that the rat's spontaneous petit mal–like seizures are not related to a deficient GABAergic control.

In conclusion, besides the *tottering* mouse, which will be described in the next section, rats with petit mal seizures are the only available petit mal model with spontaneous recurrent spike-wave discharges. Although these rats were first described in a laboratory colony in Strasbourg, such animals can also be selected from rats of commercial breeders (C. Marescaux, personal communication). Whereas in traditional petit mal models, such as the pentylenetetrazol seizure threshold test in mice, very high doses of the anti–petit mal drugs ethosuximide, valproic acid, and trimethadione are necessary to block the seizures, the active doses of these drugs in the new rat model are much lower and similar to the doses effective against absence or myoclonic seizures in man (Table 6.1). The epileptic rats not only supply data on the anticonvulsant potency of drugs but also on the duration of action, because the spike-wave discharges can be quantitated continuously after drug administration (Löscher et al., 1984). It is thus evident that spontaneous petit mal seizures in rats are a more efficient, predictive pharmacological model of petit mal epilepsy than previous models and may provide special advantages for long-term studies of anticonvulsant drugs. However, it should be noted that the elimination of antiepileptic drugs in rats is, in most instances, much more rapid than in man (Gerber et al., 1971; Farghali-Hassan et al., 1976; Klotz et al., 1976; Löscher, 1978). Thus, maintenance of active drug concentrations during prolonged treatment is often only possible by special administration techniques, such as by constant rate application via osmotic minipumps (Löscher and Nau, 1983).

Tottering Mice

The homozygous *tottering* mouse (*tg*, autosomal recessive) originally described by Green and Sidman (1962) is a presumed single-locus mutant, phenotypically characterized by spontaneous epileptic seizures (Noebels, 1979). By 3 to 4 weeks of age, affected homozygotes can be recognized by a broad-based, ataxic gait. Spontaneous focal motor seizures are observed 1 to 3 days later and occur one or more times per day throughout the normal life span of the mutant. The initial stage of each episode begins with bilateral tonic flexor spasms in the hindlegs. The second stage is marked by rapid, unilateral clonic jerks of the limbs, beginning in the hindleg and progressively spreading to affect both hindleg and foreleg simultaneously. In its final stage, the seizure terminates with prolonged clonic jerking of the forelimb alone. Generalized convulsions are never seen. In studies on *tottering* mice by Frey's group in Berlin, the focal motor seizures occurred irregularly a few times a day and 93% of the seizures lasted for 15 min or longer (Schultz, 1983). The focal seizures are not correlated with any reliable stereotyped EEG findings in surface electrocorticograms (Noebels and Sidman, 1979).

A second distinct seizure pattern in homozygous *tottering* mice are the absence, or myoclonic, seizures associated with spike-wave discharges in the EEG. As early as 32 days postnatal, bilaterally synchronous 6 to 7 sec spike-wave discharges appear as spontaneous bursts in electrocorticographic recordings (Noebels and Sidman, 1979). These spike-wave bursts, 0.3 to 10 sec in duration, occur hundreds of time per day and are in each case accompanied by a behavioral seizure, with arrest of movement, staring posture, twitching of the vibrissae, and single myoclonus jerks of the head or jaw. These seizures are present at least through 10 months of age.

The pharmacological sensitivity of the petit mal seizures and the focal motor seizures in *tottering* mice has also been studied (Table 6.1). As shown by Heller et al., (1983) in *tottering* mice with chronically implanted EEG electrodes, the spike-wave seizures can be dose-dependently blocked by ethosuximide, diazepam, and phenobarbital but not by phenytoin, which exacerbates the seizures at higher doses. These data are consistent with the results obtained by Micheletti et al., (1985a) in rats with spontaneous petit mal epilepsy (see above) and suggest that *tottering* mice with spike-wave discharges may be a valuable pharmacological model for absence and/or myoclonic epilepsy in man. In contrast to the petit mal rats, spike-wave discharges in *tottering* mice can be also effectively blocked by GABAmimetic drugs, as indicated by the anticonvulsant activity of aminooxyacetic acid, an inhibitor of GABA degradation (Heller et al., 1983).

As shown by Frey's group (Schultz, 1983), the focal motor seizures in *tottering* mice are not affected by ethosuximide and valproic acid but are potently suppressed by diazepam (Table 6.1), which is consistent with the clinical experience in focal epilepsy (Schmidt, 1981). These results thus demonstrate that in *tottering* mice the sensitivity of spike-wave absence seizures and focal motor seizures to antiepileptic drugs differ, which closely resembles absence and focal epilepsy in humans (Schultz, 1983).

In conclusion, *tottering* mutant mice are an interesting genetic animal model of chronic epilepsy that can be used to predict drugs effective against petit mal seizures, but can also be used for prediction of efficacy against focal seizures. Thus, different seizure types with different pharmacological sensitivities can be studied in the same model. The main advantage of *tottering* mice is that the seizures occur spontaneously and chronically and thus offer the opportunity to study the long-term efficacy of antiepileptic drug treatment. As in rats, it should be noted that the half-lives of most antiepileptic drugs are much shorter in mice compared to humans (cf. Löscher and Frey, 1984). However, maintenance of active drug concentrations is easily achieved in mice by use of osmotic minipumps or administration of drugs via the drinking water (Frey and Kampmann, 1965; Frey and Kretschmer, 1971; Nau and Zierer, 1982). The main disadvantage of *tottering* mice is the problem of getting enough epileptic homozygotes for pharmacological studies. At least in Frey's laboratory, breeding with homozygous mice proved to be impossible and heterozygous animals had to be used to get homozygotes with seizures for the experiments (Schultz, 1983). This procedure is very time-consuming and limits the potential of *tottering* mice as a pharmacological model for anticonvulsant drug screening. Despite this inherent drawback, *tottering* mice offer several characteristics that are not shared by other models of epilepsy and thus may contribute important new principles to our understanding of the hereditary epilepsies.

GENETIC ANIMAL MODELS OF EPILEPSY WITH REFLEX SEIZURES

In human epilepsy, only about 5% of the epileptic patients get focal or generalized seizures in response to sensory stimulation (Schmidt, 1981). About one third of these patients with "reflex" epilepsy respond to photic stimulation. Seizures in photosensitive epilepsy are mostly of the absence type, whereas grand mal, complex partial, or simple partial seizures are less frequent (Schmidt, 1981). Thus, the major drawback of all genetic animal models with reflex seizures is that this type of epilepsy is rare in humans. Furthermore, it should be noted that photomyoclonic seizures can occur in nonepileptic patients, which means that animals with photomyoclonic seizures are not necessarily models for epilepsy. The main advantage of genetic animals models with reflex seizures for anticonvulsant drug evaluation is that seizures can be easily and reproducibly evoked in these models without electrical or chemical means, and that the seizure types, at least in part, are similar in their clinical phenomenology to seizures occurring in human epilepsy. As already discussed for animal species with spontaneous seizures, several antiepileptic drugs are more rapidly eliminated in species with reflex seizures compared to man, which should especially be considered during assessment of the long-term efficacy of anticonvulsant drugs in these models.

Baboons with Photomyoclonic Seizures

A photomyoclonic syndrome in the baboon *Papio papio* was first reported by Killam et al. (1966, 1967). Since then, much work has been carried out on elucidating the characteristics of this syndrome (Naquet and Meldrum, 1972; Killam, 1979). Myoclonic responses to intermittent photic stimulation (stroboscope at 25/sec) occur in 60% to 80% of adolescent baboons (*P. papio*) from the Casamance region of Senegal, whereas seizure incidence is lower in *P. papio* from other areas. A minority (approximate 20%) of the animals show full tonic–clonic seizures. The clinical seizures are accompanied by paroxysmal discharges in the EEG, which usually consist of polyspikes or spikes and waves. The photosensitive responses, like the photoconvulsive response in humans, are age- and sex-dependent, being maximal in the adolescent female. The incidence of photosensitivity of other Papio species is generally less than 10%. Furthermore, in studies of other subhuman primates the incidence of photosensitivity has been equally low (Naquet and Meldrum, 1972). These data suggest that the photosensitivity is not a characteristic of genus or species but is a unique feature of *Papio papio* from the Casamance region. Photic stimulation is not the only effective provocative agent in *P. papio,* as hyperventilation, overexercise, the stress of capture, restraint, or heat and humidity can also induce seizures. Also, even in the absence of light, EEG paroxysms occur during sleep (Killam et al., 1967).

Photically induced seizures in *P. papio* begin with rapid, bilateral clonus of the eyelids and periocular musculature, followed by spread of diffuse clonic twitching to the face and neck, often accompanied by more intense, isolated jerks of the head. The diffuse clonic twitching may be interrupted by isolated, or groups of, clonic jerks involving the whole body, or tonic spasms of facial muscles, jaw opening, and grimaces. Finally, the entire body may become involved in violent clonic jerks with flexion of head and upper body and tonic extension of lower limbs. In some animals this is followed by marked, generalized clonus. In the most severely epileptic animals, the generalized clonus gradually shifts to a tonic spasm and finally appears again in large clonic jerks becoming more and more isolated. In the EEG, the onset of clinical seizures is accompanied by rapid spike, slower spike-waves, and slow polyspike-wave complexes (Naquet and Meldrum, 1972).

The photomyoclonic response of *P. papio* has been suggested as a model for photomyoclonic seizures and myoclonic petit mal epilepsy in humans (Killam,

1979; Naquet and Meldrum, 1972). However, the sensitivity of this response to antiepileptic drugs is only in part similar to the human syndromes. Consistent with clinical experiences, valproic acid, phenobarbital, and benzodiazepines give complete protection against the myoclonic responses in baboons, whereas phenytoin and carbamazepine are only partially active (Table 6.2). Plasma levels associated with protection are similar to human therapeutic levels for diazepam, clonazepam, and phenobarbital, whereas respective levels for valproic acid are higher than necessary for humans. However, trimethadione, which is active against myoclonic generalized seizures in man, is only weakly active in baboons. Furthermore, ethosuximide and primidone were found to have little effect on the epileptic response in *P. papio* (Table 6.2), although both drugs are highly active against myoclonic petit mal in humans (Schmidt, 1981). Investigations with higher doses of ethosuximide showed that seizures in adult baboons could be reduced or abolished at 120 to 180 mg/kg, which produced blood levels up to 130 μg/ml. However, side effects were much greater than with equieffective doses of phenobarbital and benzodiazepines (Rinne et al., 1978). *P. papio* has also been used for chronic evaluation of antiepileptic drugs. During prolonged treatment, phenytoin was much more active than after single doses, whereas in the case of diazepam and clonazepam, tolerance to the anticonvulsant effect developed (Killam et al., 1973; Stark et al., 1970).

A wide range of compounds that increase the activity of inhibitory neurotransmitters in the brain have been found to attenuate the photomyoclonic responses in baboons, including dopaminergic, serotoninergic, and GABAergic agents (Jobe and Laird, 1981; Löscher, 1985a). Furthermore, novel drugs with anticonvulsant activity have been studied in this model as part of the preclinical evaluation (e.g., Meldrum et al., 1983; Chambon et al., 1984; De Sarro et al., 1986). Interestingly, as in rats with petit mal epilepsy (see previous discussion), the GABA agonists THIP and muscimol failed to block photically induced seizures in baboons but caused paroxysmal EEG activity and myoclonic seizures (Meldrum et al., 1980). On the other hand, the GABA agonist progabide was reported to be an effective anticonvulsant in *P. papio* (Cepeda et al., 1982).

In conclusion, the baboon *Papio papio* offers a useful model of innate photomyoclonic seizures. Being the only subhuman primate model of epilepsy, it offers special advantages for the study of age-related drug effects and chronic drug treatment. However, with respect to the results with antiepileptic drugs discussed above, the predictive value of this model for drugs effective against particular types of human epilepsy still requires clarification.

Audiogenic Seizure Susceptible Mice

Audiogenic seizures are violent generalized convulsions triggered by exposure to intense auditory stimulation. These seizures were first described in mice by Studentsov in 1924 (cf. Krushinsky et al., 1970) but, with the exception of a few reports, most of the research on audiogenic seizures in mice did not begin until after World War II. Most studies on audiogenic seizure–susceptible mice have

TABLE 6.2 Anticonvulsant potency of common antiepileptic drugs in animals with reflex seizures.

	Effective doses (ED) and plasma concentrations (ED) of common antiepileptic drugs								
	Photosensitive baboons (myoclonic seiz.)		DBA/2 mice with audiogenic seiz. (clonic)	Rats with audiogenic seiz. (clonic-tonic)	Gerbils with reflex seiz. ED 50 (mg/kg p.o.)		Photosensitive fowl (tonic-clonic seiz.)	Human epilepsy: common clinical dose and therapeutic range[a]	
	ED (mg/kg i.v.)	EC (μg/ml)	ED 50 (mg/kg i.p.)	ED 50 (mg/kg p.o.)	Minor seizures	Major seizures	EC (μg/ml)	HED (mg/kg/day p.o.)	EC (μg/ml)
Phenobarbital	15	7–17	2–7	7.2	14	8.1	4–14	2–3	10–40
Primidone	>100	n.d.	n.d.	n.d.	n.d.	11	22 (as phenobarbital)	10–20	10–40 (as phenobarbital)
Phenytoin	15–50[b]	20–30	2–14	42	>40	22	8–14	5–6	3–23
Carbamazepine	40[b]	n.d.	n.d.	14	~40[b]	17	n.d. (ED = 20 mg/kg)	15–20	4–12
Ethosuximide	>100	>80	130–150	~400[b]	360	700	>430	15–20	30–120
Trimethadione	weakly active		140	195	n.d.	n.d.	210–444 (as dimethadione)	30	600–800 (as dimethadione)
Valproic acid	200	200–400	55–300	410	180	240	105–245	15–20	40–100
Diazepam	0.5–1	0.5	0.04–0.12	n.d.	0.38	0.7	n.d. (ED 50 = 0.5 mg/kg)	0.2–0.4	0.2–0.5
Clonazepam	0.15	0.025–0.04	0.005	0.1	0.02 (i.p.)	0.03 (i.p.)	n.d. (ED 50 = 0.025 mg/kg)	0.1–0.2	0.01–0.07

[a] EDs and ECs of antiepileptic drugs in human epilepsy relate to those seizure types against which the respective drugs are particularly effective, i.e., partial (focal) seizures and generalized tonic–clonic (grand mal) seizures in the case of phenobarbital, primidone, phenytoin, and carbamazepine, and generalized minor (absence, myoclonic) seizures in the case of ethosuximide, trimethadion, valproic acid, and the benzodiazepines. For myoclonic seizures, it should be noted that primidone and, less so, phenobarbital are effective against myoclonic petit mal in man (Schmidt, 1981).

[b] Only partially effective.

Data were taken from Consroe et al. (1979), Consroe and Wolkin (1977), Loscher and Meldrum (1984), and references cited therein. n.d., not determined.

been performed in the DAB/2 inbred strain of the house mouse (*Mus musculus*), which has been known since 1947 to be susceptible to sound-induced seizures (Collins, 1972; Seyfried, 1979; Chapman et al., 1984). Nearly 100% of the males and females of this strain undergo an age-dependent, often fatal, sequence of convulsions when ititially exposed to a loud mixed-frequency sound (10–120 kHz, 90–120 dB), such as a doorbell. The seizures begin with a wild running phase, followed by clonic convulsions and a tonic extension, ending in respiratory arrest (in about 60%) or full recovery. Although the clonic seizures are normally used for evaluation of anticonvulsant drug action, the wild running phase is probably the most important feature that distinguishes the syndrome from eletroconvulsive seizures and most types of chemically induced seizures. The onset of audiogenic seizure susceptibility in DBA/2 mice generally occurs over a relatively short period of time (12–17 days of age). The age of peak susceptibility is between 19 and 24 days; thereafter, susceptibility declines. By the time mice are adults (>80 day of age), they are completely resistant to auditory stimulation (Seyfried, 1979). Genetic, neurophysiological, biochemical, endocrine, and developmental studies on this strain of audiogenic rodents have been the subject of several reviews (Collins, 1972; Seyfried, 1979; Jobe and Laird, 1981; Chapman et al., 1984).

Regarding evaluation of antiepileptic drugs in this model, all of the commonly used drugs protect against clonic seizures in DBA/2 mice (Table 6.1). Anticonvulsant ED 50 values for phenobarbital, phenytoin, diazepam, and clonazepam are very similar to the doses effective in the treatment of human epilepsy, whereas effective doses for ethosuximide, trimethadione, and valproic acid are considerably higher in DBA/2 mice compared to man. However, ED 50 values of these latter three drugs in DBA/2 mice are very similar to the doses effective in the pentylenetetrazol seizure test; that is, the traditional mouse model for petit mal epilepsy. Thus, in contrast to most other known models of epilepsy, sound-induced seizures in DBA/2 mice are not particularly sensitive to a specific clinical category of antiepileptic drugs.

Besides antiepileptic drugs, various experimental anticonvulsants have proved active in audiogenic seizure–susceptible mice, especially compounds that potentiate GABAergic and monoaminergic transmission (Chapman et al., 1984). Among genetic animal models of epilepsy, DBA/2 mice with sound-induced seizures are at present certainly the most widely used model for anticonvulsant drug screening.

In conclusion, audiogenic seizure–susceptible mice are useful as a sensitive gross screening model for potential anticonvulsant drugs. However, they do not discriminate between the different clinical categories of antiepileptic drugs, and thus they cannot predict antiepileptic activity against a specific type of epilepsy. In this respect, it should be noted that sound-induced seizures are uncommon in man (Bickford and Klass, 1969) and, thus, the audiogenic seizure–susceptible mouse is obviously not a model of a particular human convulsive disorder. On the other hand, it is unlikely that a potent anticonvulsant drug would be rejected by this model. Regarding chronic evaluation of antiepileptic drugs in mice with sound-induced seizures, as already discussed above the rapid elimination of drugs in mice renders the maintenance of drug levels difficult and thus specific administration techniques are requisite.

Audiogenic Seizure Susceptible Rats

Sound-induced convulsions in rats were first described in 1924 (Donaldson, 1924). During the 1940s and 1950s selective breeding was carried out by various groups to increase the incidence and uniformity of audiogenic seizures in rats (Consroe et al., 1979). The Russian Krushinsky–Molodkina (KM) strain of Wistar descent, begun in 1947, and the University of Arizona (UAZ) strain of Sprague–Dawley descent, begun in 1958, are apparently the longest continuing and most uniform colonies currently available. Prevalence of audiogenic seizures in these colonies is 99% (KM) and 86% (UAZ), respectively.

Audiogenic seizures in rats are first observed between 17 and 21 days of age and, in contrast to audiogenic seizures in mice, the seizure susceptibility does not decline with age (Consroe et al., 1979). Sound-induced seizures in rats typically consist of the following sequence: a startle response, momentary quiescence, violent running, tonic–clinic seizure, and postictal depression (Consroe et al., 1979). In contrast to audiogenic seizure– susceptible mice, rats rarely die after the seizures. For anticonvulsant drug evaluation, the tonic–clonic component of the response pattern in rats is commonly used. Genetic and biochemical aspects of audiogenic seizure susceptibility in rats have been reviewed elsewhere (Collins, 1972; Consroe et al., 1979; Jobe and Laird, 1981; Dailey et al., 1989).

Evaluation of common antiepileptic drugs in rats with sound-induced seizures (Consroe and Wolkin, 1977; Consroe et al., 1979) has shown that drugs that are effective against generalized tonic–clonic (grand mal) and partial seizures in humans are the most potent drugs in this model, namely phenobarbital, phenytoin, carbamazepine, and clonazepam (Table 6.2). On the other hand, audiogenic seizures in rats are relatively insensitive to the anticonvulsants effective against generalized absence and myoclonic seizures in man (Table 6.2). Actually, ethosuximide failed to show a linear dose-response against audiogenic seizures and, even with high doses (600 mg/kg), no more than 60% of the animals could be protected (Consroe and Wolkin, 1977). Determination of protective indices, that is, the ratio between the median minimal neurotoxic dose (TD 50) in the rotarod test and the anticonvulsant ED 50, yielded high values for carbamazepine, phenytoin, clonazepam, and phenobarbital in audiogenic seizure–susceptible rats, thus indicating a satisfactory margin of safety for these drugs (Consroe and Wolkin, 1977; Consroe et al., 1979). Considerably lower protective indices were found for trimethadione and valproic acid, and ethosuximide was only protective in neurotoxic doses.

These data demonstrate that the audiogenic seizure test in rats is similar to the maximal electroshock seizure (MES) test in rats or mice as regards the relative effectiveness of antiepileptic drugs. Although the MES test is the standard paradigm for identifying drugs with potential activity against grand mal epilepsy in man, the audiogenic seizure–susceptible rat seems to be a valuable alternative model in this respect, because it provides important pharmacogenetic, etiological, and pathophysiological information that is not readily apparent with the use of the more traditional electroshock seizure model.

Gerbils with Reflex Seizures

Reflex epilepsy in Mongolian gerbils (*Meriones unguiculatus*) was first described by Thiessen and coworkers in 1968 (Thiessen et al., 1968), who recognized the trait in animals randomly bred in their laboratory at the University of Texas. In 27% of the gerbils in this colony seizures could be evoked by placing the gerbils in a new environment. Subsequent studies showed that seizures in gerbils can be initiated by various other precipitating environmental stimuli (e.g., onset of bright light, audiogenic stimuli, vigorous shaking of the cage) and different handling procedures (Kaplan, 1975). It has been suggested that the effectiveness of these different stimulation techniques is directly related to their stress-inducing properties (Kaplan, 1975). The incidence of seizures in gerbils could be increased to about 97% by selective breeding of seizure-sensitive animals (Loskota et al., 1974). Both in randomly bred and selectively bred gerbils seizures vary in severity from animal to animal and, in some animals, from experiment to experiment. The seizure severity can be rated by the following seven-point scale that was developed by Loskota et al. (1974): grade 0, no seizure; grade 1, twitching of vibrissae and pinnae (faster than normal), animals still moving but often "thumping" with hind legs; grade 2, motor arrest with more pronounced twitching of vibrissae, whiskers, ears, and eyelids; grade 3, motor arrest with generalized myoclonic jerks, hunched posture; after short quiescent period suddenly alertness with normal activity ("quick release"); grade 4, tonic–clonic seizures, animal remains on its feet, no "quick release"; grade 5, tonic–clonic seizures with loss of righting reflexes and often body rollover; and grade 6, seizure progressing to death (rarely seen). In selectively bred seizure-sensitive gerbils, the mean age at onset of seizures has been reported to be 57 days in males and 47 days in females (Loskota et al., 1974). In these animals, seizure severity increased with age and maximum scores were reached at 6 months and thereafter. The epileptic nature of the seizures in gerbils has been demonstrated in studies relating the motor manifestations of seizures to epilepsy-like EEG abnormalities (Loskota and Lomax, 1975). Some ictal EEG paroxysms had focal onset in parietal derivations that might be related to the stimulus-bound nature of the seizures (Loskota and Lomax, 1975). The biological, morphological, and neurochemical characteristics of epileptic gerbils have been reviewed recently (Löscher, 1992).

Despite the interesting features of the epileptic gerbil model, until recently the animals have only rarely been used for anticonvulsant drug evaluation. This is mainly due to the inherent difficulties in studies on epileptic gerbils. First, although in gerbils seizures can be triggered by different external stimuli, usually handling or change in environment, most of these stimuli are poorly defined and not quantifiable (Thiessen et al., 1968; Goldblatt et al., 1971; Kaplan, et al., 1975). Second, following a seizure gerbils exhibit a refractory period that lasts for at least several days (Consroe and Wolkin, 1977). Therefore, if a seizure is triggered without intention by handling during drug administration, evaluation of the drugs' effects is not possible. To overcome this problem, some investigators used inhalant anesthetics during drug administration (Goldblatt et al., 1971; Loskota and Lomax, 1974; Cox and Lomax, 1976), further complicating the results because of possible pharmacological contamination by the anesthetics. Third, familiarity with the experimental environment may lead to habituation of the stimulus that triggers the seizure (Loskota and Lomax, 1974). Fourth, seizures initiated by handling or change in environment can be inhibited by injection of different drug vehicles alone, including water and saline (Schonfeld and Glick, 1980; Majkowski and Kaplan, 1983).

More recently, in Frey's gerbil colony in Berlin we could overcome most of these problems by the use of a strong and quantifiable seizure stimulus, that is, by exposition of the animals to a blast of compressed air (average pressure 5 bars for 10 sec) aimed at the back of the animals (Frey et al., 1981 and 1983; Löscher et al., 1983; Löscher and Frey, 1984). By this technique, seizures could be evoked in more than 98% of randomly bred gerbils, whereas only few of the animals (<10%) showed seizures in response to handling, change in environment, or acoustic and photic stimulation (Löscher and Frey, 1984). Thus, in contrast to selectively bred seizure-sensitive gerbils, administration of drugs in these randomly bred animals was possible without precipitating seizures. Whereas randomly and selectively bred gerbils obviously differ in their seizure threshold to certain stimuli, development of seizure severity with age appears to be very similar. About all of our gerbils that convulsed upon stimulation at 7 to 10 weeks exhibited "minor" (grade 1 and 2) seizures. In most animals, these seizures occurred after two to three weekly air blast exposures. The seizure severity gradually increased with further stimulations and in most animals "major" (grade 3–5) seizures developed. Maximum seizure severity was reached at about 7 months and did not change thereafter (Löscher and Frey, 1984). Incidentally, puberty in gerbils is reached at an age of 9 to 12 weeks.

By means of the air blast technique, we established the first reliable and comparable basic data for the anticonvulsant effect of common antiepileptic drugs in gerbils (Frey et al., 1983; Frey et al., 1984; Löscher and Frey, 1984). For the evaluation of antiepileptic drugs in gerbils we divided the animals into groups with minor seizures (grades 1 and 2, facial myoclonic seizures) and major seizures (grades 3–5, generalized myoclonic and tonic–clonic seizures). Animals were not treated before they had shown a constant seizure pattern (either grades 1 and 2, or 3–5) for 3 consecutive weeks. As a consequence of the rapid increase in seizure severity with age, only a small portion of the gerbils could be used for drug testing against minor seizures, namely those animals that remained on stage 1 and 2 for at least 3 weeks. Injection of drug vehicle (water, saline, or water with 1% Tween 80) did not modify the seizure response induced by air blast. As shown in Table 6.2, clonazepam, diazepam, valproic acid, and ethosuximide were most potent against minor seizures. This type of seizure was resistant to phenytoin and could only partially be suppressed by carbamazepine. Phenobarbital was active against minor seizures but less so than against major ones. Major seizures were best suppressed by phenytoin, phenobarbital, primidone, and carbamazepine as well as by diazepam and clonazepam. The benzodiazepines, however, were less potent against major seizures than against minor ones, which is consistent with clinical experience in humans (Schmidt, 1981). Valproic acid was also less active against major seizures than it was against minor seizures, and ethosuximide protected against major seizures only at very high, sedative doses (Table 6.2). Thus, it seems possible to differentiate in the gerbil between drugs clinically useful against absence and myoclonic seizures (ethosuximide, valproic acid, benzodiazepines) and drugs particularly effective against generalized tonic–clonic seizures (phenytoin, phenobarbital, primidone, carbamazepine) in humans. At least for primidone, carbamazepine, and the benzodiazepines, anticonvulsant doses in gerbils were very similar to the doses used for clinical treatment of epilepsy (Table 6.2). Following administration of primidone, effective plasma concentrations of its active metabolite phenobarbital were in the range of 2 to 5 μg/ml (Frey et al., 1984), which is even below the "therapeutic" range in humans (Table 6.2). For clonazepam effective plasma concentrations of 10 to 20 ng/ml were determined in gerbils (W. Löscher, unpublished experiments).

The air blast technique was also used to evaluate the effects of various GABAmimetic drugs on seizures in gerbils (Löscher et al., 1983). All categories of GABAmimetics tested, that is, GABA prodrugs (cetyl GABA), GABA agonists (muscimol, THIP, progabide, and its acidic metabolite SL 75102), GABA uptake blockers [the (−)-isomer of nipecotic acid ethyl ester and (±)-*cis*-4-hydroxynipecotic methyl ester] and inhibitors of GABA degradation (amino-oxyacetic acid, gamma-acetylenic GABA) proved able to block major seizures in gerbils. Most potent were muscimol, THIP, cetyl GABA, amino-oxyacetic acid and gamma-acetylenic GABA. Actually, these five GABAmimetics proved to be clearly more potent in suppressing major seizures in gerbils than most clinically established antiepileptic drugs tested. As for minor seizures in gerbils, the $GABA_A$ receptor agonists THIP and progabide were about equally active against this seizure type as they were against major seizures, whereas the GABA prodrug cetyl GABA, which was quite potent against major seizures, failed to show a linear dose-response against minor seizures (Löscher, 1985b).

Besides GABAmimetic drugs, various other drugs that specifically manipulate inhibitory and excitatory neurotransmitter systems were studied in epileptic gerbils, and the effects obtained were compared with those determined in other genetic animal models of epilepsy (Löscher, 1985c; Löscher, 1992). These studies indicated that seizures in gerbils are strikingly more sensitive to increases in GABA and catecholamine levels in the brain than seizures in other models. Accordingly, recent data by Olson et al. (1984) have shown an impaired GABA/benzodiazepine receptor function in the substantia nigra of seizure-sensitive gerbils. On the other hand, seizures in gerbils appeared not to be sensitive to alterations in excitatory amino acid, glycine, and 5-dydroxytryptamine-mediated neurotransmission (Löscher, 1985c; 1992).

In conclusion, the unique features of the epileptic gerbil make this model extremely useful for epilepsy research. The progressive age-dependent development of seizure severity in these animals represents an interesting parallelism to human absence, or myoclonic, epilepsies in childhood, which at a later age often proceed to grand mal epilepsy (Schmidt, 1981). This parallelism is corroborated by the fact that minor and major seizures in gerbils respond differently to antiepileptic drugs. Actually, epileptic gerbils of different age can thus be used for identifying different clinical categories of antiepileptic drugs.

Unfortunately, there are some drawbacks of the epileptic gerbil model: (1) the experiments are time-consuming since, due to the long postictal refractoriness, the animals can only be tested once a week, (2) as in other rodent species the half-lives of common antiepileptic drugs in gerbils are much shorter than in man (Frey et al., 1983), which renders the maintenance of effective drug levels during chronic treatment difficult. However, this problem can be resolved by the use of special administration devices, such as osmotic minipumps (Löscher, 1986), (3) a certain skill in handling when drugs are to be administered is mandatory in order to avoid the induction of seizures, especially minor ones.

Despite these drawbacks, the gerbil may be useful in improving the understanding of the neurophysiological and neurochemical mechanisms of seizure development. It offers the opportunity of studying differential action of anticonvulsants on young and adult animals and may thus prove valuable in the evaluation of anticonvulsant drugs used or usable in human epilepsy.

Photosensitive Fowl

An autosomal recessive mutation in domestic fowl, resulting in tonic–clonic seizures in response to different stimuli, was first reported by Crawford (1969). Seizures can be induced in homozygous chickens on the day of hatching or later by stimuli that induce vigorous muscular exertion, by rhythmical auditory stimuli, by combined rhythmical auditory and optic stimulation, by intermittent photic stimulation alone, and by heat stress (Johnson et al., 1979). Seizures induced by all stimuli follow the same general course. However, the timing of events in the seizures can be determined most precisely if intermittent photic stimulation (14/sec) is used. Except for a brief period of decline in seizure susceptibility from days 3 to 7, seizures can be induced throughout the bird's life span. Twelve to 20 sec following the photic stimulation, affected chickens appear to show increased alertness and then upward and backward extension and rotation of the head and neck lasting from 5 to 15 sec, which is apparently analogous to the opisthotonus of seizures in other species. The chickens then lose control of postural muscles of the legs and assume a sitting position with wings extended and directed downward. After a few seconds in this position, running movements of the legs and flapping movements of the wings begin and the violence of these movements causes the chicken to fall and roll about the cage. This phase usually lasts from 25 to 30 sec and is followed by a period of apparent postictal depression and disorientation. Electroencephalographic recordings have shown that high voltage spikes coincide with the onset of the clinical seizures (Johnson et al., 1979). Genetic and neurochemical studies on the epileptic fowl have been reviewed elsewhere (Crawford, 1970; Johnson et al., 1979; Jobe and Laird, 1981).

Photically induced tonic–clonic seizures in epileptic fowl can be blocked by several antiepileptic drugs (Johnson and Davis, 1984). In Table 6.2, the effective plasma concentrations of antiepileptic drugs in this model are shown. Seizures can be completely abolished by pretreatment with phenobarbital, primidone, and valproic acid. Complete protection with phenytoin and trimethadione cannot be obtained because of the onset of neurological toxicity at higher doses. Ethosuximide in doses of 100 to 300 mg/kg is inactive. Carbamazepine protects against seizures at 20 mg/kg, but effective plasma concentrations have not yet been reported. Clonazepam and diazepam are also highly effective in this model (Table 6.2), but again effective plasma concentrations have not yet been reported. The effective plasma concentrations for phenobarbital, phenytoin, and primidone approximate those required in humans for the control of generalized tonic–clonic seizures (Table 6.2). Valproic acid is only active in chickens at concentrations exceeding the "therapeutic" range in human epilepsy. These data indicate that photically induced tonic–clonic seizures in chickens represent a model of grand mal epilepsy. However, the specificity of epileptic fowl in this regard is curtailed by the fact that trimethadione is highly active against tonic–clonic seizures in chickens (effetive concentrations of its metabolite dimethadione are even below the "therapeutic" range in absence epilepsy in man; see Table 6.2) although trimethadione is not effective in grand mal epilepsy in man (Schmidt, 1981).

A number of experimental anticonvulsants have been tested in epileptic fowl, including Δ^9-tetrahydrocannabinol, L-DOPA, L-tryptophan, pyridoxine, and some GABAmimetic drugs (Johnson et al., 1979; Johnson and Davis, 1984). In this respect, it should be noted that the blood–brain barrier in young chicks (1–3 days old) is less effective in excluding drugs that in adult chickens, and young chicks thus provide an excellent model for testing compounds that hardly penetrate into the brain in adults. For instance, seizure susceptibility in epileptic chicks can be reduced by systemic administration of GABA (Johnson and Davis, 1984).

In conclusion, although epileptic fowls have not yet been used extensively for the evaluation of anticonvulsant drugs, the available data suggest that they represent an interesting, sensitive model for human grand mal epilepsy. The model is simple, inexpensive, and large numbers of genetically defined subjects can be produced with ease. Epileptic chickens are thus a valuable means for large-scale testing of drugs for anticonvulsant activity and it seems likely that other uses will also be found for this unique model of epilepsy in the future.

CONCLUSIONS

The inherent epilepsies discussed in this chapter occur in species that are widely separated phylogenetically. Despite the divergence of species, there are, at least in part, similarities in the clinical manifestations of their epilepsies that may suggest similarities in basic pathophysiological mechanisms. Genetic animal models of epilepsy may thus provide information about the fundamental mechanisms involved in onset and development of the epilepsies that is inaccessible with other techniques.

One of the most important potential uses of experimental animals with inherent epilepsy is in the testing of new drugs for antiepileptic activity. In this regard, determination of the responsiveness of the seizures of each of the species to drugs having known efficacy against the major types of human epilepsy has been helpful in establishing a specific type of epilepsy for which each genetic animal model is most suited for drug testing. In reference to the seizure types and antiepileptic drug efficacies in the different genetic animal models discussed above, dogs, rats, gerbils, and chickens with generalized tonic–clonic seizures represent suitable models for grand mal epilepsy, whereas rats and *tottering* mice with spike-wave absence seizures as well as gerbils with minor seizures are unique models for petit mal epilepsy in humans. Photosensitive baboons offer a useful model of photomyoclonic seizures and possibly also myoclonic petit mal in humans. *Tottering* mice with focal motor seizures and dogs with focal seizures may prove useful and chronic models for the partial epilepsies.

The low initial and maintenance costs of the small rodents (mice, rats, gerbils) and domestic fowls, and the ease with which their seizures can be initiated, make these species particularly valuable for large-scale testing of new drugs. In contrast, the difficulties of obtaining epileptic dogs and the high prime and maintenance costs of baboons in the numbers necessary for drug testing limit the usefulness of these two species for drug evaluation. However, the canine and baboon models offer unique opportunities for testing the more subtle effects of drugs on behavior in an experimental paradigm that might permit more accurate prediction of drug effects in man.

In conclusion, the observations summarized in this chapter emphasize that genetic animal models of epilepsies are of general service to those concerned with mechanisms, processes, and treatment of the epilepsies, and offer novel approaches for the development of new, effective antiepileptic drugs. Since most of the conventional marked antiepileptic drugs were selected by one of the traditional electrical or chemical seizure models, naturally occurring models of epilepsy might be expected to identify drugs with novel mechanisms of action or greater antiepileptic specificity.

REFERENCES

Al-Tahan F, Löscher W, Frey, H-H (1984): Pharmacokinetics of clonazepam in the dog. *Arch Int Pharmacodyn* 268:180–193

Bickford RG, Klass DW (1969): Sensory precipitation and reflex mechanisms. In: *Basic Mechanisms of the Epilepsies,* Jasper HH, Ward AA Jr, Pope A, eds. Boston: Little, Brown and Company, pp 543–564

Bielfelt SW, Redman HC, McClellan RO (1971): Sire- and sex-related differences in rates of epileptiform seizures in a purebred beagle dog colony. *Am J Vet Res* 32:2039–2048

Cepeda C, Worms P, Lloyd KG, Naquet R (1982): Action of progabide in the photosensitive baboon, *Papio papio*. *Epilepsia* 23:463–470

Chambon JP, Molimard JC, Calassi R, Maruani J, Rodier D, Sigault G, Leyris R, Ronucci R, Biziere K (1984): Anticonvulsant activity of dioxo-4,9 diaza-5-10 tetradecane (CM

40142), a new GABA derivative, in mice and photosensitive baboons. *Arzneim-Forsch* (Drug Res) 34:1017–1021
Chapman AG, Croucher MJ, Meldrum BS (1984): Evaluation of anticonvulsant drugs in DBA/2 mice with sound-induced seizures. *Arzneim-Forsch* (Drug Res) 34:1261–1270
Collins RL (1972): Audiogenic seizures. In: *Experimental Models of Epilepsy—A Manual for the Laboratory Worker,* Purpura DP, Penry JK, Tower D, Woodbury DM, Walter R, eds. New York: Raven Press, pp 347–372
Consroe P, Piccioni A, Chin L (1979); Audiogenic seizure susceptible rats. *Fed Proc* 38:2411–2416
Consroe P, Wolkin A (1977): Cannabidiol—antiepileptic drug comparisons and interactions in experimentally induced seizures in rats. *J Pharmacol Exp Ther* 201: 26–32
Cox B, Lomax P (1976); Brain amines and spontaneous epileptic seizures in the Mongolian gerbil. *Pharmacol Biochem Behav* 4:263–267
Crawford RD (1969): A new mutant causing epileptic seizures in domestic fowl. *Poultry Sci* 48:1799
Crawford RD (1970): Epileptiform seizures in domestic fowl. *J Hered* 61:185–188
Croft PG (1965): Fits in dogs: A survey of 260 cases. *Vet Rec* 77:438–445
Cunningham JG (1971): Canine seizure disorders. *J Am Vet Med Assoc* 158:589–598
Dailey JW, Reigel CE, Mishra PK, Jobe PC (1989): Neurobiology of seizure predisposition in the genetically epilepsy-prone rat. *Epilepsy Res* 3:3–17
De Sarro GB, Nistico G, Meldrum BS (1986): Anticonvulsant properties of flunarizine on reflex and generalized models of epilepsy. *Neuropharmacology* 25:695–701
Donaldson HH (1924): *The Rat*. Philadelphia: Memoirs of the Wistar Institute of Anatomy and Biology 6, p 469
Edmonds HL Jr, Hegreberg GA, van Gelder NM, Sylvester DM, Clemmons RM, Chatburn CG (1979): Spontaneous convulsions in beagle dogs. *Fed Proc* 38:2424–2428
El Sayed MA, Löscher W, Frey HH (1978): Pharmacokinetics of clonazepam in the dog. *Arch Int Pharmacodyn* 234:180–192
Falco MF, Barker J, Wallace ME (1974): The genetics of epilepsy in the British alsatian. *J Small Anim Pract* 15:685–692
Farghali-Hassan BM, Bossi L, Garattini S, Gerna M, Gomeni R, Morselli PL (1976): Carbamazepine pharmacokinetics in young, adult and pregnant rats. Relation to pharmacological effects. *Arch Int Pharmacodyn* 220:125–139
Frey H-H, Göbel W, Löscher W (1979): Pharmacokinetics of primidone and its active metabolites in the dog. *Arch Int Pharmacodyn* 242:14–30
Frey H-H, Kampmann E (1965): Tolerance to anticonvulsant drugs. *Acta Pharmacol Toxicol* 22:159–171
Frey H-H, Kretschmer B-H (1971): Anticonvulsant effect of trimethadione in mice during continued treatment via the drinking water. *Arch Int Pharmacodyn* 193:181–190
Frey H-H, Löscher W (1980): Pharmacokinetics of carbamazepine in the dog. *Arch Int Pharmacodyn* 234:180–190
Frey H-H, Löscher W (1981): Clinical pharmacokinetics of phenytoin in the dog: A reevaluation. *Am J Vet Res* 41:1635–1638
Frey H-H, Löscher W, Reiche R, Schultz D (1981): Pharmacology of antiepileptic drugs in the gerbil—I. Pharmacokinetics. *Neuropharmacology* 20:769–771
Frey H-H, Löscher W, Reiche R, Schultz D (1983): Anticonvulsant potency of common antiepileptic drugs in the gerbil. *Pharmacology* 27:330–335.
Frey H-H, Löscher W, Reiche R, Schultz D (1984): Anticonvulsant effect of primidone in

the gerbil: time course and significance of the active metabolites. *Pharmacology* 28:329–335

Frey H-H, Schulz R (1970): Time course of the demethylation of trimethadione. *Acta Pharmacol Toxicol* 28:477–483

Gerber N, Weller WL, Lynn R, Rangno EE, Sweetman BJ, Busch MT (1971): Study of dose-dependent metabolism of 5,5-diphenylhydantoin in the rat using new methodology for isolation and quantitation of metabolites in vivo and in vitro. *J Pharmacol Exp Ther* 178:567–579

Goldblatt D, Konow A, Shoildson I, MacMath T (1971): Seizures in the Mongolian gerbil. *Neurology* 21:Abstract 433.

Green MC, Sidman RL (1962): Tottering—a neuromuscular mutation in the mouse. *J Hered* 53:233–237

Heller AH, Dichter MA, Sidman RL (1983): Anticonvulsant sensitivity of absence seizures in the *tottering* mutant mouse. *Epilepsia* 25:25–34

Holliday TA, Cunningham JG, Gutnick MJ (1970): Comparative clinical and eletroencephalographic studies of canine epilepsy *Epilepsia* 11:281–292

Jobe PC, Laird HE (1981): Neurotransmitter abnormalities as determinants of seizure susceptibility and intensity in the genetic models of epilepsy. *Biochem Pharmacol* 30:3137–3144

Johnson DD, Davis HL (1984): Anticonvulsants in epileptic fowl. *Arzneim-Forsch* (Drug Res) 12:1753–1757

Johnson DD, David HL, Crawford RD (1979): Pharmacological and biochemical studies in epileptic fowl. *Fed Proc* 38:2417–2423

Kaplan H (1975): What triggers seizures in the gerbil, *Meriones unguiculatus? Life Sci* 17:693–698

Killam EK (1979): Photomyoclonic seizures in the baboon, *Papio papio. Fed Proc* 38:2429–2433

Killam EK, Matsuzaki M, Killam DF (1973): Effects of chronic administration of benzodiazepines on epileptic seizures and brain electrical activity in *Papio papio*. In: *Benzodiazepines,* Garratini S, Mussini E, Randall LO, eds. New York: Raven Press, pp 443–460

Killam KF, Killam EK, Naquet RJ (1966): Mise en evidence chez certains singels d'un syndrome photomyoclonique. *Can R Acad Sci* 262:1010–1012.

Killam KF, Killam EK, Naquet RJ (1967): An animal model of light sensitivity epilepsy. *Electroencephalogr Clin Neurophysiol* 22:497–513

Klotz U, Antonin K-H, Bieck PR (1976): Pharmacokinetics and plasma binding of diazepam in man, dog, rabbit, guinea pig and rat. *J Pharmacol Exp Ther* 199: 67–73

Krushinsky LV, Molidkina LN, Fless DA, Dobrokhotova LP, Steshenko AP, Semiokhina AF, Zorina ZA, Romanova LG (1970): The functional state of the brain during sonic stimulation. In: *Physiological Effects of Noise,* Welch BL, Welch AS, eds. New York: Plenum Press, pp 159–183

Löscher W (1978): Serum protein binding and pharmacokinetics of valproate in man, dog, rat and mouse. *J Pharmacol Exp Ther* 204:255–261

Löscher W (1985a): GABAmimetics in animal models of seizure states In: *Epilepsy and GABA Receptor Agonists, Basic and Therapeutic Research,* Bartholini G, Bossi L, Lloyd KG, Morselli PL, eds. New York: Raven Press, pp 109–119

Löscher W(1985b): Anticonvulsant action of GABA agonists and prodrugs on minor and major seizures in epileptic gerbils. Comparison with mouse models of seizure states. *Arch Int Pharmacodyn* 276:263–270

Löscher W (1985c): Influence of pharmacological manipulation of inhibitory and excitatory neurotransmitter systems on seizure behavior in the Mongolian gerbil. *J Pharmacol Exp Ther* 233:204–213

Löscher W (1986): Development of tolerance to the anticonvulsant effect of GABAmimetic drugs in genetically epilepsy-prone gerbils. *Pharmacol Biochem Behav* 24:1007–1013

Löscher W (1992): The epileptic gerbil. Neuronal networks and actions of antiepileptic drugs. In: *Drugs for the Control of Epilepsy: Actions on Neuronal Networks Involved in Seizure Disorders,* Faingold CL, Fromm GH ed. Boca Raton: CRC Press pp 309-323

Löscher W, Czuczwar SJ, Wolff GL (1986): AE mice: an inbred mouse strain with interesting features for epilepsy research. *Epilepsia* 27:657–664

Löscher W, Frey H-H (1981): Pharmacokinetics of diazepam in the dog. *Arch Int Pharmacodyn* 254:180–195

Löscher W, Frey H-H (1984): Evaluation of anticonvulsant drugs in gerbils with reflex epilepsy. *Arzneim-Forsch (Drug Res)* 34:1484–1488

Löscher W, Frey H-H, Reiche R, Schultz D (1983): High anticonvulsant potency of gamma-aminobutyric acid (GABA)mimetic drugs in gerbils with genetically determined epilepsy. *J Pharmacol Exp Ther* 226:839–844

Löscher W, Meldrum BS (1984): Evaluation of anticonvulsant drugs in genetic animal models of epilepsy. *Fed Proc* 43:276–284

Löscher W, Nau H (1983): Distribution of valproic acid and its metabolites in various brain areas of dogs and rats after acute and prolonged treatment. *J Pharmacol Exp Ther* 226:845–854

Löscher W, Nau H, Marecaux C, Vergnes M (1984): Comparative evaluation of anticonvulsant and toxic potencies of valproic acid and 2-en-valproic acid in different animal models of epilepsy. *Eur J Pharmacol* 99:211–218

Löscher W, Schmidt D (1988): Which animal models should be used in the search for new antiepileptic drugs? A proposal based on experimental and clinical considerations. *Epilepsy Res* 2:145–181

Löscher W, Schwartz-Porsche D, Frey H-H, Schmidt D (1985): Evaluation of epileptic dogs as an animal model of human epilepsy. *Arzneim-Forsch (Drug Res)* 35:82–87

Loskota WJ, Lomax P (1974): The Mongolian gerbil as an animal model for the study of the epilepsies: Anticonvulsant screening. *Proc West Pharmacol Soc* 17:40–45

Loskota WJ, Lomax P (1975): The Mongolian Gerbil *(Meriones unguiculatus)* as a model for the study of the epilepsies: EEG records of seizures. *Electroencephalogr Clin Neurophysiol* 38:597–604

Loskota WJ, Lomax P, Rich ST (1974): The gerbil as a model for the study of the epilepsies. *Epilepsia* 15:109–119

Majkowski J, Kaplan H (1983): Value of Mongolian gerbils in antiepileptic drug evaluation. *Epilepsia* 24:609–615

Martinek Z, Arbeiter E (1980): Betrag zur anitkonvulsiven Wirkung von Dipropylessigsäure (DPA) bei Hunden. *Kleintier-Praxis* 25:275–280

Martinek Z, Horak F (1970): Development of so-called "genuine" epileptic seizures in dogs during emotional excitement. *Physiol Bohemoslov* 19:185–195

Maxson SC, Fine MD, Ginsburg BE, Konieck DL (1983): A mutant for spontaneous seizures in C57Bl/10Bg mice. *Epilepsia* 24:15–24

Meldrum BS, Evans MC, Braestrup C (1983): Anticonvulsant action in the photosensitive baboon, *Papio papio,* of a novel β-carboline derivative, ZK 91296. *Eur J Pharmacol* 91:255–259

Meldrum B, Pedley T, Horton R, Anlezark G, Franks A (1980): Epileptogenic and anticonvulsant effects of GABA agonists and GABA uptake inhibitors. *Brain Res Bull* 5(suppl 2): 685–690

Micheletti G, Marescaux C, Vergnes M, Rumbach L, Warter JM (1985b): Effects of GABAmimetics and GABA-antagonists on spontaneous nonconvulsive seizures in Wistar rats. In: *Epilepsy and GABA Receptor Agonists: Basic and Therapeutic Research.* Bartholini G, Bossi L, Lloyd KG, Morselli PL, eds. New York: Raven Press, pp 129–137

Micheletti G, Vergnes M, Marescaux C, Reis J, Depaulis A, Rumbach L, Warter JM (1985a): Antiepileptic drug evaluation in a new animal model: Spontaneous petit mal epilepsy in the rat. *Arzneim-Forsch (Drug Res) 35:483–490*

Naquet R, Meldrum BS (1972): Photogenic seizures in baboon. In: *Experimental Models of Epilepsy—A Manual for the Laboratory Worker.* Purpura DP, Penry JK, Tower D, Woodbury DM, Walter R, eds. New York: Raven Press, pp 373–406

Nau H, Zierer R (1982): Pharmacokinetics of valproic acid and metabolites in mouse plasma and brain following constant-rate application of the drug and its unsaturated metabolite with an osmotic delivery system. *Biopharm Drug Dispos* 3:317–328

Nellhaus G (1958): Experimental epilepsy in rabbits: development of a strain susceptible to audiogenic seizures. *Am J Physiol* 193:567–572

Noebels JL (1979): Analysis of inherited epilepsy using single locus mutations in mice. *Fed Proc* 38:2405–2410

Noebels JL, Sidman RL (1979): Inherited epilepsy: spike-wave and focal motor seizure in mutant mouse *tottering*. *Science* 204:1334–1336

Olsen RW, Wamsley JK, Lee R, Lomax P (1984): Alterations in the benzodiazepine/GABA receptor-chloride ion channel complex in the seizure-sensitive Mongolian gerbil. In: *Neurotransmitters, Seizures and Epilepsy II,* Fariello RG, Morselli PL, Lloyd KG, Quesney LF, Engel J Jr, eds. New York: Raven Press, pp 201–213

Purpura DP, Penry JK, Tower D, Woodbury DM, Walter R (1972): Experimental Models of Epilepsy—A Manual for the Laboratory Worker. New York: Raven Press

Rinne SP, Bowyer JF, Barrows EB, Killam EK (1978): EEG effects of ethosuximide in *Papio papio*. *Pharmacologist* 20:Abstract 161.

Schmidt D (1981): *Behandlung der Epilepsien*. Stuttgart: Georg Thieme Verlag

Schonfeld AR, Glick SD (1980): Neuropharmacological analysis of handling-induced seizures in gerbils. *Neuropharmacology* 19:1009–1016

Schultz D (1983): *Versuche mit einfachen Tiermodellen der Petit mal-Epilepsien,* Doctoral Thesis, Berlin: Fachbereich Veterinärmedizin

Schwartz-Porsche D, Löscher W, Frey H-H (1985): Therapeutic efficacy of phenobarbital and primidone in canine epilepsy: a comparison. *J Vet Pharmacol Ther* 8:113–119

Schweingruber R, Ketz E, Fankhauser R (1962): Zur Frage der genuinen Epilepsie beim Hund. *Psychiat Neurol (Basel)* 143:65–74

Seyfried TN (1979): Audiogenic seizures in mice. *Fed Proc* 38:2399–2404

Stark LG, Killam KF, Killam EK (1970): The anticonvulsant effects of phenobarbital, diphenylhydantoin and two benzodiazepines in the baboon, *Papio papio*. *J Pharmacol Exp Ther* 173:125–132

Thiessen DD, Lindzey G, Friend HC (1968): Spontaneous seizures in the Mongolian gerbil *(Meriones unguiculatus)*. *Psychonomic Sci* 11:227–228

Van der Velden NA (1968): Fits in Tervueren shepherd dogs: a presumed hereditary trait *J Small Anim Pract* 9:63–70

Vergnes M, Marescaux C, Micheletti G, Reis J, Depaulis Z, Rumbach L, Warter JM (1982): Spontaneous paroxysmal electroclinical patterns in rat: a model of generalized non-convulsive epilepsy. *Neurosci Lett* 33:97–101

Wallace ME (1975): Keeshond: a genetic study of epilepsy and EEG readings. *J Small Anim Pract* 16:1–10

Genetically Selected Rat Lines as Models of Alcoholism and Alcohol Intoxication

J.D. SINCLAIR, A.D. LÊ, AND K. KIIANMAA

Selectively bred lines of rodents have become increasingly important tools in alcohol research. There were none before 1950; two were started in the 1950s (the UCh-A and UCh-B lines); four in the 1960s (LS, SS, AA, ANA); six more in the next decade (AT, ANT, MA, LA, P, NP); and 14 new lines have been developed in the 1980s (HAD, LAD, HAS, LAS, WSP, WSR, SEW, MEW, SLOW, FAST, HOT, COLD, plus a pair being developed as NIAAA). These 26 lines belong to 13 pairs, with one line in each pair having a high level and the other a low level of an ethanol-related characteristic.

Most of the line development in the field has been related to two aspects: voluntary consumption of alcohol and intoxication from alcohol. It is beyond the scope of this chapter to provide comprehensive coverage of all of the experimental results obtained even with only the lines selected for these differences. For more specific information, the reader is directed to the reviews by Li et al. (1981), Phillips et al. (1989), Sinclair et al. (1989), Eriksson (1990), and the book edited by Kiianmaa et al. (1989). Instead, we will describe the historical background, the conclusions that seem to be emerging from the work, and some of the reasons for using selected lines in alcohol research.

GENETIC INFLUENCE ON ALCOHOL DRINKING

It had long been known that alcoholism tended to run in families. It was not clear, however, whether genetic factors played any part in this relationship or if it was caused solely by environmental factors. The importance of inheritance was first demonstrated in research with animals. Initially, it was found that different inbred strains of mice differ in their alcohol preference (McClearn and Rodgers, 1959). Then, experiments in Chile (Mardones, 1972), Finland (Eriksson, 1968, 1969), and Indiana (Li et al., 1981) showed that the offspring of heavier drinking rats consumed more ethanol than the offspring of rats that avoided alcohol. This research resulted in the establishment of pairs of rat lines with high and low voluntary selection of ethanol: the UCh-B/UCh-A pair in Chile, the Alko Alcohol/Alko Non-Alcohol (AA/ANA) pair in Helsinki, and the Preferring/Non-

Preferring (P/NP) and High Alcohol Drinking/Low Alcohol Drinking (HAD/LAD) pairs in Indianapolis (see Sinclair et al., 1989).

If all variation in alcohol drinking were caused by differences in the environment and experiences, selective breeding of high and low alcohol drinkers would not cause any divergence between the lines. Thus the fact that it was possible to develop lines differing in their voluntary alcohol drinking proved that inheritance is important in alcohol selection and the reinforcement from alcohol in animals. This in turn has helped to stimulate studies that have established that genetic factors also play a role in human alcohol drinking and abuse (e.g., Goodwin et al., 1973; Cadoret and Gath, 1978; Schuckit, 1985; Tarter et al., 1985; Topel, 1985; Cloninger, 1987; Finn and Pihl, 1987; Heath and Martin, 1988) and to establish the goal of finding the underlying mechanisms and genetic markers for a predisposition to develop alcoholism.

Alcohol Intake and Motivation

The behavior selected with these rat lines was the amount of alcohol the animals consume when given a free choice between an unflavored alcohol solution (e.g., 10% v/v) in one bottle and water in a second bottle. The consumption measures used in the selection are obtained only after the animals have been given several weeks of prior free access in which to learn either to drink or avoid the alcohol solution. This consumption also tends to be more stable and uniform within each line than that during the first few days of access.

Each of the lines selected for heavy drinking has eventually come to consume large quantities, and the ones selected for low drinking to avoid practically all alcohol. For example, in the F_{55} generation, the AA male rats consumed an average of 7.6 g/kg/day of ethanol, whereas the ANA males drank only 0.4 g/kg/day. It should be noted that in the development of the P/NP lines, the nonpreferring NPs stabilized after fewer generations than the preferring rats, suggesting that fewer genes may be controlling the aversion to ethanol (Li et al., 1981).

The AA rats learn to perform an operant task such as lever pressing in order to obtain alcohol, but ANA rats appear to get little reinforcement from alcohol (George, 1987; Kalant, 1988; Ritz et al., 1986; Sinclair, 1974). The AAs learn it despite free access to food and water, no shaping or training, and no prior ethanol experience (Hyytiä and Sinclair, 1989). Similar results are generally obtained with P and NP rats. The P rats have also been shown to work for intragastric infusions of ethanol (Waller et al., 1984), thus indicating that the reinforcement is at least partly postabsorptive. It has been concluded that the AA rats (Kalant, 1988) and P rats (Froehlich et al., 1990) generally meet the requirements for animal models of alcoholism.

Correlates of Differences in Alcohol Drinking

A wide variety of behavioral and physiological characteristics have been examined in both the AA/ANA and P/NP pairs in the hope of discovering factors and mechanisms underlying the differences in alcohol drinking. In some cases, however, factors identified in one pair have not been confirmed in the other. For example, the ANAs develop much higher levels of acetaldehyde accumulating after ethanol injection than the AAs (cf. Hilakivi et al., 1984; Koivula et al., 1975), but preliminary results suggest that this difference is not present in the P/NP pair (P. Eriksson, personal communication).

Nevertheless, other factors have been found to differentiate the two pairs similarly. For instance, both the AA and P rats tend to show a larger stimulatory effect from low doses of ethanol than do ANA and NP animals (cf. Hilakivi et al., 1984; Waller et al., 1986). Similarly, the alcohol-preferring lines both develop tolerance to ethanol better than do the nonpreferring lines (Lê and Kiianmaa, 1989; Gatto et al., 1987).

Another similarity recently observed is that both the AA and P rats, but not Wistar animals, show a "short-alcohol-deprivation effect" (Sinclair and Li, 1989): their alcohol drinking is increased after only a few hours of abstinence. Interestingly, severely dependent human alcoholics had earlier also been found to show an increase in craving and rate of alcohol drinking induced by only 3 hr of abstinence (Rankin et al., 1979).

It is easy to imagine how the amount of reinforcement produced by alcohol, the accumulation of acetaldehyde from ethanol metabolism, the development of tolerance, and the rate at which craving grows during abstinence could promote or limit alcohol drinking. The work with selected rat lines has demonstrated that each of these factors is influenced by genetics. Thus, one can hypothesize that these factors might be among the genetically determined characteristics influencing the development of alcoholism in humans. There are, however, many other differences found in rat lines selected for differential alcohol drinking that may help to generate less intuitively obvious hypotheses for genetically determined mechanisms underlying alcoholism.

GENETIC INFLUENCE ON INTOXICATION

The second area of alcohol research to use selectively bred lines of rodents was the study of intoxication. First, it was shown that existing inbred strains of mice differ in their sensitivity to inhaled alcohol (McClearn, 1962; McClearn and Rodgers, 1959), their ability to metabolize alcohol (Rodgers et al., 1963), and the length of time that their righting reflex was lost after an injection of alcohol (Kakihana et al., 1966; Kakihana and Butte, 1980). A selective-breeding program started in 1962 at the University of California in Berkeley, with eight inbred strains crossed to produce a heterogeneous foundation stock, succeeded in producing the Long Sleep (LS) and Short Sleep (SS) mouse lines, which differ greatly in the duration of righting reflex loss produced by ethanol (e.g., Collins, 1981; McClearn, 1979; McClearn and Kakihana, 1981). The lines were moved later to the University of Colorado, Boulder, where they are still being maintained and studied today.

Intoxication is not, however, a single, unitary phenomenon. The hypnotic effect of high doses examined in the LS and SS mice may be influenced by different mechanisms than the stimulatory effect at low doses, the lethal effect of very high doses, or the various effects at moderate doses such as hypothermia and motor incoordination. These other components have been examined in more recent selective breeding programs. The HOT and COLD mice lines have been developed for differences in their hypothermic effects from ethanol, and the FAST and SLOW lines for differences in the locomotor stimulatory effects of mice (see Phillips et al., 1989).

The motor-impairment effects are of particular practical importance because much of the total harm from alcohol in human society comes from the decrease in coordination. Consequently, a program of selective breeding of rat lines for differences in ethanol-induced motor impairment was begun at the Alko laboratories in Helsinki (cf. Eriksson and Rusi, 1981). A heterogeneous base stock, obtained by crossing Wistar, Sprague–Dawley, and Long–Evans derived animals, was tested for the decrease produced by ethanol (usually 2 g/kg) in the angle when they begin sliding on a tilting plane (Arvola et al., 1958). The blood-alcohol levels were also measured and the breeders selected so as to minimize any difference in ethanol elimination. The program resulted in the AT and ANT lines, which differ greatly in their motor impairment from alcohol measured on the tilting plane test, an accelerating Rotorod test (Eriksson and Rusi, 1981), and the horizontal wire test (Hellevuo and Korpi, 1988).

In the most recently tested generation, F_{36}, the mean ($\pm$SE) differences in sliding angles on the tilting plane for male and female ATs were $7.6° \pm 9.2°$ and $5.5° \pm 7.1°$, respectively, in comparison to $25.3° \pm 4.9°$ and $24.2° \pm 5.9°$ for ANT animals. Another selective breeding program (Riley et al., 1976, 1978) produced the MA (Most Affected) and LA (Least Affected) rat lines developed for differences in the suppressive effects of ethanol on locomotor activity. It might be noted that although this difference in sliding angle is easier to understand, the difference in the sines of the sliding angles is theoretically a better measure of the motor impairment because the sine of the angle is proportional to the amount of force the animal must exert to remain on the plane.

Line pairs developed for a difference on one component of alcohol intoxication do not necessarily differ on other components. There is no difference between the AT and ANT rat lines or between the FAST and SLOW mouse lines in their hypothermic reaction to ethanol (Eriksson and Sarviharju, 1984; Lê and Kiianmaa, 1989; Philips et al., 1989) and the FAST/SLOW lines do not differ in their hypnotic reactions. Similarly, the LS and SS mice show only a small difference in hypothermic effect (Tabakoff et al., 1980), and this may to some extent be a reflection of the faster ethanol elimination rate in the SS mice (Smolen and Smolen, 1989).

One conclusion from the work with the AT/ANT lines may be of particular interest. So far there has been little evidence for neural mechanisms that are more sensitive to the effects of ethanol in the ANT line, which shows more behavioral impairment. Instead, two instances have been found that suggest that the AT animals have better "control mechanisms" than the ANTs for counteracting the effects of ethanol: a GABA inhibitory system more responsive to ethanol (Hellevuo and Kiianmaa, 1989) and a higher rate of acquiring tolerance (Lê and

Kiianmaa, 1990). This emphasizes the fact that organisms possess many homeostatic mechanisms at various levels, all operating to oppose change, as is illustrated for the case of alcohol in Figure 7.1. Consequently, when lines are developed for high and low levels of effects from a drug, it is possible and, indeed, perhaps more likely that differences will be found in the homeostatic control mechanisms of the lines rather than in the sensitivity per se of some target for the drug. Furthermore, since tolerance is the ability to counteract an effect, it suggests that the names for our lines, AT = "Alcohol Tolerant", and ANT = "Alcohol Nontolerant," may actually be rather accurate.

USES OF SELECTED LINES IN ALCOHOL RESEARCH

The initial purpose for selective breeding programs is to determine whether genetic factors actually have an influence on a particular feature. The ability to develop lines differing in alcohol preference showed that genetic influences help determine how much alcohol a rat drinks. Similarly, the successful breeding programs related to intoxication have demonstrated that inheritance has an important role in determining the amount of stimulatory effect, motor impairment, hypothermic effect, and hypnosis produced by ethanol. There has been little research with humans on this question, but the animal results suggest that inheritance may help to determine the susceptibility to ethanol intoxication in people, too.

The most common use of the lines, once they have been established, has been the search for correlates of the selected trait, that is, of alcohol preference or sensitivity. The purpose is to find the underlying factors and mechanisms. A problem with this research, however, is that genetic drift may cause spurious line differences unrelated to the selected trait to appear. The problem increases the longer the lines are maintained and increases much more rapidly if a smaller number of breeding animals is used.

Maintaining replicate lines, for example, two or more high lines and two or more low ones, has been seen as a solution to this problem. This does give an independent estimate of the amount of genetic drift that has occurred, which is useful statistically (Henderson, 1989), but it does not solve the problem and may be misleading. In earlier generations, you can be no more certain that a difference is really related to the selected trait if you find it in all five replicate lines maintained with four breeders each than if you find it in a single line maintained with 20 breeders. (The probability of an allele being lost—or being fixed—in all k replicate lines of n breeders is approximately the same as the probability of losing it in a single large line with k times n breeders.)

As indicated by computer simulations (Sinclair et al., 1987), there is a more practical and powerful solution that minimizes the noise introduced by genetic drift. For the traits selected in our lines, a high degree of separation was found by about the fifth generation, with only a relatively small additional separation thereafter. Genetic drift, in contrast, is likely to begin producing spurious line

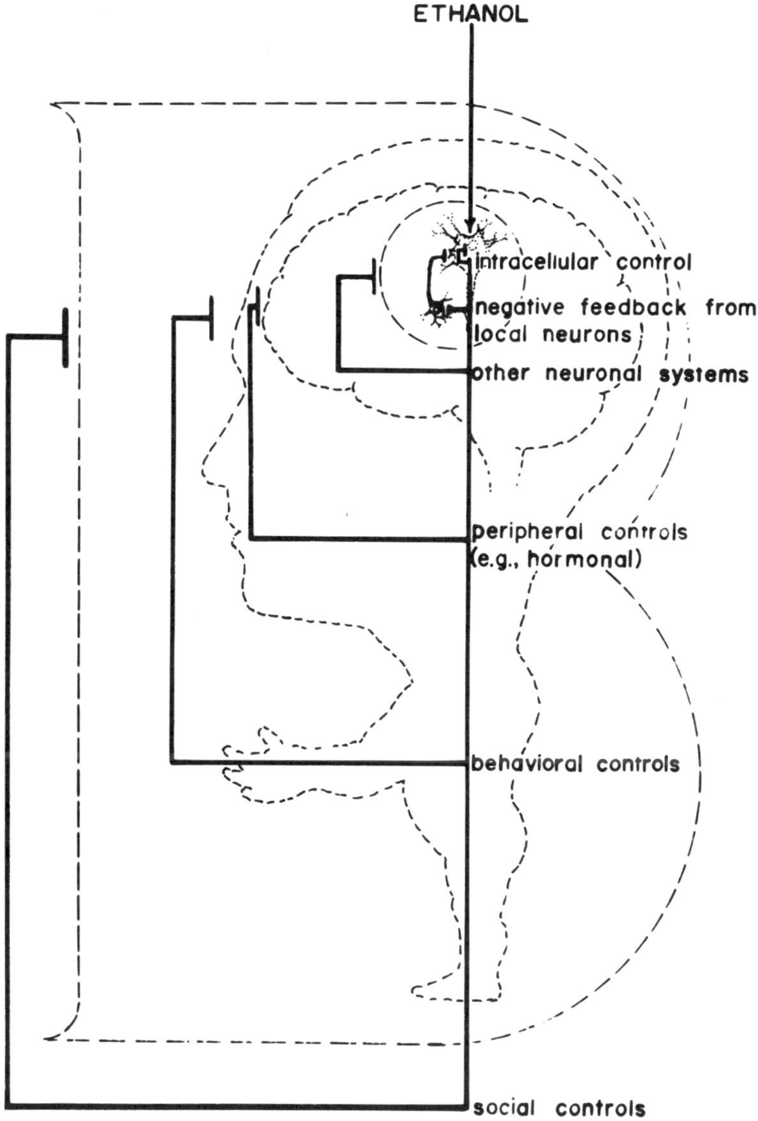

FIGURE 7.1. Some of the "control mechanisms" that may influence ethanol intoxication. From the level within a single cell to that of social factors, there are homeostatic control mechanisms opposing change of all kinds, including the changes involved in the production of intoxication. Genetic factors that influence intoxication need not be restricted to influencing the sensitivity of the alcohol target; instead, they may operate through such control mechanisms to make some individuals better able to tolerate ethanol.

differences after only about 10 generations, if 10 to 20 breeders are used per line. Consequently, there is a "window" present from about the fifth to tenth generation during which any line difference found is quite likely to be related to the selected train. By perpetually restarting the lines about every fifth generation, either from the same base population or from a crossing of some individuals in the separated lines, but at the same time maintaining each selected line for study until about the tenth generation, it is possible to have animals within this window always available.

Another reason for using selected lines in alcohol research is that they provide a convenient means for obtaining animals with extreme reactions for use in other research. For example, a study on the effect of some drug on heavily drinking rats (e.g., ones consuming 6 g/kg/day) could be more easily conducted with AA or P rats than by screening hundreds of rats from nonselected strains in order to obtain a large enough sample showing this behavior.

The existence of such lines provides the only means for conducting certain types of studies involving factors that could be influenced by prior ethanol exposure. The process of screening animals from nonselected strains for alcohol preference or alcohol intoxication sensitivity would itself confound any results obtained with such factors. The selected lines allow one to examine animals that would show high or low alcohol drinking, or high or low sensitivity, without ever exposing the animals to alcohol.

Finally, as will be discussed elsewhere (Sinclair, 1991), lines selected for differences in the amount of intoxication they display provide a unique means for localizing the primary sites where genetic factors first interact with ethanol intoxication.

REFERENCES

Arvola A, Sammalisto L, Wallgren H (1958): A test for the level of alcohol intoxication in the rat. *Q J Stud Alcohol* 19:563–572

Cadoret RJ, Gath A (1978): Inheritance of alcoholism in adoptees. *Br J Psychiatry* 132:252–258

Cloninger CR (1987): Neurogenetic adaptive mechanisms in alcoholism. *Science* 236:410–416

Collins AC (1981): A review of research using Short-Sleep and Long-Sleep mice. In: *Development of Animal Models as Pharmacogenetic Tools, NIAAA Research Monograph 6,* McClearn GE, Deitrich RA, Erwin VG, eds. Washington, DC: U.S. Government Printing Office.

Eriksson CJP (1990): Finnish selective breeding studies for initial sensitivity to ethanol: update 1988 on the AT and ANT rat lines. In: *Initial Sensitivity to Alcohol, NIAAA Research Monograph 20,* DHHS Pub. No. ADM 90-1611, Deitrich RA, Pawlowski AA, eds. Washington, DC: U.S. Government Printing Office

Eriksson CJP, Sarviharju M (1984): Motor impairment, narcosis and hypothermia by ethanol: separate genetic mechanisms. *Alcohol* 1:59–62

Eriksson K (1968): Genetic selection for voluntary alcohol consumption in the albino rat. *Science* 159:739–741

Eriksson K (1969): Factors affecting voluntary alcohol consumption in the albino rat. *Ann Zool Fenn* 6:227–265
Eriksson K, Rusi M (1981): Finnish selection studies on alcohol-related behaviors: general outline. In: *Development of Animal Models as Pharmacogenetic Tools, NIAAA Research Monograph 6,* McClearn GE, Deitrich RA, Erwin VG, eds. Washington, DC: U.S. Government Printing Office
Finn PR, Pihl RO (1987): Men at high risk for alcoholism: the effect of alcohol on cardiovascular response to unavoidable shock. *J Abnorm Psychol* 96:230–236
Froehlich JC, Li T-K (1990): Enkephalinergic involvement in voluntary alcohol drinking. In: *Opioids, Bulimia, and Alcohol Abuse & Alcoholism,* Reid LD, ed. New York: Springer-Verlag
Gatto GJ, Murphy JM, Waller MB, McBride WJ, Lumeng L, Li T-K (1987): Persistence of tolerance to a single dose of ethanol in the electively-bred alcohol-preferring rat. *Pharmacol Biochem Behav* 28:105–110
George FR (1987): Genetic and environmental factors in ethanol self-administration. *Pharmacol Biochem Behav* 27:379–384
Goodwin DW, Schulsinger F, Hermansen L, Guze SB, Winokur G (1973): Alcohol problems in adoptees raised apart from alcoholic biological parents. *Arch Gen Psychiatry* 28:238–243
Heath AC, Martin NG (1988): Teenage alcohol use in the Australian twin register: genetic and social determinants of starting to drink. *Alcoholism: Clin Exp Res* 12:735–741
Hellevuo K, Kiianmaa K (1989): GABA turnover in the brain of rat lines developed for differential ethanol-induced motor impairment. *Pharmacol Biochem Behav,* 34:905–909
Hellevuo H, Korpi ER (1988): Failure of Ro 15-4513 to antagonize ethanol in rat lines selected for differential sensitivity to ethanol and in Wistar rats. *Pharmacol Biochem Behav* 30:183–188
Henderson ND (1989): Interpreting studies that compare high and low-selected lines on new characters. *Behav Genetics* 19:473–502
Hilakivi L, Eriksson CJP, Sarviharju M, Sinclair JD (1984): Revitalization of the AA and ANA rat lines: effects on some line characteristics. *Alcohol* 1:59–62
Hyytiä P, Sinclair JD (1989): Demonstration of lever pressing for oral ethanol by rats with no prior training or ethanol experience. *Alcohol* 6:161–164
Kakihana R, Brown DR, McClearn GE, Tabershaw IR (1966): Brain sensitivity to alcohol in inbred mouse strains. *Science* 154:1574–1575
Kakihana R, Butte JC (1980): Biochemical correlates of inherited drinking in laboratory animals. In: *Animal Models in Alcohol Research,* Eriksson K, Sinclair JD, Kiianmaa K, eds. London: Academic Press
Kalant H (1988): Free choice consumption of ethanol by AA and ANA rats in an operant model. In: *Problems of Drug Dependence, 1987, NIDA Research Monograph No. 81,* Harris LS, ed. DHHS Publication No. (ADM)88-1564, Washington, DC: United States Government Printing Office
Koivula T, Koivusalo M, Lindros KO (1975): Liver aldehyde and alcohol dehydrogenase activities in rat strains genetically selected for their ethanol preference. *Biochem Pharmacol,* 24:1807–1811
Kiianmaa K, Tabakoff B, Saito T, eds. (1989): *Genetic Aspects of Alcoholism.* Helsinki: Finnish Foundation for Alcohol Studies
Lê AD, Kiianmaa K (1989): Initial sensitivity and the development of acute and rapid tolerance to ethanol in the AT and ANT rats. In: *Genetic Aspects of Alcoholism,*

Kiianmaa K, Tabakoff B, Saito T, eds. Helsinki: Finnish Foundation for Alcohol Studies

Lê AD, Kiianmaa K (1990): Role of initial sensitivity and genetic factors in the development of tolerance to ethanol in the AT and ANT rats. *Psychopharmacology,* 102:11–16

Li T-K, Lumeng L, McBride WJ, Waller MB (1981): Indiana selection studies on alcohol-related behaviors. In: *Development of Animal Models as Pharmacogenetic Tools, NIAAA Research Monograph 6,* McClearn GE, Deitrich RA, Erwin VG, eds. Washington, DC: U.S. Government Printing Office

Mardones J (1972): Experimentally induced changes in alcohol appetite. In: *Biological Aspects of Alcohol Consumption,* Forsander OA, Eriksson K, eds. Helsinki: Finnish Foundation for Alcohol Studies

McClearn GE (1962): Genetic differences in the effect of alcohol upon behavior of mice. In: *Proceedings of the Third International Conference on Alcohol and Road Traffic,* Havard JDJ, ed. London: British Medical Association

McClearn GE (1979): Influence of genetic variables on means, variances, anc covariances of behavioral response to toxicological and pharmacological substances. *J Toxicol Environ Health* 5:145–156

McClearn GE, Kakihana R (1981): Selective breeding for ethanol sensitivity: short-sleep and long-sleep mice. In: *Development of Animal Models as Pharmacogenetic Tools, NIAAA Research Monograph 6,* McClearn GE, Deitrich RA, Erwin VG, eds. Washington, DC: U.S. Government Printing Office

McClearn GE, Rodgers DA (1959): Differences in alcohol preference among inbred strains of mice. *Q J Stud Alcohol* 20:691–695

Phillips TJ, Feller DJ, Crabbe JC (1989): Selected mouse lines, alcohol and behavior. *Experientia* 45:805–838

Rankin H, Hodgson R, Stockwell T (1979): The concept of craving and its measurement. *Behav Res Ther* 17:389–396

Riley EP, Freed EX, Lester D (1976): Selective breeding of rats for differences in reactivity to alcohol: an approach to an animal model of alcoholism. *J Stud Alcohol* 37:1535–1547

Riley EP, Lochry EA, Freed EX (1978): Differential tolerance to pentobarbital in rats bred for differences in alcohol sensitivity. *Psychopharmacology* 58:167–170

Ritz MC, George FR, deFiebre CM, Meisch RA (1986): Genetic differences in the establishment of ethanol as a reinforcer. *Pharmacol Biochem Behav* 24:1089–1094

Rodgers DA, McClearn GE, Bennet EL, Hebert M (1963): Alcohol preference as a function of its caloric utility in mice. *J Comp Physiol Psychol* 56:666–672

Schuckit MA (1985): Genetics and the risk for alcoholism. *JAMA* 253:2614–2617

Sinclair JD (1974): Rats learning to work for alcohol. *Nature* 2:409–412

Sinclair JD (1991): Localization of the primary sites of genetic influence. *Behav Genet* (in press)

Sinclair JD, Lê AD, Kiianmaa K (1989): The AA and ANA rat lines, selected for differences in voluntary alcohol consumption. *Experientia* 45:798–805

Sinclair JD, Li T-K (1989): Long and short alcohol deprivation: effects on AA and P alcohol-preferring rats. *Alcohol* 6:505–509

Sinclair JD, Viitamaa T, Hyytiä P (1987): Behavioral and color variations betweeen rat lines developed for differential alcohol sensitivity. *Alcohol Alcoholism* (suppl 1):449–453

Smolen A, Smolen TN (1989): Reproducibility of ethanol elimination rates in long-sleep and short-sleep mice. *J Stud Alcohol* 50:519–524

Tabakoff B, Ritzmann RF, Raju TS, Deitrich RA (1980): Characterization of acute and chronic tolerance in mice selected for inherent differences in sensitivity to ethanol. *Alcoholism: Clin Exp Res* 4:70–73

Tarter RE, Alterman AI, Edwards KL (1985): Vulnerability to alcoholism in men: a behavior-genetic perspective. *J Stud Alcohol* 46:329–356

Topel H (1985): Biochemical basis of alcoholism: statements and hypotheses of present research. *Alcohol* 2:711–788

Waller MB, McBride WJ, Gatto GJ, Lumeng L, Li T-K (1984): Intragastric self-infusion of ethanol by ethanol-preferring and -nonpreferring lines of rats. *Science* 225:78–79

Waller MB, Murphy JM, McBride WJ, Lumeng L, Li T-K (1986): Effect of low dose ethanol on spontaneous motor activity in alcohol-preferring and -nonpreferring lines of rats. *Pharmacol Biochem Behav* 24:617–623

Genetic Studies of Nicotine and Nicotine/Alcohol Reactivity in Humans and Animals

ALLAN C. COLLINS AND MICHAEL J. MARKS

Although psychoactive substances are readily available in today's society, not everyone uses or abuses these agents. Some individuals may choose not to use a drug because of environmental influences or societal attitudes, but even when drugs are readily available and the societal attitudes permissive or supportive not everyone will use a drug and not everyone who experiments with the drug becomes dependent. Whenever individual differences of this sort are observed, potential genetic influences on the trait in question must be considered. This chapter will briefly review the literature that suggests that genetic factors contribute to the use and abuse of alcohol and will attempt to provide a comprehensive review of the literature that indicates that genetic factors may influence tobacco use. Similarly, a cursory review of the studies done in animals that have attempted to explain and identify the genetic underpinnings of human alcohol use will be presented along with a thorough review of the animal studies that have dealt with the genetics of nicotine-induced behavioral and physiological effects. Some emphasis will be placed on potential genetic regulation of the interactions between alcohol and nicotine.

GENETIC INFLUENCES ON PSYCHOACTIVE DRUG USE BY HUMANS

The idea that genetic factors might predispose individuals to misuse and abuse psychoactive agents is hardly a revolutionary concept. The ancient Greeks were concerned with this issue, as evidenced by comments made by Plutarch and Aristotle. Plutarch suggested that "one drunk begets another" and Aristotle argued that "drunken women bring forth children like themselves." Although these ancient philosophers were most concerned with the possibility that alcohol use at the time of conception might have a deleterious effect on the "seed," it seems likely that even in ancient Greece the familial incidence of alcoholism was evident. In more recent times, Davenport (1912) noted in a pedigree analysis of a number of behavioral traits that narcotism seemed to have a heritable basis. Despite this long-standing suspicion that genetic factors may influence substance use and abuse, we still have only fragmentary knowledge concerning the details of this genetic control.

Genetics of Alcoholism

Without question, we know more about potential genetic influences on alcohol use and abuse than any other drug. The evidence that alcoholism has a heritable basis has come from studies of twins, of families, and of adoptees.

Twin Studies

Two types of twins exist, monozygotic (MZ, identical) and dizygotic (DZ, fraternal). The MZ twins are identical at all loci whereas DZ twins are no more similar than any other sibling pair; on average, DZ twins are the same at 50% of all genetic loci. Twins are often compared for traits of interest, and are judged to be concordant if both members of a twin pair express the trait. It has been argued that the trait of interest is probably regulated by genetic factors if MZ twins are more concordant than are the DZ twins.

Several studies have determined the concordance for alcoholism in MZ–DZ twins, and the results do not provide unequivocal support for a genetic theory of alcoholism. Although Kaij (1960) found the concordance for alcoholism in Swedish MZ twins was 58% whereas the concordance was only 28% in DZ twins, subsequent studies by Partanen et al., (1977) and Murray et al., (1983) failed to replicate this finding in studies of Finnish and English twins, respectively. The Finnish study detected an MZ–DZ difference in concordance among younger twins, but the overall sample failed to detect a difference. The English study also failed to detect an MZ–DZ difference. However, a study of American twins obtained from the United States Veteran's Administration found that the concordance for alcoholism in MZ twins was considerably higher than was the concordance in DZ twins (Hrubec and Omenn, 1981). Thus, based on twin studies that do not include an adoption analysis, it is not clear whether alcoholism has a heritable basis.

Family Studies

Studies of the familial incidence of alcoholism have produced more consistent results. Approximately 10 years ago, Cotton (1979) summarized the data obtained in 39 studies that assessed alcoholism in the families of 6251 alcoholics. The overall average from these 39 studies indicated that almost one third of all the alcoholics had at least one parent with an alcohol abuse problem. Recent studies of the familial incidence of alcoholism have only served to reinforce the conclusion that alcoholism "runs in families."

Although most researchers would agree that alcoholism has a heritable basis, there is not universal agreement concerning all alcoholics or forms of alcoholism. Goodwin (1985) has argued that there may be two types of alcoholism, familial and nonfamilial. He argues that the familial form is heritable and is characterized by early onset and is generally more severe than is the nonfamilial form of alcoholism. Goodwin has designated the nonfamilial form as an "environmentally induced phenocopy." Cloninger (1987) has recently taken issue with the suggestion that an environmentally induced form of alcoholism exists. He argues that there are regular relationships between average per capita consumption of alcohol and the proportion of heavy drinkers and suggests that alcoholism increases quadratically with average consumption. He also argues that it may be that the lesser evidence for a genetic influence on alcoholism in Goodwin's nonfamilial alcoholics may simply reflect contemporary age group–related differences in average consumption.

Adoption Studies

Twins and families often share very similar environments. This similar environment could explain the increased incidence of alcoholism seen in certain families. Adoption studies can minimize this problem. If an adoptee resembles his/her biological parents, genetic influence is presumed to be the dominant factor in influencing the trait of interest; if the adoptee resembles the adopting parents, environmental influences likely dominate. The adoption method has been used many times in alcoholism studies (see Schuckit et al., 1972; Bohman, 1978; Cadoret et al., 1979; Goodwin, 1985 for examples), and the results obtained are in general agreement; the alcohol consumption of biological sons of alcoholic fathers resembles that of their biological father, not their adoptive parents. The results concerning female alcoholism are equivocal in that the incidence of alcoholism in adopted daughters of alcoholics is not different from the population average (Goodwin et al., 1977).

Individual Differences in Sensitivity to Ethanol

Although many studies indicate that genetic factors seem to regulate the development of alcoholism, little progress has been made in identifying precisely what factors related to alcohol use are heritable. Nearly 20 years ago, Omenn and Motulsky (1972) suggested that alcoholism might be regulated by five "simpler" factors: sensitivity to acute effects, metabolism of ethanol, central nervous system adaptation to chronic ethanol intake (addictability), predisposing personality factors, and sensitivity to medical and behavioral complications. Although these factors sound reasonable, only a handful of studies have attempted to ascertain whether any of these suggestions are valid and most of these studies have dealt with individual differences in sensitivity to acute intoxicating effects. Two recent studies examined the sensitivities to ethanol of MZ and DZ twins (Wilson and Plomin, 1984; Martin et al., 1985). Both studies reported that MZ twins were more similar in response to ethanol than were DZ twins. Because neither of these two studies examined ethanol sensitivity of naive subjects, it is not known whether the MZ–DZ differences represent differences in innate sensitivity or differences in the development of tolerance to ethanol.

Other investigators have examined the sensitivity to ethanol of individuals who have family histories that differ in alcoholism. Schuckit and Gold (1988) reported that males from families with a positive history for alcoholism are less sensitive to alcohol's intoxicating actions and alcohol-induced prolactin release than are males from families that have no history of alcoholism, and argued that individuals who are less sensitive to the actions of alcohol can become alcoholics because they can consume larger doses of alcohol. Nagoshi and Wilson (1987) also reported on the relative sensitivities to alcohol of young, adult family history positive and negative males. These investigators assessed different parameters than did Schuckit and Gold (1988) and obtained different results; no mean group differences were detected. Surprisingly, however, the family history positive males perceived themselves to be more impaired by alcohol than did the family history negative males. As was the case with twin studies of alcohol sensitivity, neither of these studies used alcohol-naive subjects and neither controlled for previous intake. Consequently, no conclusions can be drawn with regard to the relationship between acute sensitivity to alcohol and the development of alcoholism, but this factor may be very important since Samson et al. (1989) have noted that the negative effects of early experiences with ethanol are predictive for abstinence in both sexes. Apparently those individuals who experience excess toxic responses to ethanol refrain from further use. This may mean that genetic factors that regulate sensitivity to toxic actions of ethanol will serve to reduce the probability of alcoholism.

Genetics of Smoking

Although it is not nearly as well studied as alcoholism, tobacco use may also be regulated by genetic factors. Most of the studies of tobacco use have used the twin method and most of these studies have assumed that if an individual is a smoker that they are like an alcoholic, addicted to their drug. Despite this potential problem, the data currently at hand indicate that genetic factors do, indeed, regulate tobacco use.

Twin Studies

More than 30 years ago, Fisher (1958a,b) reported that concordance for smoking (whether both members of a twin pair were smokers or nonsmokers) was higher in MZ twins than in DZ twins. This result was obtained in several subsequent studies (for example, see Raaschou-Nielsen, 1960; Shields, 1962; Conterio and Chiarelli, 1962; Dies and Reznikoff, 1969). Two of these twin studies (Fisher, 1958b; Shields, 1962) included studies of MZ twins that had been raised in separate families. Neither study detected any effect of rearing status on concordance for smoking, which indicates that family environment is not an important factor that influences tobacco use. In addition to higher concordance for ever/never smoking, MZ twins are also more similar to one another, than are DZ twins, for amount of tobacco used (Raaschou-Nielsen, 1960).

Heritability (h^2) has been reported for ever/never smoking in some of these studies; heritability estimates can range between 0 and 1.0 and represent an estimate of the total phenotypic variance that is due to additive genetic variance. If the h^2 value is 0 the trait in question is largely regulated by environment whereas values approaching 1.0 indicate that genetic control is dominant. Traits such as height or weight commonly have h^2 values in the 0.9 to 0.95 range. Pedersen (1981) reported an h^2 value of 0.84 for smoking status based on her study of Swedish twins whereas Kaprio et al. (1981) reported an overall heritability of 0.45 in a study of Finnish twins. The Finnish study included an estimate of heritability for different age groups. Heritability values of 0.55 were obtained for smoking status in 18- to 29-year-olds. This value declined to 0.17 for those over 60 years of age. We have calculated the heritabilities for two of the older studies (Shields, 1962; Dies and Reznikoff, 1969) and obtained values of 0.78 and 0.45, respectively. These results indicate that genetic factors are very important in regulating smoking status, but they indicate an important role for environmental influences as well.

FAMILY STUDIES

Eysenck (1980) carried out the only major study of potential genetic influences on smoking that involved more than a twin analysis; his study involved a combined twin and family approach. Potential influences of assortative mating, age/gene interactions, and sex differences were explored. Overall heritabilities of 0.22 for age of onset of smoking and 0.38 for average amount of tobacco consumed were obtained. Eysenck argued that these are modest heritabilities and that this indicates that environment is the most important factor in influencing initiation of smoking and amount consumed. However, when Eysenck analyzed data concerning smoking persistence he found that 68% of the variance was ascribed to genetic factors. Gurling et al. (1985) have suggested that Eysenck's observation that genetic factors are very important in influencing persistence of smoking may mean that nonsmokers have a strong biological reason for never having smoked or for giving it up quickly. This could arise if nonsmokers are more sensitive to nicotine's noxious actions or if they develop tolerance to these undesirable actions slowly, or not at all. Alternatively, initial sensitivity to the toxic actions of nicotine may be similar in smokers and nonsmokers, but nonsmokers may fail to detect any reinforcing effects from tobacco.

Individual Differences in Sensitivity to Nicotine

One of the most frequently encountered models that has been developed to explain smoking is the so-called nicotine titration model. This model states that smokers attempt to maintain blood nicotine within some specifically defined window. If levels are too high toxic actions of nicotine develop that serve to limit smoking, whereas if levels are too low abstinence symptoms develop and smokers "light up" to avoid or ameliorate their withdrawal symptoms. Kozlowski and Herman (1984) have suggested that individuals differ in these upper and lower boundaries because of differences in sensitivity to nicotine and that as long as blood nicotine levels remain between these boundaries, smoking is influenced by psychosocial factors.

Jones (1986) is one of the few investigators who has actually looked for individual differences in sensitivity to nicotine. He studied the effects of smoking on the heart rate of 124 smokers. Some individuals showed an increase in heart rate (responsive) whereas others did not (nonresponsive). Jones noted that the responsive individuals had smoking characteristics of light smokers whereas the nonresponsive individuals seemed to be heavier smokers. It is not known whether Jones' results are due to individual differences in first dose sensitivity to nicotine that have persisted or to differences in tolerance development. Along these lines, a recent report by Shiffman (1989) noted that light smokers, who he referred to as "chippers," had long-standing differences from heavy smokers. The chippers reported considerably fewer toxic symptoms following initial attempts at smoking than did the dependent (heavy) smokers. The heavy smokers apparently developed tolerance to nicotine actions to the point that they were equisensitive to the chippers. This observation suggests that individual differences in tolerance development exist and also implies that those individuals who develop tolerance to nicotine also become dependent.

Unfortunately, from a scientific viewpoint, no studies of individual differences in sensitivity to nicotine have used nonsmokers or nicotine-naive subjects. Consequently, it is impossible to determine whether differences among smokers in responsiveness to smoking or nicotine are due to pre-existing differences, to differences in smoking history, or to differences in tolerance development. Similarly, it is not known whether the observation made by Johnston (1942) that smokers perceived intravenous nicotine infusions as pleasant, whereas nonsmokers experienced toxic symptoms, is due to a pre-existing difference or to a difference that emerged after chronic nicotine exposures.

ANIMAL STUDIES OF DRUG ACTIONS

Many of the questions regarding genetic influences on psychoactive drug use and abuse cannot be ethically answered in studies involving humans. This is especially true if chronic drug treatment is desired or if mechanistic studies are contemplated. In addition, if genetic explanations are sought, genetic factors are not readily controlled in humans. Animal studies can minimize such problems, provided that adequate controls are used and realistic expectations are set. The remainder of this chapter will discuss the genetic methods that have been used in animal studies as well as some of the most important of the results obtained from these animal studies. Special emphasis will be placed on nicotine-related studies.

Techniques Used in Animal Studies

Presumably, a major goal of an animal study is to model the human condition(s). One way of assuring that animal studies can be generalized to the human condition is to increase the number of genetic stocks that are studied. Several different species might be studied or, if available, several genetically defined stocks within a species could be used. The use of several genetically defined stocks within a species has the advantage that cross-breeding experiments can be done to test further any genetic relationships that seem evident.

Inbred Strains

One of the most frequently encountered genetic techniques is the use of the inbred strain. Properly developed inbred strains are obtained by brother–sister mating for 20 or more generations. Such animals are like MZ twins in that all animals of an inbred strain are identical at all genetic loci. The precise genetic configuration in a given inbred strain is achieved by chance, with the exception of those genes that are highly related to fitness. More than 100 inbred mouse strains and a smaller number of inbred rat strains have been developed. A screening of inbred strains can be done quite readily and, if the strains differ in a trait of interest, unequivocal evidence will have been obtained that indicates that the trait being measured is regulated by genetic factors and that genetic variance exists.

Inbred strains offer advantages for ascertaining whether a given trait is regulated by genetic factors, but they have disadvantages as well. Inbred strain comparisons do not readily allow an estimate of the number of genes that might influence a trait of interest nor do they provide any information regarding dominance relationships. Perhaps of greater importance, inbred strain comparisons have major problems for correlational studies. Two traits might appear to be correlated with one another in a potential cause-effect relationship in an inbred strain analysis when, in fact, the genes that regulate both traits are actually closely linked on the same chromosome.

Diallel Crosses

Problems with inbred strain analyses have led researchers to develop several alternative strategies. Various breeding techniques have been developed and some of these have proven to be useful in understanding drug actions. A seldom used, but very informative cross-breeding technique involves systematically breeding several different inbred strains. This crossing technique, called a diallel cross, often involves the generation of reciprocal crosses; that is, females from each test strain are mated with males from each of the test strains in a Latin square design. This will result, for example, in 36 test generations if a diallel cross is developed from six inbred strains.

The diallel cross method involves only crosses between the parental inbreds. Consequently, only first generation hybrids (F1 hybrids) will be developed. The F1 hybrids are heterozygous at all loci where the parental inbred strains differ. The principal value of a diallel cross is that it provides information concerning dominance relationships and a preliminary estimate of the number of genes that regulate a trait may be obtained. However, linkage relationships are not broken down. Therefore, the diallel analysis is only slightly better than is an inbred strain analysis with regard to a correlational study.

Classical Crosses

The classical, or Mendelian cross, is often useful because linkage relationships are broken down. The classical cross usually consists of mating two inbred strains of interest to produce the F1 hybrid. The F1 hybrids are mated back to the parental inbreds to produce the so-called backcross generations. In addition, the F1 hybrids may be mated together to produce the F2 hybrids. It is in the production of the backcross and F2 hybrids that significant reshuffling of the genes occurs and linkage is broken down. The classical cross is useful for assessment of dominance relationships and may also be useful in estimating the number of genes that regulate a trait of interest. Correlational studies are best done in an F2 or even an F3 generation. Accurate correlations can be obtained in F2 or F3 animals, provided the measurement of one trait does not affect the other trait of interest.

Selectively Bred Stocks

Selective breeding normally involves testing a large number of heterogeneous stock or F2–F3 animals for a trait of interest (see DeFries, 1981, for a thorough discussion of the method). Animals with high values are bred with other high value animals and low value animals are bred with low value animals. Control animals are bred together by ignoring their values. This procedure is used in every generation. Inbreeding is minimized by avoiding the mating of animals who share a common grandparent or great grandparent. Assuming that selection is successful, the high line will eventually become homozygous for all of the alleles that regulate high values for the trait of interest and the low line will be homozygous for all alleles that contribute to low scores. Note that high values and low values are not necessarily caused by alternative alleles at the same locus. If outbreeding is maintained during the selection, the selectively bred animals will maintain normal levels of heterozygosity at loci that do not influence the selected-for trait. This results in reduced problems with linkage and explains why selectively bred animals are very valuable for identifying potential biochemical mechanisms that may regulate the selected-for trait.

Other Methods

Several other methods are available to the animal geneticist. These include the use of specifically outbred heterogeneous stocks and recombinant inbred strains. These will not be discussed here because neither of these methods have been used, to date, in an analysis of nicotine actions.

GENETIC STUDIES OF ALCOHOL-RELATED BEHAVIORS

Genetic studies of alcohol actions in animals have been reviewed extensively by Crabbe (1989) and by Phillips et al. (1989) and will receive only cursory treatment here. Until recently, the dominant approaches used in animal studies of alcohol actions have been inbred strain comparisons and selective breeding. These methods have unequivocally demonstrated that nearly every alcohol-related behavior is regulated by genetic factors. For example, inbred strains differ in initial sensitivity to alcohol, the development of tolerance, severity of withdrawal, and ethanol self-selection or preference (see Belknap, 1980, and Deitrich and Spuhler, 1984, for reviews of these studies), and selective breeding has been used successfully to obtain mouse and rat lines that differ in sensitivity to a first dose of ethanol, in ethanol self-selection (preference), and in severity of withdrawal following chronic treatment (reviewed in Crabbe, 1989). These studies clearly demonstrate genetic influences on ethanol-related behaviors. Furthermore, the animal models that have been developed have been useful in testing relevant hypotheses concerning potential mechanisms of ethanol actions.

GENETIC STUDIES OF NICOTINE-RELATED BEHAVIOR

Although many papers have been published that describe studies of behavioral or biochemical actions of nicotine in animals, only a small fraction of these studies have attempted to ascertain whether genetic factors influence nicotine actions. Nonetheless, the data currently available suggest that genetic factors regulate at least some aspects of nicotine actions. Most of these studies have used the rat or mouse, and almost all of the genetic studies have been concerned with potential genetic influences on initial (first dose) sensitivity to one or more of the behavioral or physiological actions of nicotine; a limited number of studies have dealt with potential genetic influences on the development of tolerance to nicotine.

Studies With the Rat

Only a handful of studies have attempted to determine whether genetically defined rat stocks differ in one or more responses to nicotine. Those studies that have been reported have used only inbred and selectively bred strains. Consequently, only limited genetic information has been observed.

Nicotine-Induced Alterations in Locomotor Activities

The first suggestion that the locomotor-altering actions of nicotine are regulated by genetic factors came from a study of the effects of nicotine on four categories of behavior (rearing, moving, grooming, and immobility) in three rat strains (Morrison and Lee, 1968). Lister, Sprague–Dawley, and Wistar rats were exposed to the test chamber, a $60 \times 30 \times 30$ cm box, for 5 min four times on a biweekly basis. Five min before the fifth test the animals were injected with 0.4 mg/kg nicotine. Strain differences in control activity were found and the 0.4 mg/kg dose of nicotine increased the activity scores of the inactive Wistar strain whereas it decreased activity scores in the Lister rats.

Several studies of nicotine effects on locomotor activities have used selectively bred rat strains (inbred) or lines (outbred). Rat strains, or lines, that differ in response to one or more of nicotine's effects have not been developed. However, nicotine actions, primarily nicotine effects on locomotor activity, have been measured in the Maudsley Reactive (MR) and Nonreactive (MNR) strains, which were selectively bred for differences in "emotionality" as assessed by defecation scores when placed in an open-field arena (Broadhurst, 1960) and in the Roman High Avoidance (RHA) and Low Avoidance (RLA) lines, which were selectively bred for differences in conditioned avoidance response (Bignami, 1965). Since the Maudsley and Roman rats are known to differ in locomotor activity (for a review, see Driscoll and Battig, 1982), it may be that included among those genes that influence defecation scores and acquisition of a conditioned avoidance response are genes that also influence locomotor activity; that is, one or more of the "emotionality" genes exert a pleiotropic effect on locomotor activity. Thus, a study of nicotine actions on locomotor activity of the Maudsley and Roman rats is a study of pleiotropic gene action.

Differences in nicotine effects on locomotor activity of the Roman lines and Maudsley strains have been reported several times (Garg, 1968, 1969a,b; Battig et al., 1976; Fleming and Broadhurst, 1975), but these studies must be viewed as being incomplete. None of these studies, except for the study of Fleming and Broadhurst (1975), tested more than one nicotine dose or one time after administration. Therefore, it is not clear whether either the Roman lines or Maudsley strains differ qualitatively or quantitatively in their sensitivities to nicotine. Complete dose-response curves and time-course studies will be required to resolve this question. Also, many of the rat studies used multiple nicotine injections and, as a result, it is not clear whether the genetic differences that were reported represent differences in sensitivity to a first dose of nicotine or differences in the development of tolerance.

The studies using the rat also did not attempt to provide an estimate of the number of genes that regulate the differences that were detected, nor did these studies attempt to identify potential biochemical explanations for the genetic differences. We do not know, for example, whether the strains and lines differ in the rate of metabolism of nicotine or whether they differ in central nervous system sensitivity to the drug. All that can be said at this time is that rats differ genetically in response to nicotine's actions, particularly in effects on locomotor activity.

Studies With the Mouse

Nicotine affects a wide variety of behavioral and physiological processes, and many of these actions have been studied from a genetic perspective in the mouse. The progress that has been made in characterizing the genetic regulation of a single challenge dose of nicotine on learning, locomotor activities, body temperature, respiration rate, heart rate, acoustic startle response, seizures, and adrenal hormone release will be discussed in the remainder of this chapter. In addition, the progress that has been made in characterizing genetic influences on the development of tolerance to nicotine's effects will be analyzed.

Nicotine Effects on Learning and Memory

Only one study has attempted to characterize the potential genetic influence on nicotine effects on a learning task in the mouse. Bovet et al. (1967) studied the effects of nicotine on a conditioned avoidance response in nine inbred mouse strains. In this study, the mice were exposed to a conditioning stimulus (turning on a 10-watt light) that preceded a mild shock by 5 sec. The mice could avoid the shock by running into an adjacent compartment of the test apparatus. The mice were given 100 trials for each of 5 consecutive days. Control animals were injected with saline before each test session whereas nicotine-treated animals were injected with a 0.5 mg/kg dose. Clear differences were seen in the saline-injected controls. Some mouse strains, such as the DBA, A/J, and C57BR strains, learned the task very well and would avoid 70% to 80% of the shocks by the fifth test day. Other strains, such as the C3H, C57BL/6, and CBA strains, never learned to avoid the shock. Nicotine treatment affected the various mouse strains differently; six of the strains showed a nicotine-induced enhancement of learning, but this enhancement varied from strain to strain. Two of the strains, the C57BL/10 and C57BR/cd strains, were impaired by nicotine. Established conditioned behavior was not affected in any strain by nicotine. Thus, genetic factors influence both the magnitude and the direction of nicotine's effects on conditioned avoidance learning in the mouse. Unfortunately, no one seems to have pursued these studies any further. Consequently, we know only that genotype influences the effects of five injections with nicotine on acquisition of a conditioned avoidance response.

Genetic Influences on Acute Sensitivities to Nicotine

Because nicotine's effects on locomotor activities appeared to be regulated by genetic factors in the rat, Hatchell and Collins (1980) assessed the effects of nicotine on locomotor activity in both sexes of three inbred mouse strains (C57BL/6, DBA/2, C3H). The animals were injected with a 1.0-mg/kg nicotine dose and locomotor activity was measured for a 3-min time period in a symmetrical Y-maze starting 2.5, 5, 10, 20, and 40 min after injection (i.e., a time course of nicotine action was determined for each strain). All of the animals were naive to injection and to the Y-maze at the time of testing. Therefore, these experiments assessed first-dose sensitivity (acute sensitivity) to nicotine. For two of the strains, the C57BL and DBA, nicotine elicited clear-cut decreases in locomotor activity at the early time points following injection. These effects were transient, however, since both sexes of the C57BL strain and the females of the DBA strain regained control levels, or slightly greater than control levels, of activity at 20 and 40 min after injection. The C3H female mice were virtually unaffected, at all testing times, by the 1.0-mg/kg nicotine dose whereas the males exhibited significant decreases in activity at the 2.5-min test time only. The data from the sexes were combined to obtain an overall strain comparison: DBA and C57BL mice were equally responsive to nicotine and were much more affected than were the C3H mice.

Strain differences in sensitivity to a behavioral effect of a drug could arise because of differences in drug absorption or metabolism. Therefore, the nicotine content in brain and liver from 1 to 20 min postinjection in both sexes of all three strains was measured following injection with the 1-mg/kg test dose. In general, females eliminated nicotine from the liver more quickly than did the males. Only modest strain differences in rate of nicotine elimination from brain and liver were detected, and no consistent association between brain nicotine content and behavioral response to nicotine was observed. Petersen et al. (1984) have also measured nicotine metabolism in these three mouse strains using a sophisticated capillary gas chromatography-mass spectroscopy method to measure the blood, brain, and liver content of nicotine, cotinine, and nicotine-N-oxide. This study indicated that the three strains differ modestly in the rate of nicotine absorption; the C3H mice absorbed nicotine slightly more quickly than did the other two strains. No differences in nicotine metabolism were evident; the half-lives of nicotine disappearance from the blood were 5.87, 5.97, and 6.85 min for the three strains. These results suggest that the strain differences in nicotine effects on Y-maze activity seen in C57BL, DBA, and C3H mice reported by Hatchell and Collins (1980) are a consequence of differences in CNS sensitivity to nicotine.

The metabolism studies described above yielded an unexpected result: the half-lives of nicotine in the three mouse strains were less than 10 min. This compares with an apparent half-life of more than 30 min for the rat (Rosecrans and Schechter, 1972). This finding more than likely explains why the nicotine doses used in rats to elicit measurable behavioral effects are considerably less than those used in the mouse.

The observation that strains differ in response to nicotine, as measured by Y-maze activity, was expanded in a subsequent study that measured the effects of several nicotine doses (0.5, 1.0, 1.5, and 2.0 mg/kg) on open field activity in C57BL/6, DBA/2, C3H and BALB/c mice (Marks et al. 1983a). Activity was measured for 5 minutes starting 2 minutes after injection. The open field test was selected because activity in this test apparatus is probably influenced by anxiety; recall that activity in the open-field arena is probably a stressful environment for the rat and mouse (Broadhurst, 1962; DeFries et al., 1966). Both C57BL/6 and DBA/2 mice showed a dose-dependent decrease in activity; the 1.0-mg/kg nicotine dose eliminated activity in both strains. BALB/c mice were unaffected by the lowest nicotine doses, but activity was ultimately decreased to 0 at the 1.5-mg/kg dose. C3H mice were unique in that at the 0.5-, 0.75-, and 1.0-mg/kg doses they were activated; at 1.5 mg/kg activity was decreased to about one third of the control activity levels. These results indicate that inbred mouse strains differ in sensitivity to nicotine both quantitatively (dose required to elicit a standard response) and qualitatively (stimulation or depression at a prescribed dose).

This study also involved an analysis of nicotine effects on four other measures: rotarod performance, acoustic startle response, respiration rate, and body temperature. These tests were selected in order to obtain a broader assessment of potential genetic influences on sensitivity to nicotine. The rotarod test is a test of motor coordination and the acoustic startle test is a measure of central nervous system (CNS) excitability. The respiration and body temperature tests were selected because smoking affects these measures in humans. The time courses of nicotine effects on each of these measures were determined and, subsequently, dose-response curves were constructed for each test. Animals were tested for the dose-response studies at the time of maximal nicotine effect. Strain differences were found for each of these tests; both quantitative and qualitative differences were observed. The strains differed in the doses required to elicit standard effects such as a 50% reduction in locomotor activity and, for some of the tests, they also differed in direction of response. In general, the C3H strain seemed to be uniquely sensitive to stimulant effects of nicotine with the exception of the effects of nicotine on respiration rate. The rank order of the strains also varied from test to test. This result would be expected if nicotine elicits its effects on these tests via different mechanisms or via actions on different brain regions.

An analysis of the binding of DL-[^{3}H]-nicotine and [^{125}I]-α-bungarotoxin to seven brain regions was also included in this study. These ligands were chosen because earlier studies from our laboratory indicated that they bind to high and low affinity, respectively, nicotinic receptors in mouse brain (Marks and Collins, 1982). No strain differences in nicotine binding were observed in the seven brain regions but strain differences in α-bungarotoxin binding were detected in the hippocampus. We tentatively concluded that differences in the number or affinity of nicotinic receptors did not underlie strain differences in sensitivity to nicotine. As will be documented below, this conclusion was incorrect.

The experiments described to this point indicate only that genetic factors influence several responses to nicotine. However, these studies provide little information concerning the genetic architecture that regulates nicotine's effects. Consequently, diallel analyses of the effects of nicotine on body temperature (Marks et al., 1984) and open-field activity (Marks et al., 1986a) were conducted. Both of these studies used five inbred strains (A, BALB, C3H, C57BL/6, and DBA/2). A five-by-five cross was generated from these strains; that is, all 25 of the F1 hybrids were developed from these five inbred strains. Such a cross allows an estimate of maternal effects. The hypothermia study included an analysis of the effects of 0.75- and 1.5-mg/kg nicotine doses on body temperature. The analysis indicated that dominance effects and additivity influence nicotine actions. Dominance was toward a more intense response to nicotine. Epistatic effects (one gene modifying the effect of other genes) were seen in all of the C57BL crosses. Both narrow-sense and broad-sense heritabilities were estimated from this study. These were 0.30 and 0.35, respectively, for the 0.75-mg/kg dose and 0.19 and 0.20 for the 1.5-mg/kg dose. The lower heritability estimates at the higher nicotine dose may be due to a floor effect; the maximal effect of nicotine on body temperature is influenced by the ambient temperature. These results, coupled with the observation that both dominance and additive genetic variance were detected, suggest that nicotine's effects on body temperature are regulated by several genes.

The diallel analysis of open-field activity (Marks et al., 1986a) used only the 0.75-mg/kg dose. Higher and lower doses gave confounding results due to ceiling and floor effects. This study also detected both additive and dominance components. The dominance was toward a more intense response to nicotine. Narrow- and broad-sense heritabilities were calculated for the control data; these were 0.59 and 0.72, respectively. The heritabilities for the nicotine data were lower, 0.18 and 0.58. The finding that heritabilities decrease following nicotine treatment suggests that environmental factors become more important in regulating open-field activity following nicotine administration. We also calculated the number of genes that influence nicotine's effects on open-field activity (approximately seven). Thus, this response to nicotine is not mediated by a simple genetic system.

In all of the mouse studies that have been described to this point, individual animals were injected with saline or nicotine and performance on one test was measured at some time point after injection, usually the time of maximal nicotine effect. Another strategy that has proved to be valuable in determining responsiveness to nicotine has been developed. This strategy involves the measurement of several responses in each animal (Marks et al., 1985a). Included among these responses are the measurement of nicotine effects on respiration rate, Y-maze crossing and rearing activities (the number of entries into different sections of a Y-maze and the number of times an animal stands up on its hind legs are measured), acoustic startle response, heart rate, and body temperature. It was found that the results obtained when each of these tests was administered to each animal, at the time of maximal nicotine effect, were identical to the results obtained when the tests were administered individually; administering one test did not affect the responses in the other tests. Nicotine dose-response curves were constructed for this test battery in four inbred strains (C57BL/6, DBA/2, BALB/c, and C3H). Consistent with the results discussed above, substantial quantitative and qualitative strain differences in sensitivity to nicotine were observed.

More recently, this test battery has been used to evaluate sensitivity to an acute dose of nicotine using 19 inbred strains (Marks et al., 1989a). Dose-response curves were constructed for each of the tests in each of the 19 inbred strains. The 19 inbred mouse strains differed dramatically in their sensitivities to nicotine. Parameters such as ED_{50} values (the nicotine dose that elicits a 50% change in a measure) were calculated for each test; the strains differed in sensitivity for each of the tests by a factor of 2 to 3, as assessed by the ED values.

Increasing the number of strains allowed correlational assessments. Correlations were calculated for the ED values for each test in all of the strains. Correlations among the measures differed widely. For example, the Y-maze crosses and Y-maze rears tests were highly correlated ($r = 0.93$), but nicotine effects on heart rate and respiratory rate were not correlated ($r = 0.04$). Other factors showed intermediate correlation (e.g., body temperature–Y-maze crosses, $r = 0.72$; body temperature–respiratory rate, $r = 0.33$).

These results were also subjected to factor analysis in an attempt to ascertain whether a simplified relationship among the variables exists. Principal component, maximum likelihood, and general least-square factor analyses were used to analyze the data set. The same general results were obtained: two factors accounted for 69.8% of the variance. Three of the factors (Y-maze crosses, Y-maze rears, and body temperature) loaded almost exclusively on factor 1. Heart rate also loaded exclusively on factor 1, but the loading was less. Respiratory rate and acoustic startle loaded on both factors 1 and 2. Two measures of sensitivity to nicotine-induced seizures were also included in this analysis. These measures loaded exclusively on factor 2. These results suggest that a similar set of genes regulate nicotine's effects on Y-maze crosses, rears, body temperature, and heart rate and another set of genes regulate the seizure responses. Nicotine effects on respiratory rate and acoustic startle presumably are mediated by genes from both the "activity-temperature" and seizure-sensitivity gene pools.

The binding of nicotine and α-bungarotoxin was also measured in eight brain regions obtained from these 19 inbred strains (Marks et al., 1989b). In this study, both affinity (K_D) and maximal number of binding sites (B_{max}) were determined. The K_D values did not vary among the brain regions or strains but variability in B_{max} was observed between regions and among the strains within a region. In general, those strains with high or low nicotine binding in one brain region had high or low nicotine binding in all of the other brain regions except for cortex and cerebellum that did not vary among the strains. Strong differences in α-bungarotoxin binding were observed in several of the brain regions, and those strains with high binding in one region also had high binding in the other brain regions. Nicotine and α-bungarotoxin binding were not correlated. Thus, genetic factors influence the number of each of these receptors; affinity apparently does not vary.

We also examined the correlation between nicotine and α-bungarotoxin binding and the sensitivities of the 19 strains to nicotine's effects on the test battery and nicotine-induced seizures. Nicotine binding was significantly correlated with nicotine effects on Y-maze rears and crosses and on body temperature ($r = -0.62$). α-Bungarotoxin binding was not correlated with those measures ($r = 0.10$), but a significant correlation with seizures was obtained ($r = -0.63$). These results suggest that several low-dose effects of nicotine (locomotor activities, body temperature) are regulated by the high affinity nicotine binding site and also indicate that the seizure response that is elicited by high nicotine doses is mediated by the lower affinity (for nicotine) α-bungarotoxin binding sites. Nicotinic receptor numbers account for approximately 36% of the variance in sensitivity to nicotine.

GENETICS OF NICOTINE-INDUCED SEIZURES.

Several other studies have obtained results that indicate that genetic factors influence sensitivity to nicotine-induced seizures. Tepper et al. (1979) were the first to investigate potential genetic influences on this response to nicotine. C57BL/6 and DBA/2 mice of both sexes, 21 and 42 days of age, were injected with nicotine doses ranging between 0.75 and 13 mg/kg. Animals were evaluated for the appearance of tremor or clonic seizures and latency to death, if it occurred. The strains did not differ in any measure at 21 days of age, but at 42 days a small but statistically significant difference in sensitivity to seizures and lethality was found; the C57BL/6 mice were slightly more sensitive to nicotine. Blood and brain nicotine levels were also determined in this study by measuring ^{14}C-nicotine content 5 min after injection. No strain differences were observed in the 21-day-old animals, but at 42 days strain differences were detected. The C57BL/6 mice had slightly higher levels of nicotine in both blood and brain than did the DBA mice, which may explain the modest strain differences in seizure sensitivity that were detected.

Several studies from our laboratory have used C3H and DBA mice. These strains were used because we had previously noted that these two mouse strains differed maximally in the number of α-bungarotoxin binding sites in the hippocampus. We reasoned that if the α-bungarotoxin binding site is a low affinity nicotine receptor it might mediate a high dose nicotine effect such as seizures. In order to increase the power of the analysis, we also tested mice obtained from a classical cross of these two strains [i.e., F1, F2, and backcross (F1-by-DBA, and F1-by-C3H) generations were tested (Miner et al., 1984].

Animals from each genetic stock were injected with a single nicotine dose ranging between 2.0 and 7.0 mg/kg and observed for 3 min. Whether a clonic seizure occurred, and latency to seizure, was recorded for each animal. The ED_{50} dose, the nicotine dose that elicited seizures in 50% of the animals, was calculated for each genetic stock. The ED_{50} values for the C3H and DBA mice were 3.13 ± 0.20 and 5.21 ± 0.27 mg/kg. The F1 and F2 crosses had ED_{50} values that were nearly identical to the DBA values. The ED_{50} values of the F1-by-DBA backcross generation was also similar to that of the DBA strain. Thus, dominance toward the DBA phenotype was seen. The F1-by-C3H value was midway between that of the parental C3H and DBA strains. This is the pattern that would be expected if nicotine-induced seizures are regulated by a relatively simple system. Several models were tested and satisfactory fit was obtained for a single-locus, two-allele model with dominance toward low susceptibility to seizures.

This study also included several biochemical analyses. No differences were detected among the various groups in brain nicotine levels, which indicates that the various generations differ in brain sensitivity to nicotine-induced seizures. Nicotine and α-bungarotoxin binding were also measured in cortex, midbrain, and hippocampus in each of the groups. No differences in nicotine binding were found whereas differences in midbrain and hippocampal α-bungarotoxin binding were detected. The DBA, F1-by-DBA, F1, and F2 groups were identical whereas the C3H strain was significantly higher. The F1-by-C3H group was intermediate between the F1 and C3H groups. These data also fit a single-locus, two-allele dominance model. These results suggest that nicotine-induced seizures are regulated, at least in part, by the number of α-bungarotoxin receptors, especially hippocampal receptors. The correlation between seizure ED_{50} and midbrain and hippocampal α-bungarotoxin was significant ($r = -0.69$ for midbrain and -0.80 for hippocampus). These results suggest that those animals with greater numbers of α-bungarotoxin binding sites are more sensitive to nicotine-induced seizures (ED_{50} value is lower).

Further support for this conclusion was obtained in a study that used only the genetically segregating generations derived from a DBA-by-C3H cross, that is, the F1-by-DBA, F1-by-C3H, and F2 generations (Miner et al., 1985). Animals from these generations were injected with a single 4.0-mg/kg nicotine dose. Only some of the animals from each generation seized within 3 min after receiving this dose. Immediately after behavioral testing, the animals were sacrificed and α-bungarotoxin binding was measured in cortex, midbrain, and hippocampus. Those animals that seized following nicotine had significantly greater numbers of bungarotoxin binding sites in midbrain and hippocampus.

The relationship between seizure sensitivity and bungarotoxin binding was examined in yet another way in the C3H-DBA classical cross generations (Miner et al., 1986) by infusing nicotine intravenously, at a rate of 2 mg/kg/min, until a seizure occurred. The latency to seizures was determined. Bungarotoxin binding was also determined in cortex, midbrain, and hippocampus. The results obtained were slightly different from those seen following intraperitoneal (i.p.) nicotine injection, but the conclusion obtained was the same: increased sensitivity to nicotine-induced seizures correlates highly with the number of bungarotoxin binding sites.

Most recently, nicotine-induced seizure sensitivity was determined following i.p. injection and intravenous (i.v.) infusion in 19 inbred mouse strains (Miner and Collins, 1989). α-Bungarotoxin binding was measured in cortex, midbrain, and hippocampus. The 19 strains differed dramatically in seizure sensitivity; the ED_{50} values ranged between 2.34 mg/kg for mice of the ST/b strain to 6.16 mg/kg for DBA/1 mice. Significant correlations were obtained between the ED_{50} values following i.p. nicotine injection and latency to seizure following i.v. nicotine infusion ($r = 0.52$, $p < .05$). Consistent with all of the previously described studies, significant correlations were obtained between seizure sensitivity and bungarotoxin binding. These correlations were highest for seizure latency following i.v. nicotine ($r = -0.55$ for midbrain and -0.64 for hippocampus). These results clearly argue that the number of bungarotoxin binding sites in midbrain and hippocampus are genetically influenced and that this biochemical measure plays a major role in regulating sensitivity to nicotine-induced seizures.

Further support for the argument that C3H and DBA/2 mice differ in sensitivity to seizures that originate in the hippocampus comes from an electrophysiological study (Freund and Wehner, 1987). The addition of nicotine to a hippocampal slice resulted in an increase in population spike amplitude and in the generation of epileptiform activity (multiple population spikes). Slices from DBA mice were affected by lower concentrations of nicotine. Because these effects were blocked by nicotinic antagonists such as mecamylamine, the involvement of nicotinic receptors is indicated.

Correlations Between Sensitivity to Ethanol and Nicotine

A high correlation between smoking and alcohol consumption exists in humans. Alcoholics are almost invariably heavy smokers (Ayers et al., 1976) and experimental studies indicate that providing an alcoholic with ethanol results in an increased use of tobacco (Griffiths et al., 1976; Henningfield et al., 1983). These observations led us to assess the relative sensitivities to nicotine of two mouse lines that have been selectively bred for differences in duration of ethanol-induced sleep time or anesthesia (McClearn and Kakihana, 1973). The two lines are referred to as long-sleep (LS) and short-sleep (SS). Our first study of potential genetically determined interactions between ethanol and nicotine (de Fiebre et al., 1987) consisted of an examination of the relative sensitivities of the LS and SS mice to nicotine as monitored by the test battery (respiration rate, Y-maze crosses and rears, acoustic startle, heart rate, and body temperature) and by nicotine-induced seizures. Dose-response curves were constructed for each of these measures. The LS and SS mice did not differ in sensitivity to the effects of nicotine on respiration rate and acoustic startle response, but differences were observed for all of the other measures. The alcohol-sensitive LS mice were sensitive to nicotine; the SS mice were resistant. The ED_{50} and ED_{50}-like values of the SS mice were two- to threefold greater than the LS values. When compared to the results obtained in our 19 inbred strains analysis (Marks et al., 1989a), the LS are among the most sensitive and the SS among the most resistant. The two mouse lines were not nearly as different for the seizure test. The nicotine dose that elicited seizures in 50% of the LS animals was 3.34 mg/kg; the SS ED_{50} was 4.91 mg/kg. These results suggest that selective breeding for differences in sensitivity to a high dose of ethanol also results in differential sensitivity to nicotine. Such a result might be obtained if ethanol exerts its actions, at least in part, on the same systems as does nicotine.

We also studied nicotine metabolism and distribution to brain in this study. No differences were found that could account for the differences in sensitivity to nicotine. Thus, the LS and SS mice differ in CNS sensitivity to some, but not all, of nicotine's effects. The binding of nicotine and α-bungarotoxin to eight brain regions was also measured. Surprisingly, given our results with the inbred strains, no differences in binding were detected. Because ethanol seems to alter the desensitization of nicotinic receptors (Young and Sigman, 1981), we are currently investigating whether the nicotinic receptors of the LS and SS mice differ in the desensitization process.

Strain Differences in Nicotine–Corticosterone Interactions

A number of studies have demonstrated that nicotine treatment results in an increase in plasma cortisol levels in humans and in plasma corticosterone levels in animals. Two recent studies have demonstrated that inbred mouse strains differ in sensitivity to the corticosterone-releasing effects of nicotine (Freund et al., 1988; Martin and Wehner, 1988). Both of these studies examined the effects of various doses of nicotine in plasma corticosterone levels. Freund et al. (1988) examined the effects of saline and nicotine (0.5–2.0 mg/kg) injection in C57BL/6, DBA/2, C3H, and A mice. Strain differences were found; the C57BL mice were significantly less sensitive to nicotine's effects. This was specific to nicotine since cold water stress was equieffective in inducing corticosterone release in the C57BL and DBA strains.

Martin and Wehner (1988) were more concerned with the issue of whether pretreatment with a low dose of nicotine alters the corticosterone release elicited by a subsequent challenge dose of nicotine. These investigators did not detect any acute tolerance to nicotine's hormone-releasing actions, which is in contrast to observations made in the rat (Balfour, 1980; Sharp and Beyer, 1986), but they did replicate the observation that the DBA/2, C3H, and A strains are equisensitive to nicotine's corticosterone-releasing actions.

More recent studies have addressed the issue of whether the release of corticosterone influences behavioral sensitivity to nicotine. The first study in this area (Pauly et al., 1988) assessed the effects of adrenalectomy (ADX) and subsequent corticosterone replacement on behavioral sensitivity to nicotine in C3H mice. Adrenalectomy resulted in an increased sensitivity to nicotine; dose-response curves for nicotine effects on Y-maze crosses and rears, heart rate, and body temperature were shifted to the left; that is, ADX animals were more sensitive to nicotine. This shift was reversed by treating ADX animals chronically with corticosterone. Adrenalectomy did not result in altered nicotine metabolism. This report also noted that ADX did not alter nicotine or α-bungarotoxin binding, but this study did not involve measurements of binding in the hippocampus (see below).

A subsequent study (Pauly et al., 1990) examined the effects of ADX on sensitivity to nicotine in four inbred strains (C3H, C57BL/6, DBA/2, BUB) and in the selectively bred LS and SS mice. The C3H and LS mice showed an increased sensitivity to nicotine, following ADX, for all of the following measures: startle response, Y-maze rears and crosses, heart rate, and body temperature. The other strains showed altered sensitivity to nicotine, but generally only for one or two of the measures. Effects of ADX on nicotine and α-bungarotoxin binding were also determined in eight brain regions in four of the strains; ADX resulted in an increase in α-bungarotoxin binding in hippocampus only in all four strains. These results suggest that corticosterone, or a metabolite, has antinicotine actions and that these actions are variably expressed in different mouse strains.

Tolerance to Nicotine

Chronic drug administration is frequently accompanied by a reduction in intensity of response to the drug. This reduction in response may occur in as little as one dose, in which case it is often referred to as acute tolerance. Tolerance that develops following longer term treatment is usually called chronic tolerance. The development of tolerance to nicotine's noxious effects may allow the increase in consumption that is necessary to convert a tobacco user into a habitual smoker.

Strain Differences in Acute Tolerance

Acute tolerance to behavioral or physiological effects of nicotine has received minimal attention even though many studies have demonstrated that the nicotinic receptors found in *Torpedo California* and at the neuromuscular junction desensitize. If brain nicotinic receptors desensitize, a decreased behavioral sensitivity to nicotine should result.

Three studies from our laboratory have demonstrated that acute tolerance develops to the seizure-inducing effects of nicotine and that this tolerance is seen in some mouse strains, but not in others. These studies involved measuring the effects of pretreatment with a subseizure dose of nicotine on nicotine-induced seizures. In the first of these studies (Miner and Collins, 1988a), the effects of nicotine pretreatment (1 and 2 mg/kg) on nicotine-induced seizures were determined in DBA/2 and C3H mice. Nicotine pretreatment shifted the dose-response curves for intraperitoneally injected nicotine to the right (ED_{50} values were doubled) in DBA mice; minimal effects were seen in the C3H strain. This shift in the dose-response curves was maximal when the interval between preinjection and challenge with the seizure level dose was 15 min. Nicotine pretreatment was without effect when the pretreatment preceded challenge by 60 min. Comparable results were obtained if the nicotine seizures were elicited by i.v. administration of nicotine. However, nicotine pretreatment did not modify sensitivity to another seizure-inducing agent, flurothyl.

An identical experiment using the lines of mice that were bred for differential alcohol sensitivity, the LS and SS mice, gave comparable results (de Fiebre and Collins, 1988). Nicotine-pretreated LS mice showed a greater shift to the right of seizure dose-response curves than did the SS mice. A related study (de Fiebre and Collins, 1989) examined the effects of ethanol treatment on acute tolerance to nicotine. Pretreatment with low doses of ethanol augmented the desensitizing effects of nicotine pretreatment in both the LS and SS mice, but this effect was seen at lower doses and was more pronounced in LS mice.

These experiments provide evidence that nicotine pretreatment will produce acute tolerance or behavioral desensitization in some mouse strains but not in others. Results of this sort might be obtained if mouse strains have nicotinic receptors that differ in the desensitization process. The LS and SS differences in behavioral desensitization could result if these mouse lines have nicotinic receptors that differ in sensitivity to an ethanol effect on the desensitization process. We are currently exploring these hypotheses in biochemical studies of receptor desensitization.

Strain Differences in Chronic Tolerance

Although chronic tolerance to nicotine has been studied extensively by many investigators, only a few studies have attempted to ascertain whether genetic factors influence chronic tolerance development. However, results from our laboratory indicate that mouse strains differ in tolerance development. The first evidence that we obtained that supports this contention comes from a study of the effects of chronic nicotine injection in DBA/2 and C57BL/6 mice (Hatchell and Collins, 1977). Animals of both sexes from both strains were injected three times daily with saline (controls) or nicotine (1 mg/kg) for 2, 4, or 7 days. The morning following the last treatment dose the animals were challenged with a 1-mg/kg dose of nicotine and locomotor activity was measured for 3 min starting 5 min after injection. Both sex and strain influenced tolerance development. Females of both strains developed more tolerance than did the males; this effect may be explained by sex differences in nicotine metabolism that were detected in this study. When collapsed across the sexes, the C57BL/6 mice developed tolerance more quickly and possibly to a greater degree than did the DBA/2 mice.

Although this first study of nicotine tolerance suggested strain differences, we were not particularly satisfied with the procedures that were used. Chronic injections, even with saline, were clearly very stressful to the mice; the animals became progressively more hostile as chronic injections continued. Consequently, we were not certain whether the strain differences in Y-maze response to nicotine were true indicators of nicotine tolerance. Therefore, we developed an alternative method for the chronic administration of nicotine, chronic i.v. infusion. For most of our studies we have infused mice continuously (24 hr/day) for 7 to 10 days. Tolerance is assessed by challenging the animals with varying doses of nicotine (dose-response curves are constructed) 2 hr after infusion was stopped. Nicotine effects on respiration rate, Y-maze crosses and rears, acoustic startle, heart rate, and body temperature are measured using our "test battery" procedures. Early studies with DBA/2 mice examined effects of various infusion doses (Marks et al., 1983b) and different times of infusion and withdrawal (Marks et al., 1985b) on sensitivity to nicotine and on brain nicotinic receptors. These studies indicated that tolerance to nicotine develops for virtually every measure, in a dose-dependent fashion. Nicotine infusion also resulted in a dose-dependent increase in nicotine binding that was also maximal at the 3-mg/kg/hr dose. α-Bungarotoxin binding also increased following chronic nicotine treatment but this change did not parallel tolerance development, as measured by the test battery. The time course study (Marks et al., 1985b) also produced evidence that suggested that maximal tolerance to nicotine's effects on the test battery developed within 2 to 3 days of infusion and was lost within 8 days after infusion was stopped. Tolerance acquisition and loss were tightly coupled to the increase and return to control of nicotine binding. The acquisition and loss of tolerance to nicotine's seizure-inducing effects are paralleled by changes in α-bungarotoxin binding (Miner and Collins, 1988b).

The finding that nicotine tolerance is accompanied by increases in the number of brain nicotinic receptors was unexpected, especially since studies of initial sensitivity to nicotine suggested that greater numbers of receptors correlated with increased sensitivity to nicotine. We have suggested that chronic nicotine treatment may result in a short-term stimulation of brain nicotinic receptors followed by a prolonged desensitization or inactivation. Behavioral tolerance is seen, even though there are more total receptors, because the vast majority of the receptors are desensitized.

In three recent studies we have explored potential genetic influences on tolerance development. In the first of these studies, DBA/2 and C3H mice were infused with saline or 2, 4, or 6 mg/kg/hr of nicotine for 10 days (Marks et al.,

1986c). Two hr after infusion was stopped, the animals were challenged with nicotine and drug effects on the test battery were measured. The DBA/2 mice developed a dose-dependent tolerance to nicotine whereas the C3H strain did not develop tolerance to any measure until the 4-mg/kg/hr infusion dose. Even though the two strains differed dramatically in tolerance development, both strains showed similar effects of nicotine infusion on nicotine and α-bungarotoxin binding. In a second study (Marks et al., 1986b) we examined the effects of one infusion dose (3 mg/kg/hr) on tolerance development in four inbred strains (DBA/2, C3H, C57BL/6, and BALB). All of these strains, except the C3H, developed some tolerance to nicotine at this infusion dose. Comparable changes in nicotine and α-bungarotoxin binding were observed in the four strains.

Most recently, we reported results of a study of tolerance development that has used seven mouse strains: DBA/2, C3H, C57BL/6, A, BUB, LS, and SS (Collins and Marks, 1989). These strains were selected because they differ widely in sensitivities to a first dose of nicotine. The mice were infused chronically with saline or nicotine (0.5, 1, 2, 4, or 6 mg/kg/hr) for 10 days. Tolerance was determined 2 hr after infusion was stopped by constructing dose-response curves for nicotine effects in each strain. Some mouse strains, such as C57BL/L and DBA, developed tolerance at the lowest infusion doses whereas others, such as the BUB and SS, developed tolerance only at the highest doses. The strains also differed in the maximal amount of tolerance. For example, C57BL/6 mice showed fourfold changes in ED_{50} values whereas DBA mice showed a twofold change in these values. The effects of chronic treatment with several doses of nicotine or nicotine and α-bungarotoxin binding have been measured in eight brain regions (Marks et al., 1991). Nicotine infusion elicited dose-dependent increases in nicotine binding in all brain regions in all strains. This change correlated highly with the development of tolerance to nicotine in C57BL/6 and DBA/2 mice, but did not parallel tolerance development in other strains such as the C3H. Surprisingly, tolerance to nicotine paralleled changes in α-bungarotoxin binding in C3H mice. In addition, a relationship between initial sensitivity to nicotine and tolerance development was detected. A high correlation ($r = 0.86$) was observed between initial sensitivity to nicotine's effects on body temperature and the minimum infusion dose that elicited a detectable shift in the hypothermia dose-response curve. This high correlation suggests that initial sensitivity may predict tolerance development. The most sensitive animals (to a first dose of nicotine) may develop tolerance more readily.

Summary of the Mouse Studies

Several methods have been used to study the genetics of acute (first dose) sensitivity to nicotine. The results obtained from inbred strain and derived crosses clearly demonstrate that genetic factors regulate several different responses to nicotine. At least two major classes of response have been identified; one of these classes includes nicotine effects on locomotor activity and body temperature and the other class is characterized by nicotine-induced seizures. Diallel analyses indicate that nicotine's effects on locomotor activity and body temperature are polygenically determined, but a major gene may have been identified. This gene regulates the number of brain nicotine binding sites. Seizures induced by nicotine appear to be regulated by a simpler genetic system, perhaps a single gene. This response to nicotine may be regulated by the number of α-bungarotoxin binding sites in the hippocampus. Although receptor numbers seem to explain some of the variance in response to nicotine, it must be emphasized that only 35% to 40% of the variation in sensitivity correlates with receptor numbers.

The finding that the LS and SS mice, which were selectively bred for differences in sensitivity to ethanol (McClearn and Kakihana, 1973), also differ in sensitivity to several effects of nicotine suggests that ethanol's effects may be mediated, in part, by nicotinic systems in the brain. However, LS–SS differences might arise by coincidence because some inbreeding occurred as selection proceeded. We are encouraged that the LS–SS differences are not coincidental because we have recently found (de Fiebre et al. 1991) that rat lines that are being selectively bred for differences in sensitivity to ethanol also differ in sensitivity to nicotine. This finding increases the probability that some overlap exists between those genes that regulate sensitivity to ethanol and those genes that influence sensitivity to nicotine. The most parsimonious explanation is that ethanol elicits some of its effects by modifying the activity of brain nicotinic systems.

Mouse strains differing in the development of acute and chronic tolerance to nicotine have also been observed, but the mechanisms that underlie these strain differences have not been identified. There does, however, seem to be a relationship between sensitivity to a first dose of nicotine and the propensity for developing chronic tolerance to nicotine; more sensitive mouse strains develop tolerance to nicotine at lower chronic treatment doses than do strains that are less sensitive to a first dose of nicotine. This, of course, should be expected if tolerance develops only after an effect is elicited. Tolerance occurs to the effects elicited by the drug, not to the drug itself. Although work from our laboratory has demonstrated that chronic nicotine infusion results in receptor upregulation and that this upregulation parallels tolerance development in some mouse strains, particularly the DBA/2 (Marks et al., 1983b, 1985b), we must emphasize that concluding that strain differences in tolerance development are due to differences in receptor upregulation is probably premature. We suspect that strain differences in tolerance development are due to factors other than strain differences in receptor upregulation.

CONCLUSIONS

Studies of alcohol and tobacco use by humans indicate that genetic factors are important in regulating the abuse of these substances as is evidenced by twin, family, and, for alcohol, adoption studies. Animal studies have clearly demonstrated that genetic factors regulate several "simpler" factors related to the actions of these drugs, such as initial sensitivity and the development of tolerance.

We suspect that many of the responses to ethanol and nicotine that have been measured in animal studies are not directly related to why people use and abuse these substances. Nonetheless, if acute sensitivity to rewarding and noxious effects of these drugs, or in the development of tolerance to these effects, varies in humans as they do in animals, an explanation for individual differences in alcohol consumption and tobacco use may be at hand. Overlap between those genes that influence ethanol and nicotine sensitivity might explain why alcoholics are almost invariably heavy smokers. The long-term value of unraveling the potential genetic regulation and biochemical underpinnings of drug sensitivity is that better individualized approaches can be developed that will be useful in helping alcohol and tobacco users/abusers stop their self-medication.

Acknowledgments. Work from the authors' laboratory was supported by grants from the National Institute on Drug Abuse (DA-03194 and DA-05131), the National Institute on Alcoholism and Alcohol Abuse (AA-06391), and the RJ Reynolds Tobacco Company. Dr. Collins is the recipient of a Research Scientist Development Award (DA-00116).

REFERENCES

Ayers J, Ruff CF, Templer DI (1976): Alcoholism, cigarette smoking, coffee drinking and extraversion. *J Stud Alcohol* 37:983–985

Balfour DJK (1980): Studies on the biochemical and behavioral effects of oral nicotine. *Arch Int Pharmacodyn* 245:95–103

Battig K, Driscoll P, Schlatter J, Uster HJ (1976): Effects of nicotine on the exploratory locomotion patterns of female Roman high and low avoidance rats. *Pharmac Biochem Behav* 4:435–439

Belknap JK (1980): Genetic factors in the effects of alcohol: neurosensitivity, functional tolerance, and physical dependence. In: *Alcohol Tolerance and Dependence,* Rigter H, Crabbe J, eds. Amsterdam: Elsevier, pp 157–180

Bignami G (1965): Selection for high rates and low rates of avoidance conditioning in the rat. *Anim Behav* 13:221–227

Bohman M (1978): Some genetic aspects of alcoholism and criminality: a population of adoptees. *Arch Gen Psychiatry* 35:269–276

Bovet D, Bovet-Nitti F, Oliverio A (1967): Action of nicotine on spontaneous and acquired behavior in rats and mice. *Ann NY Acad Science* 142:261–267

Broadhurst PL (1960): Experiments in psychogenetics: applications of biometrical genetics to behavior. In: *Experiments in Personality, vol. 1, Psychogenetics and Psychopharmacology,* Eysenck HJ, ed. London: Routledge and Keenan Paul, pp 1–102

Broadhurst PL (1962): A note on further progress in a psychogenetic selection experiment. *Psych Rep* 10:65–66

Cadoret RJ, Cain CA, Grove WM (1979): Development of alcoholism in adoptees raised apart from alcoholic biologic relatives. *Arch Gen Psychiatry* 37:561–563

Cloninger CR (1987): Recent advances in family studies of alcoholism. In: *Genetics of Alcoholism,* Goedde HW, Agarwal DP, eds. New York: Alan R. Liss, pp 47–60

Collins AC, Marks MJ (1989): Chronic nicotine exposure and brain nicotinic receptors—influence of genetic factors. *Prog Brain Res* 79:137–146

Conterio F, Chiarelli B (1962): Study of the inheritance of some daily habits. *Heredity* 17:347–359

Cotton NS (1979): The familial incidence of alcoholism: a review. *J Stud Alcohol* 40:89–116

Crabbe JC (1989): Genetic animal models in the study of alcoholism. *Alcohol Clin Exp Res* 13:120–127

Davenport CB (1912): *Heredity in Relation to Eugenics*. London: Williams and Norgate

de Fiebre CM, Collins AC (1988): Decreased sensitivity to nicotine-induced seizures as a consequence of nicotine pretreatment in long-sleep and short sleep mice. *Alcohol* 5:55–61

de Fiebre CM, Collins AC (1989): Behavioral desensitization to nicotine is enhanced differentially by ethanol in long-sleep and short-sleep mice. *Alcohol* 6:45–51

de Fiebre CM, Medhurst LJ, Collins AC (1987): Nicotine response and nicotinic receptors in long-sleep and short-sleep mice. *Alcohol* 4:493–501

de Fiebre CM, Romm E, Collins J, Draski LJ, Deitrich RA, Collins AC (1991) Responses to cholinergic agonists of rats selectively bred for differential sensitivity to ethanol Alcohol Clin Exp Res 15:270-276

DeFries JC (1981): Selective breeding for behavioral and pharmacological responses in laboratory mice. In: *Genetic Research Strategies for Psychobiology and Psychiatry,* Gershon ES, Matthysse S, Breakefield XO, Ciaranello RD, eds. Boxwood Press, Pacific Grove, CA

DeFries JC, Hegmann JP, Weir MW (1966): Open-field behavior in mice: Evidence for a major gene effect mediated by the visual system. *Science* 154:1577–1579

Deitrich RA, Spuhler K (1984): Genetics of alcoholism and alcohol actions. *Res Adv Alcohol Drug Prob* 8:47–98

Dies R, Reznikoff M (1969): Personality and smoking patterns in a twin population. *J Pro Tech Pers Assess* 33:457–463

Driscoll P, Battig K (1982): Behavioral, emotional and neurochemical profiles of rats selected for extreme differences in active, two-way avoidance performance. In: *Genetics of the Brain,* Liblich, I, Amsterdam: Elsevier, pp 95–123

Eysenck HJ (1980): *The Causes and Effects of Smoking*. London: Tempie Smith

Fisher RA (1958a): Lung cancer and cigarettes? *Nature* 182:108

Fisher RA (1958b): Cancer and smoking. *Nature* 182:596

Fleming JC, Broadhurst PL (1975): The effects of nicotine on two-way avoidance conditioning in bi-directionally selected strains of rats. *Psychopharmacologia* 42:147–152

Freund RK, Martin BJ, Jungschaffer DA, Ullman EA, Collins AC (1988): Genetic differences in plasma corticosterone levels in response to nicotine injection. *Pharmacol Biochem Behav* 30:1059–1064

Freund RK, Wehner JM (1987): Strain-selective effects of nicotine on electrophysiological responses evoked in hippocampus from DBA/2Ibg and C3H/2Ibg mice. *J Neurogenet* 4:75–86

Garg M (1968): The effect of nicotine on rearing in two strains of rats. *Life Sci* 7 (Part II):421–429

Garg M (1969a): The effects of some central nervous system stimulant and depressant drugs on rearing activity in rats. *Psychopharmacologia* 14:150–156

Garg M (1969b): Variation in effects of nicotine in four strains of rats. *Psychopharmacologia* 14:432–438

Goodwin DW (1985): Alcoholism and genetics: the sins of the fathers. *Arch Gen Psychiatry* 42:171–178

Goodwin DW, Schulsinger F, Knop J (1977): Alcoholism and depression in adopted-out daughters of alcoholics. *Arch Gen Psychiatry* 34:751–755

Griffiths RR, Bigelow GE, Liebson I (1976): Facilitation of human tobacco self-administration by ethanol: a behavioral analysis. *J Exp Anal Behav* 25:279–292

Gurling HMD, Grant S, Dangl J (1985): The genetic and cultural transmission of alcohol use, alcoholism, cigarette smoking and coffee drinking: a review and an example using a log linear cultural transmission model. *Br J Addiction* 80:269–279

Hatchell PC, Collins AC (1977): Influences of genotype and sex on behavioral tolerance to nicotine in mice. *Pharmacol Biochem Behav* 6:25–30

Hatchell PC, Collins AC (1980): The influence of genotype and sex on behavioral sensitivity to nicotine in mice. *Psychopharmacology* 71:45–49

Henningfield JE, Chait LD, Griffiths RR (1983): Cigarette smoking and subjective response in alcoholics: effects of pentobarbital. *Clin Pharmacol Ther* 31:806–812

Hrubec Z, Omenn GS (1981): Evidence of genetic predisposition to alcoholic cirrhosis and psychosis: twin concordances for alcoholism and its biological end points by zygosity among male veterans. *Alcoholism: Clin Exp Res* 5:207–215

Johnston LM (1942): Tobacco smoking and nicotine. *Lancet* 2:742

Jones RA (1986): Individual differences in nicotine sensitivity. *Addict Behav* 11:435–438

Kaij L (1960): *Studies of the Etiology and Sequels of Abuse of Alcohol*. Lund, Sweden: University of Lund

Kaprio J, Koskenvuo M, Sarna S (1981): Cigarette smoking, use of alcohol, and leisure-time physical activity among same-sexed adult male twins. In: *Twin Research 3: Epidemiological and Clinical Studies,* Gedda L, Parisi P, Nance WE, eds. New York: Alan R. Liss pp 37-46

Kozlowski LT, Herman CP (1984): The interaction of psychosocial and biological determinants of tobacco use: more on the boundary model. *J Appl Soc Psychol* 14:244–256

McClearn GE, Kakihana R (1973): Selective breeding for ethanol sensitivity in mice. *Behav Genet* 3:409–410

Marks MJ, Burch JB, Collins AC (1983a): Genetics of nicotine response in four inbred strains of mice. *J Pharmacol Exp Ther* 226:291–302

Marks MJ, Burch JB, Collins AC (1983b): Effects of chronic nicotine infusion on tolerance development and cholinergic receptors. *J Pharmacol Exp Ther* 226:806–816

Marks MJ, Collins AC (1982): Characterization of nicotine binding in mouse brain and comparison with the binding of α-bungarotoxin and quinuclidinyl benzilate. *Mol Pharmacol* 22:554–564

Marks MJ, Miner L, Burch JB, Fulker DW, Collins AC (1984): A diallel analysis of nicotine-induced hypothermia. *Pharmacol Biochem Behav* 21:953–959

Marks MJ, Miner LL, Cole-Harding S, Burch JB, Collins AC (1986a): A genetic analysis of nicotine effects on open field activity. *Pharmacol Biochem Behav* 24:743–749

Marks MJ, Romm E, Bealer SM, Collins AC (1985a): A test battery for measuring nicotine effects in mice. *Pharmacol Biochem and Behav* 23:325–330

Marks MJ, Romm E, Campbell SM, Collins AC (1989b): Variation of nicotinic binding sites among inbred strains. *Pharmacol Biochem Behav* 33:679–689

Marks MJ, Romm E, Gaffney DK, Collins AC (1986b): Nicotine-induced tolerance and receptor changes in four mouse strains. *J Pharmacol Exp Ther* 237:809–819

Marks MJ, Stitzel JA, Collins AC (1985b): Time course study of the effects of chronic nicotine infusion on drug response and brain receptors. *J Pharmacol Exp Ther* 235:619–628

Marks MJ, Stitzel JA, Collins AC (1986c): Dose-response analysis of nicotine tolerance and receptor changes in two inbred mouse strains. *J Pharmacol Exp Ther* 239:358–364

Marks MJ, Stitzel JA, Collins AC (1989a): Genetic influences on nicotine responses. *Pharmacol Biochem Behav* 33:667–678

Marks MJ, Campbell SM, Romm E, Collins AC (1991) Genotype influences the development of tolerance to nicotine in the mouse. *J Pharmacol Exp Ther* 259:392–402

Martin BJ, Wehner JM (1988): Influence of genotype on nicotine-induced increases of plasma corticosterone in mice as a result of acute nicotine pretreatment. *Pharmacol Biochem Behav* 30:1065–1070

Martin NG, Oakeshott JG, Gibson JB, Starmer GA, Perl J, Wilks AV (1985): A twin study of psychomotor and physiological responses to an acute dose of alcohol. *Behav Genetics* 15:305–347

Miner LL, Collins AC (1988a): Effect of nicotine pretreatment on nicotine-induced seizures. *Pharmacol Biochem Behav* 29:375–380

Miner LL, Collins AC (1988b): The effect of chronic nicotine treatment on nicotine-induced seizures. *Psychopharmacology* 95:52–55

Miner LL, Collins AC (1989): Strain comparison of nicotine-induced seizure sensitivity and nicotinic receptors. *Pharmacol Biochem Behav* 33:469–475

Miner LL, Marks MJ, Collins AC (1984): Classical genetic analysis of nicotine-induced seizures and nicotinic receptors. *J Pharmacol Exp Ther* 231:545–554

Miner LL, Marks MJ, Collins AC (1985): Relationship between nicotine-induced seizures and hippocampal nicotinic receptors. *Life Sci* 37:75–83

Miner LL, Marks MJ, Collins AC (1986): Genetic analysis of nicotine-induced seizures and hippocampal nicotinic receptors in the mouse. *J Pharmacol Exp Ther* 239:853–860

Morrison CF, Lee PN (1968): A comparison of the effects of nicotine and physostigmine on a measure of activity in the rat. *Psychopharmacologia* 13:210–221

Murray RM, Clifford C, Gurling HMD (1983): Twin and alcoholism studies. In: *Recent Developments in Alcoholism,* Galanter M, ed. New York: Gardner Press, pp 25–48

Nagoshi CT, Wilson JR (1987): Influence of family alcoholism history on alcohol metabolism, sensitivity, and tolerance. *Alcoholism: Clin Exp Res* 11:392–398

Omenn GS, Motulsky AG (1972): A biochemical and genetic approach to alcoholism. *Ann NY Acad Sci* 197:16–23

Pauly JR, Ullman EA, Collins AC (1988): Adrenocortical hormone regulation of nicotine sensitivity in mice. *Physiol Behav* 44:109–116

Partanen J, Bruun K, Markkanen T (1977): Inheritance of drinking behaviour: a study of intelligence, personality, and use of alcohol of adult twins. In: *Emerging Concepts of Alcohol Dependence*. Pattison EM, Sobell MB, Sobell LC (eds) New York: Springer, pp. 74–159

Pauly JR, Ullman EA, Collins AC (1988): Adrenocortical hormone regulation of nicotine sensitivity in mice. *Physiol Behav* 44:109–116

Pauly JR, Ullman EA, Collins AC (1990): Strain differences in adrenalectomy-induced alterations in nicotine sensitivity in the mouse. *Pharmacol Biochem Behav,* 35:171–179

Pedersen N (1981): Twin similarity for usage of common drugs. In: *Twin Research 3: Epidemiological and Clinical Studies,* Gedda L, Parisi P, Nance WE, eds. New York: Alan R. Liss pp 53–59

Petersen DR, Norris KJ, Thompson JT (1984): A comparative study of the disposition of nicotine and its metabolites in three inbred strains of mice *Drug Metab Dispos* 12:725–731

Phillips TJ, Feller DJ, Crabbe JC (1989): Selected mouse lines, alcohol and behavior. *Experientia* 45:805–827

Raaschou-Nielson E (1960): Smoking habits in twins. *Dan Med Bull* 3:82–88

Rosecrans JA, Schechter MD (1972): Brain area nicotine levels in male and female rats of two strains. *Arch Int Pharmacodyn* 196:46–54

Samson HH, Maxwell CO, Doyle TF (1989): The relation of initial alcohol experiences to current alcohol consumption in a college population. *J Stud Alcohol* 50:254–260

Schuckit MA, Gold EO (1988): A simultaneous evaluation of mutiple markers of ethanol/placebo challenges in sons of alcoholics and controls. *Arch Gen Psychiatry* 45:211–216

Schuckit MA, Goodwin DW, Winokur G (1972): A half-sibling study of alcoholism. *Am J Psychiatry* 128:1132–1136

Sharp BM, Beyer BS (1986): Rapid desensitization of the acute stimulatory effects of nicotine on the rat plasma adrenocorticotropin and prolactin. *J Pharmacol Exp Ther* 238:486–491

Shields J (1962): *Monozygotic Twins*. London: Oxford University Press, pp 121–124

Shiffman S (1989): Tobacco "chippers"—individual differences in tobacco dependence. *Psychopharmacology* 97:539–547

Tepper JM, Wilson JR, Schlesinger K (1979): Relations between nicotine-induced convulsive behavior and blood and brain levels of nicotine as a function of sex and age in two inbred strains of mice. *Pharmacol Biochem Behav* 10:349–353

Wilson JR, Plomin R (1984): Individual differences in sensitivity and tolerance to alcohol. *Soc Biol* 32:162–184

Young AP, Sigman DS (1981): Allosteric effects of volatile anesthetics on the membrane-bound acetylcholine receptor. II. Alteration of α-bungarotoxin binding kinetics. *Mol Pharmacol* 20:506–510

Potential Genetic Models of Aggression and Violence in Males

STEPHEN CLARK MAXSON

In 1989 and 1990, the National Research Council sponsored workshops on the roots of interpersonal violence in humans (Barnes, 1989), which will result in a report from the National Academy of Sciences reviewing the present state of our knowledge with regard to biological and behavioral aspects of aggression and violence, and recommending directions for future research in these areas. Research on genetic variants for different types of aggression in animals, such as the laboratory mouse, is relevant, at least to some degree, to issues concerning human aggression and violence (Carey, 1989; Scott, 1989). In this chapter, potential genetic models of four types of aggression in male mice will be reviewed. The types of aggression are offense, defense, infanticide, and predation.

It should be recognized, however, that any animal behavior will always be, at best, both similar and different in comparison to that of humans. For this reason, Scott (1984, 1989) suggested that no animal species can serve as an exact model for human aggression, but rather that some aspects of animal behavior in this or that species will appear as being relevant to human aggression. Consequently, he proposed that information should be accumulated on the various kinds of aggression in a wide range of animal species. Scott has carried out part of this program by consideration of the advantages and disadvantages of dogs (1984) and of mice (1989) as potential animal models for some aspects of human aggression. Blanchard and Blanchard (1984, 1989) have taken a similar approach in their research program on animal models of aggression. Theirs is a two-part strategy. The first is to describe, examine, and analyze in detail aggressive behavior(s) in the context of a species' ecology and evolution. The second is a systematic comparison of the human behavior with the animal model to determine the extent of point-by-point correspondence. They have done this for two types of aggression, offense and defense, in rats and humans, as well as to some degree in several other mammalian species.

TYPES OF GENETIC MODELS

With regard to behavior, there are three types of genetic models. These are models at the genic, biological, or behavioral levels.

Genic models are based on the identification of variants in an animal species for one or more genes that are homologous in that animal species and in humans (for examples, see Erickson, 1989; Hardy, 1988). Many (more than 250) gene homologies are known for mice and humans (Nadeau and Reiner, 1989). The number of known homologous genes for other animal species and humans is far less than that for mice and humans (O'Brien et al., 1988). Unfortunately, individual genes with effects on one or more types of aggression in mice, other animal species, or humans have not been identified (Carey, 1989; Maxson, 1991). Therefore, it is not possible at the time to consider potential genetic models of aggression at the genic level. However, there may be an effect of genes in the *t* complex of chromosome 17 on aggression in male and female mice (Lenington, 1991).

Biological models are based on the hypothesis that genetic variants for anatomical, physiological, or biochemical systems may have similar or related effects on animal and human behavior. For example, androgens have effects on male aggression in animals (Barfield, 1984; Brain, 1983), and they may have similar or related effects on male aggression in humans (Brain, 1984; Mazur, 1983). Many genetic variants for androgen metabolism in mice have been identified and investigated in relation to one or more types of male aggression (Carlier et al., 1990; Maxson et al., 1983; Olsen, 1983; Selmanoff and Ginsburg, 1981; Simon, 1983). These may be useful in developing and testing hypotheses for mice and other species, including humans, as to the effects of androgens on predispositions for aggressive responses in particular contexts (Brain, 1991).

Behavioral models are based on genetic variants detected in tests of one or more types of aggression in mice or other animals. For example, selective breeding has been used to establish strains of mice that differ in one or more measures of aggression of males (Cairns et al., 1983; Lagerspetz and Lagerspetz, 1974; van Oortmerssen and Bakker, 1981). Also, many inbred strains of mice have been compared for one or more measures of aggression by males (see reviews by Maxson, 1981; Michard and Carlier, 1985; Simon, 1979). Many of these selected or inbred strains are the bases for potential genetic models at the behavioral level, some of which will be considered in this chapter.

TYPES OF BEHAVIORAL MODELS

There are many definitions for aggression (for examples, see Archer, 1988; Huntingford and Turner, 1987; Johnson, 1972), and there are many types of aggression (for examples, see Brain, 1979; Miczek, 1987; Moyer, 1976). In mice, four types of aggression are offense, defense, infanticide, and predation, and in mice, variants for each type of aggression may be potential genetic models.

Offense and defense are agonistic behaviors not only of mice but also other species, including humans. Scott (1966, p. 683) defined agonistic behavior as ". . . behavior patterns having the common function of adaptation to situations involving physical conflict between members of the same species." Blanchard and

Blanchard (1984, 1988, 1989) have cogently argued that across species, including mice and humans, (1) offense and defense motor patterns differ, (2) defensive attack is more likely to be seriously injurious than is offensive attack, and (3) offense and defense serve different functions. They suggest that defensive behaviors serve the function of self-protection from injury by others, and that offensive behaviors serve the function of obtaining and retaining survival and reproductive resources. Furthermore, it has been proposed that each of these are motivational systems with potential neural homologies across related species (Adams, 1979, 1980, 1983; Blanchard and Blanchard, 1988).

Infanticide is, by some definitions, a type of aggressive behavior, and this type of aggression is the killing of conspecific young by adults. It is widespread among vertebrate species (Hausfater and Hrdy, 1984), and there are genetic variants for infanticide among laboratory mice (Svare et al., 1984). Unfortunately, infanticide occurs among not only animals but also humans (Scrimshaw, 1984), and it has been proposed that some aspect of genetic variants of infanticide in mice may be relevant to this type of aggression in humans (Svare et al., 1984; vom Saal, 1984).

Predation (interspecific or intraspecific) is considered by some to be a type of aggression (Brain, 1979; Moyer, 1976). Also, it has been suggested that interspecific predatory aggression in animals may be models for predatory-like intraspecific aggression in humans. Harris (1989) has proposed that under conditions of protein (meat) insufficiency, there are anthropological and historic records indicative of intraspecific predation of humans, and Eibl-Eibesfeldt (1979) has argued that such intraspecific predation may occur in humans when our conspecifics are so dehumanized by cultural means as to be considered, at some level, another species and therefore as potential prey. Also, Brain (1991) has suggested that some aggressive behaviors of psychopaths resemble predation more than offense or defense; the similarity is in efficient killing without visible signs of emotional arousal. For these reasons, genetic variants of interspecific predation in mice should also be considered as potential models for some types of human aggression.

GENETIC MODELS OF MOUSE AGGRESSION

Variants of four types of male mouse aggression will be presented as potential genetic models. These are variants for male mouse offense, defense, infanticide, and predation. There are two reasons for focusing on male rather than female mouse aggression. First, a large proportion of all homicides and interpersonal violence occurs between acquainted, young men (Daly and Wilson, 1989; Roberts, 1990; Wilson and Daly, 1985). Second, genetic models of mouse aggression are more fully developed for males than females (see review by Michard and Carlier, 1985).

Neither of these statements should be taken as support for the old and outdated myth that males but not females are aggressive in most species, including mice and humans. Rather, there is considerable evidence showing that females in many species, including mice (Ebert, 1983; Floody, 1983) and humans (Hrdy, 1981; Langevin, 1991), can be and are aggressive. There are some genetic studies of offense, defense, infanticide, and predation in female mice; reviews of these will be cited in the relevant sections of this chapter. It is my belief that the genetics of aggression should certainly be studied in both male and female mice, and that findings and hypotheses in each sex of mouse should be related to each other and to those in each sex for other species, including humans.

Offense

Offense in mice is characterized by unique motor patterns and unique bite targets (Adams, 1980). The motor patterns are chase, sideways offensive posture, upright offensive posture, and bite-and-kick attacks; the bite targets are primarily the tail, rump, and back. The genetics of offense for male mice has been reviewed by Hewitt and Broadhurst (1983), Maxson (1981, 1991), Michard and Carlier (1985), Scott (1966), and Simon (1979); and the genetics of offense in female mice has been reviewed by Ebert (1983), Hood and Cairns (1988), Hyde (1983), Lenington (1991), and Parmigiani (1989). Here, the focus will be on Y chromosomal genes and offense in males.

The mouse Y chromosome consists of two major regions (Maxson, 1990). One of these can recombine with the X chromosome and can be transmitted to either son or daughter. This is the pseudoautosomal region (PAR), which is at the telomeric end of the long arm of the mouse Y chromosome (Bishop et al., 1988; Keiteges et al., 1987; Soriano et al., 1987). The other region is perpetually monosomic and is always transmitted from father to son. This is the nonpseudoautosomal region (NPAR). The NPAR consists of a short arm, centromere, and the long arm from centromere to PAR (McLaren et al., 1988; Roberts et al., 1988). There is evidence that there is one or more genes in the NPAR Y chromosome (Carlier et al., 1990, Maxson and Monahan, 1991) and in the PAR of the X and Y chromosomes (Carlier et al., 1990, van Oortmerssen and Busser, 1989) with effects on offense of male mice.

There is a differential effect on measures of offense of three pairs of Y chromosomes (NPAR). The pairs of Y chromosomes (NPAR) are by strain of origin: DBA1 versus C57BL10 (Didier-Erickson et al., 1989; Maxson et al., 1979, 1989), CBA/Fa versus C57BL6Fa (Stewart et al., 1980), and CBA/H versus NZB (Carlier and Roubertoux, 1986; Carlier et al., 1990, 1991). Furthermore, for each of these, there appears to be an epistatic interaction between Y chromosomal haplotype and autosomal genes. For example, the DBA1 Y chromosome (NPAR) incremental effect on measures of offense occurs when 100% or 50% but not 0% of the autosomes are DBA1 in origin.

A hypothesis has been developed to account, at the behavioral and neural levels, for the effect of this epistatic interaction on offensive behavior of male mice (Carlier et al., 1990; Maxson and Monahan, 1991). Recent research results support the hypothesis that these genetic effects on this murine social behavior are mediated by chemosignaling, chemoperceptual, and motivational processes

(Monahan and Maxson, 1990). Furthermore, these genetic effects on chemosignaling and chemoperceptual processes may be mediated, at least in part, by serum testosterone levels and target organ sensitivity to testosterone. There is evidence in mice for a Y chromosomal gene with effects on serum levels of testosterone and for another Y chromosomal gene with effects on target organ sensitivity to testosterone (Jutley and Stewart, 1985). With regard to chemosignaling, the target organ could be the preputial gland, which releases testosterone-dependent elicitors of offense into urine (Ingersoll et al., 1986), or it could act on another target tissue that affects the levels in bladder urine of other testosterone-dependent elicitors of offense (Ingersoll, 1986). With regard to chemoperception, the target organ(s) may be neural structures in the main or vomeronasal olfactory systems (Simerly, 1990). Also, these genetic effects on motivational processes for offense may not involve testosterone metabolism but rather may be the direct consequence of Y chromosomal transcription in neural structures such as the ventromedial tegmentum (Adams, 1987). Y chromosomal DNA is transcribed in brain (Nallaseth and Dewey, 1986). Maxson (1990) has recently considered whether there might be a homologous gene(s) on the human Y chromosome.

There also appears to be a differential effect on offense of two pairs of X and Y chromosomes (PAR). The pairs of heterosomes (PAR) are by strain of origin: CBA/H versus NZB (Carlier et al., 1990) and SAL versus LAL (van Oortmerssen and Bakker, 1981; van Oortmerssen and Busser, 1989). The genes on the NPAR and PAR Y chromosomes have additive effects on measures of offense.

It has been proposed that one or more genes in the NPAR heterosomes affect general responsiveness, both physiological and behavioral, to change in the mouse's environment and thereby affect offensive behavior (van Oortmerssen and Busser, 1989). The SAL mice appear to have an active coping strategy, and the LAL mice appear to have a passive coping strategy. SAL mice function best in familiar environments, where behavioral persistence pays off. LAL mice function best in changing environments, where behavioral flexibility pays off. A persistent coping strategy facilitates aggressiveness as a way to manipulate the conspecific environment; a flexible coping strategy facilitates pacificity as a way of responding to the conspecific environment. These differences in coping strategies and thereby in offensive behavior appear to be due to effects of one or more PAR genes on neural sensitivity to testosterone (van Oortmerssen et al., 1987) and on brain serotonin turnover (Olivier et al., 1988). Maxson (1990) also considered the possible homology of mouse and human genes in the heterosome PAR. Interestingly, some putative genes with behavioral effects appear to have been mapped to the human heterosomal PAR (Collinge et al., 1989; Crow, 1989; Crow et al., 1989).

These genetic models for male offensive aggression are well developed. For offense, much is known with regard to gene mapping, gene function (at the behavioral level), gene action, and potential gene homology. Also, there is considerable information on the endocrinology (Brain, 1979, 1983, 1991) and pharmacology (Miczek, 1983, 1987; Miczek et al., 1991) of offense in male mice.

Unfortunately, there is a paucity of information on the neural basis of offense in mice or other species (Blanchard and Blanchard, 1988).

Defense

Defense in mice is also characterized by unique motor patterns and unique bite targets (Adams, 1980). The motor patterns are flight, sideways defensive posture, upright defensive posture, lunge-and-bite attacks, and freezing or immobility. The bite targets are primarily the face and shoulders. No genetic models for defense per se in male or female mice have yet been reported. However, there is a set of selected strains that may be useful in research on the genetics of defense in mice, and there are strain differences for interrelations of pain and defense systems in male mice. Also, there is considerable information for mice and other mammalian species on the neural control of defense (Blanchard and Blanchard, 1988; Siegel and Brutus, 1990), on the endocrinology of defense (Blanchard and Blanchard, 1988; Leshner, 1983), and on the pharmacology of defense and defense analgesia (Miczek, 1987; Rodgers and Randall, 1987; Siegfried et al., 1990).

Mice were selectively bred bidirectionally for two measures of attack (frequency and latency) by Cairns et al. (1983). For the high line, there was little or no change in the index measures of aggression across the generations of selection, whereas for the low line, there was, across the generations, a rapid decrease in mean number of attacks and a rapid increase in mean latency to attack. Thus, most, if not all, of the effects of selective breeding were on the low line. This cross-generational decrement in mean attacks and increment in attack latency for the low line appears to be primarily due to an increment in behavioral immobility (Gariepy et al., 1988). Since behavioral immobility in dyadic encounters of male mice may be an indicator of defensive motivation, it could be that selective breeding in this study was for genes with effects on the defensive rather than the offensive system. Thus, this set of selected strains may be a potential genetic model for defense in male mice. They may also be a potential genetic model for defense in female mice; there was a correlated response for measures of female aggression in these selected lines (Hood and Cairns, 1988).

With regard to interactions of pain and defensive systems in mice, it has been proposed that stimuli eliciting the defensive response of flight activate a nonopioid analgesia (GABA and 5HT) and of tonic immobility activate an opioid analgesia (Rodgers, 1989; Rodgers and Randall, 1987). In both cases, it is argued that this analgesia facilitates effective defensive responses to conspecific offensive attacks. In the context of this or similar hypotheses, the following strain differences for males have been reported (see review by Siegfried et al., 1990): (1) DBA2 but not C57BL6 mice show analgesia to bites received during social conflict. This difference occurs even though as a consequence of these bites, the same decrement (possibly indicating utilization) in levels of β-endorphin occurs in structures of DBA2 and C57BL6 brain. Also, DBA2 but not C57BL6 mice show analgesia to morphine, and DBA2 mice show higher naloxone binding to μ-type receptors in the hippocampus and show lower naloxone binding to μ-type receptors in the striatum than do C57BL6 mice; and (2) at low levels of conflict or pain (10 bites from conspecifics), DBA2 mice show more escape than C57BL6 mice, whereas C57BL6 mice show more defensive sideways and upright postures as well as crouch and immobility than do DBA2 mice. The same pattern occurs for higher levels of conflict or pain (30–50 bites from conspecifics). These findings suggest that inhibition of both pain systems by social conflict bites may be associated with escape in DBA2 mice, and that although inhibition of neither pain system by social conflict bites occurs in C57BL6 mice, defensive sideways or upright postures as well as immobility are not blocked by these bites. Thus, these strains may provide variants for probing the relationship between pain and defense systems.

Without further research, neither the selected strains nor inbred strains mentioned in the previous two paragraphs should be considered as potential genetic models for some aspect(s) of defense. Differences between mice of selected or inbred strains may be due not only to nuclear genes but also to maternal factors (mitochondrial DNA, ovarian cytoplasm, uterine environment, and postnatal maternal environment; see Roubertoux et al., 1990, for a discussion of these issues and of methods to resolve them). Genetic models are based on differences in nuclear or chromosomal genes and not maternal factors. However, the difference between DBA2 and C57BL6 males with regard to morphine analgesia appears to be due to nuclear genes rather than maternal factors (Castellano and Oliverio, 1975). Thus, these strains are potential genetic models for some aspect(s) of the pain system.

Infanticide

Until recently, mouse infanticide was measured in most experiments by placing a 1- to 3-day-old mouse pup in a cage with an adult mouse for a limited period, usually 15 min (Svare et al., 1984). Adult behavior was classified into three categories: (1) retrieving the pup, (2) killing the pup, and (3) ignoring the pup. Data were reported as frequency of each behavior. Fortunately, a humane test has now been developed (Perrigo et al., 1989). For this, there is a minor modification to the above procedure. The pup is placed in a tube made of 1.5 mm^2 wire mesh screen. In the tube, the pup is protected from injury. Biting or attacking the tube is taken as a measure of infanticidal propensity of an adult mouse.

There are strain differences in male infanticide. From 25 to 65 days of age, a higher proportion of C57BL6 males are infanticidal in comparison to DBA2 males (Svare and Mann, 1981). However, after 2 months of age, the same high incidence of infanticide occurs for males of both strains. These differences occur regardless of the strain, sex, or age of the stimulus pup or the duration of the test (Svare and Broida, 1982). Results from studies with reciprocal F_1's and with cross-fostering of these strains are consistent with this strain difference in male infanticide being due to variants of one or more nuclear genes (Svare and Broida, 1982). These genetic differences on infanticide are not mediated by adult differences in serum testosterone levels, but may be mediated by target organ sensitivity to testosterone

(Svare and Mann, 1981). For equal levels of exogenous testosterone in castrates, the incidence of infanticide is still higher in C57BL6 than in DBA2 males. It has been suggested that the strain difference in target organ sensitivity to testosterone with regard to infanticide may be due to effects of prenatal steroids. This hypothesis is consistent with reports that there is an inverse relationship between fetal levels of testosterone and infanticide in male mice (vom Saal, 1989). Also, strain differences for female infanticide have been observed (Mann et al. 1983), and Svare et al. (1985) have proposed that strain differences in infanticide of males and females may be due to interactions between effects of genotype and of prenatal hormone metabolism.

There are extensive comparative data on infanticide in mammals and other vertebrates (Hausfater and Hrdy, 1984), and there is considerable speculation on the adaptive function(s) of infanticide in male mice (Parmigiani, 1989; vom Saal, 1989). However, there is little or no information on the neural basis or pharmacology of infanticide in mice or other species.

Predation

Predatory behavior is a directed attack against a prey animal. Locusts, crickets, and other insects are prey animals for laboratory and wild mice (Brain and Al-Maliki, 1978; Butler, 1973; Lynds, 1980). Mice will kill (mostly by bites to head and thorax of prey) and consume cockroaches (*Periplaneta americana* and *Balberus craniifer*), locusts *(Locusta migratoria),* and crickets *(Acheta domesticus).* There are strain differences in this predatory behavior (killing) by male and by female mice (Butler, 1973; Corning et al., 1977; Ebert and Green, 1984; Jones and Brain, 1987; Yanai and Ginsburg, 1976).

Unfortunately, it is not known whether these strain differences are due to nuclear genes or maternal factors (see discussion under Defense). Also, there is little, if any, information on the neural basis, the endocrinology, or the pharmacology of this murine predatory behavior. However, it has been suggested that different biological mechanisms may be involved in the killing and eating of prey by mice (Ebert and Green, 1984; Whelton and O'Boyle, 1977). In contrast, there is extensive information on the neural basis, endocrinology, and pharmacology of predatory behavior in rats (Karli, 1981; Karli et al., 1969, 1974; Olivier et al., 1989) and in cats (Mirsky and Siegel, 1991; Siegel and Brutus, 1990).

PROSPECTUS

None of the potential genetic models of male aggression are fully developed. In addition to determining the point-by-point comparisons of the different types of aggression across species, including humans, much remains to be done on the genetics of each type of mouse aggression. For each type of aggression, individual genes should be mapped chromosomally. Then, each gene's function(s), at all levels, and each gene's action (from DNA sequence to behavior) should be determined. Methods and reasons for doing this have been presented elsewhere (Maxson, 1991; Maxson and Monahan, 1991). In closing this chapter, two advantages will be considered here for further development of these genetic models of male aggressive behaviors in mice.

First, there has been considerable interest and argument as to whether or not the same genes are involved in variation and development of more than one type of mouse aggression both within a sex and between sexes. The usual approach to this issue has been to determine the between-strain correlations for two or more types of aggression within males or females or to determine the between-sex correlation for a type of aggression across several strains (for examples, see Corning et al., 1977; Ebert, 1983; Hood and Cairns, 1988; Jones and Brain, 1987). There are two methodological problems with this approach. First, strain correlations may be due to variants of nuclear genes and/or of maternal factors. Second, the sample of strains is usually small (no more than six). If rank ordering is used to assess the magnitude of the correlation (as it must be for between sex comparisons), the correlation must be very high ($r \geqslant 0.8$) to be significant for such a small number of strains. A better approach would be to determine whether strain differences for two or more types of aggression within the same sex, or for a single type of aggression between sexes, were due to allelic variants for the same and/or different loci. For each trait, this can be done by chromosomally mapping the segregating loci derived from a pair of inbred strains. Methods for doing this are discussed by Maxson (1991).

Second, mapping individual genes with effects on each type of aggression within each sex shifts the model(s) from the behavioral to the genic level (see section on Types of Genetic Models). There are two advantages to doing this. These are benefits, as presented below, to developmental and comparative studies of each type of aggression.

With regard to developmental research, mapping a gene to its chromosomal position can be an initial step in studying the gene's DNA sequence, its transcription, its translation, and its cellular or temporal regulation. It may then be possible with transgenic mice and *in situ* hybridization to relate these molecular and cellular aspects to the gene's physiological and behavioral function(s). Also, by investigating a single, mapped locus, exact questions can be asked and answered about the biological effects of the locus and the dependence of those effects on various aspects of the environment.

With regard to comparative research, the chromosomal mapping in mice of a gene with effects on one or more types of aggression may aid in identifying and mapping the same (homologous) gene in other mammals, including humans. Many syntenies between pairs of loci are conserved between mice and other mammals, including humans (O'Brien et al., 1988). Also, this gene may have at some levels, but not necessarily at all levels, some functional similarities in mice and other mammals, including humans (Erickson, 1989). This may assist in identifying the aspects of an animal model that are relevant to human behavior.

In conclusion, it is likely that only an approach based on the chromosomal mapping of individual genes with effects on one or more types of aggression will contribute to the turning of potential genetic models into actual genetic models of aggression and violence in males.

Acknowledgment. Supported in part by NATO grant No. 0235/89.

REFERENCES

Adams DB (1979): Brain mechanisms for offense, defense, and submission. *Behav Brain Sci* 2:210–241.

Adams DB (1980): Motivational systems of agonistic behavior in muroid rodents: a comparative review and neural model. *Aggress Behav* 6:295–346

Adams DB (1983): Hormone-brain interactions and their influence on agonistic behavior. In: *Hormones and Aggressive Behavior,* Svare BB, ed. New York: Plenum Publishing Corp

Adams DB (1987) Ventromedial tegmental lesions abolish offense without disturbing predation or defense. *Physiol Behav* 38:165–168

Archer J (1988): *The Behavioural Biology of Aggression*. Cambridge, UK: Cambridge University Press

Barfield RJ (1984): Reproductive hormones and aggressive behavior. In: *Biological Perspectives on Aggression,* Flannelly KJ, Blanchard RJ, Blanchard DC, eds. New York: Alan R. Liss

Barnes DM (1989): The origins of violent behavior: what research strategies will emerge? *J NIH Res* 1:27–29

Bishop CE, Weith A, Mattei M-G, Roberts C (1988): Molecular aspects of sex determination in mice: an alternative model for the origin of the Sxr region. *Phil Trans Roy Soc Lond B* 322:119–124

Blanchard DC, Blanchard RJ (1984): Affect and aggression: an animal model applied to human behavior. In: *Advances in the Study of Aggression,* Vol. 1, Blanchard RJ, Blanchard DC eds. New York: Academic Press

Blanchard DC, Blanchard RJ (1988): Ethoexperimental approaches to the biology of emotion. *Annu Rev Psychol* 39:43–68

Blanchard DC, Blanchard RJ (1989): Experimental animal models of aggression: what do they say about human behaviour? In: *Human Aggression: Naturalistic Approaches,* Archer J, Browne K, eds. New York: Routledge

Brain PF (1979): *Hormones, Drugs, and Aggression,* Vol. 3. Exeter: Eden Press

Brain PF (1983): Pituitary-gonadal influences on social aggression. In: *Hormones and Aggressive Behavior,* Svare BB, ed. New York: Plenum Press

Brain PF (1984): Biological explanations of human aggression and the resulting therapies offered by such approaches: a critical evaluation. In: *Advances in the Study of Aggression,* Vol. 1, Blanchard RJ, Blanchard DC, eds. New York: Academic Press

Brain PF (1991): Hormonal aspects of aggression and violence. Commissioned paper for the panel on the Understanding and Control of Violent Behavior. National Academy of Sciences/National Research Council

Brain PF, Al-Maliki S (1978): A comparison of 'intermale fighting' in 'standard opponent' test and attack directed toward locust by 'TO' strain mice: effects of simple experimental manipulations. *Anim Behav* 26:723–737

Butler K (1973): Predatory behavior in laboratory mice. *J Comp Physiol Psychol* 85:243–249

Cairns RB, MacCombie DJ, Hood KE (1983): A developmental–genetic analysis of aggressive behavior in mice: I. Behavioral outcomes. *J Comp Psychol* 87:69–89

Carey G (1989): Genetics and violence: human studies. Initial Draft for commissioned paper for Panel on the Understanding and Control of Violent Behavior. National Academy of Sciences/National Research Council

Carlier M, Roubertoux P (1986): Differences between CBA/H and NZB mice on intermale aggression. In: *Genetic Approaches to Behaviour,* Medioni J, Vaysse G, eds. Toulouse: I.E.C. (Privat)

Carlier M, Roubertoux PL, Kottler ML, Degrelle H (1990): Y chromosome and aggression in strains of laboratory mice. *Behav Genet* 20:137–156

Carlier M, Roubertoux PL, Pastoret C (1991): The Y-chromosome effect on intermale aggression in mice is dependent upon the maternal environment. *Genetics* 129:231–236

Castellano C, Oliverio A (1975): A genetic analysis of morphine-induced running and analgesia in the mouse. *Psychopharmacology* 41:197–200

Collinge J, Boccio A, DeLesi L, Johnstone E, Lofthouse R, Owen F, Poulter M, Risby D, Shah T, Crow TJ (1989): Evidence for a pseudoautosomal locus for schizophrenia. *Cytogenet Cell Genet* 51:978

Corning PA, Kessler S, Kakihana R (1977): Three kinds of aggressive behavior in laboratory mice. *Behav Genet* 7:51–52

Crow TJ (1989) Pseudoautosomal locus for the cerebral dominance gene. *Lancet* 2:339–340

Crow TJ, DeLisi LE, Johnstone EC (1989): Concordance by sex of sibling pairs with schizophrenia is paternally inherited. Evidence for a pseudoautosomal locus. *Br J Psychiatry* 155:92–97

Daly M, Wilson M (1989): Homicide and cultural evolution. *Ethol Sociobiol* 10:99–110

Didier-Erickson A, Maxson SC, Ogawa S (1989): Differential effect of the DBA1 and C57BL10 Y chromosomes on the response to social or other stimuli for offense. *Behav Genet* 19:675–683

Ebert PD (1983): Selection for aggression in a natural population. In: *Aggressive Behavior: Genetic and Neural Approaches,* Simmel EC, Hahn ME, Walters JK, eds. Hillsdale, NJ: Lawrence Erlbaum Associates

Ebert PD, Green YV (1984): Predatory aggression in lines of wild mice selected for interfemale aggression. *Aggress Behav* 10:21–26

Eibl-Eibesfeldt I (1979): *The Biology of Peace and War.* New York: The Viking Press

Erickson RP (1989): Why isn't a mouse more like a man? *Trends Genet* 5:1–3

Floody OR (1983): Hormones and aggression in female mammals. In: *Hormones and Aggressive Behavior,* Svare BB, ed. New York: Plenum Press

Gariepy J-L, Hood KE, Cairns RB (1988): A developmental-genetic analysis of aggressive behavior in mice *(Mus musculus)*. III. Behavioral mediation by heightened reactivity or immobility. *J Comp Psychol* 102:392–399

Hardy J (1988): Mouse models of human neurogenetic disorders. *Trends Neurosci* 11:89–90

Harris M (1989): *Our Kind.* New York: Harper Perennial

Hausfater G, Hrdy SB (1984): *Infanticide: Comparative and Evolutionary Perspectives.* New York: Aldine Publishing Co

Hewitt JK, Broadhurst PL (1983): Genetic architecture and the evolution of aggressive behavior. In: *Aggressive Behavior: Genetic and Neural Approaches,* Simmel EC, Hahn ME, Walters JK, eds. Hillsdale, NJ: Lawrence Erlbaum Associates

Hood KE, Cairns RB (1988): A developmental-genetic analysis of aggressive behavior in mice. II. Cross-sex inheritance. *Behav Genet* 18:605–619

Hrdy SB (1981): *The Woman That Never Evolved.* Cambridge, MA: Harvard University Press

Huntingford F, Turner A (1987): *Animal Conflict.* London: Chapman and Hall

Hyde JS (1983): The genetics of agonistic and sexual behavior. In: *Behavior Genetics: Principles and Applications,* Fuller JL, Simmel WC, eds. Hillsdale, NJ: Lawrence Erlbaum Associates

Ingersoll DW (1986): Latent aggression-promoting properties of mouse bladder urine activated by heat. *Behav Neurosci* 100:783–789

Ingersoll DW, Morley KT, Benvenga M, Hands C (1986): An accessory sex gland aggression-promoting chemosignal in mice. *Behav Neurosci* 100:777–782

Johnson RN (1972): *Aggression in Man and Animals.* Philadelphia: WB Saunders

Jones SE, Brain PF (1987): Performance of inbred and outbred laboratory mice in putative tests of aggression. *Behav Genet* 17:87–96

Jutley JK, Stewart AD (1985): Genetic analysis of the Y-chromosome of the mouse: evidence for two loci affecting androgen metabolism. *Genet Res Camb* 47:29–34

Karli P (1981): Conceptual and methodological problems associated with the study of brain mechanisms underlying aggressive behaviour. In: *The Biology of Aggression,* Brain PF, Benton D, eds. Rockville, MD: Sijthoff and Noordhoff

Karli P, Eclancher F, Vergnes M, Chaurand JP, Schmitt P (1974): Emotional responsiveness and interspecific aggressiveness in the rat: interaction between genetic and experiential determinants. In: *The Genetics of Behaviour,* van Abeelen JHF, ed. New York: American Elsevier Publishing Co

Karli P, Vergnes M, Didiergeorges F (1969): Rat-mouse interspecific aggressive behavior and its manipulation by brain ablation and by brain stimulation. In: *Aggressive Behaviour,* Garttini S, Sigg EB, eds. New York: John Wiley & Sons

Keiteges EA, Schorderet DF, Gartler SM (1987): Linkage of the steroid sulfatase gene to the sex-reversed mutation in the mouse. *Genetics* 116:465–468

Lagerspetz KMJ, Lagerspetz KYH (1974): Genetic determination of aggressive behaviour. In: *The Genetics of Behaviour,* van Abeelen JHF, eds. New York: American Elsevier Publishing Co

Langevin R (1991): Biological and psychological factors in human aggression. In: *Heterotypical Behaviour in Man and Animals,* Haug M, Brain PF, Aron C, eds. London: Chapman and Hall

Lenington S (1991): The *t* complex: a story of genes, behavior and populations. *Adv Study Behav* 20:51–86

Leshner AL (1983): The hormonal responses to competition and their behavioral significance. In: *Hormones and Aggressive Behavior,* Svare BB, ed. New York: Plenum Press

Lynds PG (1980): Intermale fighting and predatory behavior in house mouse: an analysis of behavioral content. *Aggress Behav* 6:139–147

Mann MA, Kinsley C, Broida J, Svare B (1983): Infanticide exhibited by female mice: genetic, developmental and hormonal influences. *Physiol Behav* 30:697–702

Maxson SC (1981): The genetics of aggression in vertebrates. In: *The Biology of Aggression,* Brain PF, Benton D, eds. Rockville, MD: Sijthoff and Noordhoff International Publishers

Maxson SC (1990): The evolution of the mammalian Y chromosome. *Behav Genet* 20:109–126

Maxson SC (1991): Methodological issues in genetic analyses of an agonistic behavior (offense) in male mice. In: *Techniques for the Genetic Analysis of Brain and Behavior: Focus on the Mouse,* Goldowitz D, Wimer RE, Wahlsten D, eds. Amsterdam, Elsevier Science Publishers

Maxson SC, Didier-Erickson A, Ogawa S (1989): The Y chromosome, social signals, and offense in mice. *Behav Neural Biol* 52:251–259

Maxson SC, Monahan E (1991): The Y chromosome and an agonistic behavior (offense) in mice. In: *Genetic Disorders Related to Brain Function,* Guastivino J-M, Jallon J-M, Archer T, eds. Hillsdale, NJ: Lawrence Erlbaum Associates

Maxson SC, Shrenker P, Vigue LC (1983): Genetics, hormones and aggression. In: *Hormones and Aggressive Behavior,* Svare BB, ed. New York: Plenum Publishing Corp

Mazur A (1983): Hormones, aggression, and dominance in humans. In: *Hormones and Aggressive Behavior,* Svare BB, ed. New York: Plenum Publishing Corp

McLaren A, Simpson E, Epplen JT, Studer R, Koopman P, Evans EP, Burgoyne PS (1988): Location of the genes controlling H-Y antigen expression and testis determination on the mouse Y chromosome. *Proc Natl Acad Sci USA* 85:6442–6445

Michard C, Carlier M (1985): Les conduites d'agression intraspecifique chez la souris domestique: differences individuelles et analyses genetiques. *Biol Behav* 10:123–146

Miczek KA (1983): Ethopharmacology of aggression, defense, and defeat. In: *Aggressive Behavior: Genetic and Neural Approaches,* Simmel EC, Hahn ME, Walters JK, eds. Hillsdale, NJ: Lawrence Erlbaum Associates

Miczek KA (1987): The psychopharmacology of aggression. In: *Handbook of Psychopharmacology, Vol. 19, New Directions in Behavioral Pharmacology,* Iversen LL, Iversen SD, Snyder SH, eds. New York: Plenum Press

Miczek KA, Weerts E, Haney M (1991): Neurobiological mechanisms controlling aggression: preclinical developments for pharmacotherapeutic interventions. *Neurosci Biobehav Res,* in press

Mirsky AF, Siegel A (1991): The neurobiology of violence and aggression. Commissioned paper for the Panel on Understanding and Control of Violent Behavior. National Academy of Sciences/National Research Council

Monahan E, Maxson SC (1990): Chemosignal independent and dependent effects of Y chromosomal gene(s) on offense in mice. *Behav Genet* 20:735

Moyer KE (1976): *The Psychobiology of Aggression.* New York: Harper and Row

Nadeau JH, Reiner AH (1989): Linkage and synteny homologies in mouse and man. In: *Genetic Variants and Strains of the Laboratory Mouse* 2nd ed, Lyon MF, Searle AG, eds. Oxford: Oxford University Press

Nallaseth FS, Dewey MJ (1986): Moderately repeated mouse Y chromosomal sequence families present distinct types of organization and evolutionary change. *Nuc Acid Res* 14:5295–5307

O'Brien SJ, Seuanez HN, Womack JE (1988): Mammalian genome organization: an evolutionary view. *Annu Rev Genet* 22:323–351

Olivier B, Mos J, Tulp M, Schipper J, Den Daas S, van Oortmerssen GA (1988): Serotonergic involvement in aggressive behavior in animals. In: *Monoaminergic Regulation of Aggressive Behavior and Impulse Control,* van Praag HM, ed. A.C.N.P. Monographs

Olivier B, Mos J, van Oorschot R (1989): Etho-experimental studies of similarities and differences in male and female agonistic behavior. In: *Ethoexperimental Approaches to the Study of Behavior,* Blanchard RJ, Brain PF, Blanchard DC, Parmigiani S, eds. Dodrecht: Kluwer Academic Publishers

Olsen KL (1983): Genetic determinants of sexual behavior. In: *Hormones and Behaviour in Higher Vertebrates,* Balthazart J, Prove E, Gilles E, eds. New York: Springer–Verlag

Parmigiani S (1989): Maternal aggression and infanticide in the house mouse: consequences on the social dynamics. In: *House Mouse Aggression: A Model for Understanding the Evolution of Social Behavior,* Brain PF, Mainardi D, Parmigiani S, eds. New York: Harwood Academic Publishers

Perrigo G, Bryant WC, Belvin L, vom Saal FS (1989): The use of live pups in a humane, injury-free test for infanticidal behaviour in male mice. *Anim Behav* 38:897–904

Roberts C, Weith A, Passage E, Michot JL, Mattei MG, Bishop CE (1988): Molecular and cytogenetic evidence for the location of Tdy and Hya on the mouse Y chromosome short arm. *Proc Natl Acad Sci USA* 85:6446–6449

Roberts SS (1990): Murder, mayhem, and other joys of youth. *J NIH Res* 2:67–72

Rodgers RJ (1989): Physiological consequences of defeat. In: *House Mouse Aggression: A Model for Understanding the Evolution of Social Behavior,* Brain PF, Mainardi D, Parmigiani S, eds. New York: Harwood Academic Press

Rodgers RJ, Randall JI (1987): Defensive analgesia in rats and mice. *Psychol Rec* 37:335–347

Roubertoux PL, Nosten-Bertrand M, Carlier M (1990): Additive and interactive effects between genotype and maternal environments, concepts and facts. *Adv Stud Behav* 19:205–247

Scott JP (1966): Agonistic behavior of mice and rats: a review. *Am Zool* 6:683–701

Scott JP (1984): The dog as a model for human aggression. In: *Biological Perspectives on Aggression,* Flannelly KJ, Blanchard RJ, Blanchard DC, eds. New York: Alan R. Liss

Scott JP (1989): *The Evolution of Social Systems,* New York, Gordon and Breach Science Publishers

Scrimshaw SCM (1984): Infanticide in human populations: societal and individual concerns. In: *Infanticide: Comparative and Evolutionary Perspectives,* Hausfater G, Hrdy SB, eds. New York: Aldine Publishing Co

Selmanoff M, Ginsburg BE (1981): Genetic variability in aggression and endocrine functions in inbred strains of mice. In: *Multidisciplinary Approaches to Aggression Research,* Brain PF, Benton D, eds. Amsterdam: Elsevier/North-Holland Biomedical Press

Siegel A, Brutus M (1990): Neural substrates of aggression and rage in the cat. In: *Progress in Psychobiology and Physiological Psychology,* Vol. 14, Epstein AN, Morrison AR, eds. New York: Academic Press

Siegfried B, Frischknecht H-R, De Souza RLN (1990): An ethological model for the study of activation and interaction of pain, memory and defensive systems in the attacked mouse. Role of endogenous opioids. *Neurosci Biobehav Rev* 14:481–491

Simerly RB (1990): Hormonal control of neuropeptide gene expression in sexually dimorphic olfactory pathways. *Trends Neurosci* 13:104–110

Simon NG (1979): The genetics of intermale aggressive behavior in mice: recent research and alternative strategies. *Neurosci Biobehav Rev* 3:97–106

Simon NG (1983): New strategies for aggression research. In: *Aggressive Behavior: Genetic and Neural Approaches,* Simmel EC, Hahn ME, Walters JK, eds. Hillsdale, NJ: Lawrence Erlbaum Associates

Soriano P, Keiteges EA, Schorderet DF, Harbers K, Gartler SM, Jaenisch R (1987): High rate of recombination and double crossover in the mouse pseudoautosomal region during male meiosis. *Proc Natl Acad Sci USA* 84:7218–7220

Stewart AD, Manning A, Batty J (1980): Effects of Y-chromosome variants on the male behaviour of the mouse *Mus musculus. Genet Res Camb* 35:261–268

Svare B, Broida J (1982): Genotypic influences on infanticide in mice: environmental, situational and experiential determinants. *Physiol Behav* 28:171–175

Svare B, Broida J, Kinsley C, Mann M (1984): Psychobiological determinants underlying infanticide in mice. In: *Infanticide: Comparative and Evolutionary Perspectives,* Hausfater G, Hrdy SB, eds. New York: Aldine Publishing Co

Svare B, Kinsley CH, Mann M, Broida J (1985): Infanticide: accounting for genetic variation in mice. *Physiol Behav* 32:137–152

Svare B, Mann M (1981): Infanticide: genetic, developmental and hormonal influences in mice. *Physiol Behav* 27:921–927

van Oortmerssen GA, Bakker ThCM (1981): Artificial selection for high and low attack latencies in wild *Mus musculus domesticus*. *Behav Genet* 11:115–126

van Oortmerssen GA, Busser J (1989): Studies in wild house mouse. 3. Disruptive selection of aggression as a possible force in evolution. In: *House Mouse Aggression: A Model for Understanding the Evolution of Social Behaviour,* Brain PF, Mainardi D, Parmigiani S, eds. New York: Harwood Academic Publishers

van Oortmerssen GA, Dijk DJ, Schurman T (1987): Studies in wild house mice II. Testosterone and aggression. *Horm Behav* 21:139–152

vom Saal FS (1984): Proximate and ultimate causes of infanticide and parental behavior in male house mice. In: *Infanticide: Comparative and Evolutionary Perspectives,* Hausfater G, Hrdy SB, eds. New York: Aldine Publishing Co

vom Saal FS (1989) Perinatal testosterone exposure has opposite effects on adult intermale aggression and infanticide in mice. In: *House Mouse Aggression: A Model for Understanding the Evolution of Social Behavior,* Brain PF, Mainardi D, Parmigiani S, eds. New York: Harwood Academic Publishers

Whelton JP, O'Boyle M (1977): Early experience and the development of predatory aggression and interspecific aggression in mice. *Anim Learn Behav* 5:291–296

Wilson M, Daly M (1985): Competitiveness, risk taking, and violence: the young male syndrome. *Ethol Sociobiol* 6:59–73

Yanai J, Ginsburg BE (1976): Long-term effects of early ethanol on predatory behavior in inbred mice. *Physiol Psychol* 4:409–411

Cerebral Laterality as a Determinant of Behavioral Function and Dysfunction

JEFFREY N. CARLSON AND STANLEY D. GLICK

The human brain has been shown to be laterally specialized with regard to a variety of functions. The genetic determinants of this specialization have also been explored. There has been some emphasis placed on individual differences in the degree of this specialization and the determination of its significance for lending a predisposition toward behavioral dysfunction and psychiatric disease. Whereas many animal models of human behavioral dysfunction have been developed, there has been relatively little interest and consequently relatively little data bearing on the issue of how differences in brain asymmetry in animals may be important parameters in determining the modeled dysfunction. However, over the past 15 years, data have accumulated to suggest that morphologic, chemical, and behavioral indices of brain asymmetry are present in animals. It has also been suggested that variations in the magnitude and direction of these indices are determined by a complex interaction of genetic, hormonal, and experiential factors. Individual differences in cerebral laterality have been shown to covary with or predict individual differences in drug responses, spatial behavior, and stressor reactivity as well as susceptibility to stressor-evoked behavior pathology and drug abuse predilection. Such findings suggest that it is possible to study genetically influenced individual differences in lateralized brain function as they exist in animals and to relate these differences to those seen in humans.

In this chapter we will review evidence bearing on cerebral laterality in animals and its implications for similar processes in humans. We will attempt to explore its origins and determinants with respect to heritability and its effects on the modulation of behavioral processes such as learning and the effects of stress. Finally, we will suggest how variation in cerebral asymmetry in animals is associated with susceptibility to various behavioral dysfunctions and consider how they model similar processes seen in humans.

CEREBRAL LATERALITY IN HUMANS AND IN ANIMALS

The phenomenon of cerebral dominance in humans has been known since Broca's original discovery that lesions that resulted in language disorders were, in most cases, located in the left cerebral hemisphere (Broca, 1861). The human temporal lobe manifests clear asymmetries in gross anatomy, especially on its upper surface (Galaburda et al., 1978). A temporal lobe asymmetry was described by Heschl in 1878 and subsequently Pfeifer reported asymmetries in the region of cortex between Heschl's gyrus and the posterior edge of the Sylvian fossa (Galaburda et al., 1978; Pfeifer, 1936), the planum temporale. Pfeifer's discovery was later confirmed and more fully characterized by Geschwind and Levitsky (1968). Newer radiological and computerized methodology has made it possible to demonstrate cerebral asymmetries in living subjects. Using computerized axial tomography, the asymmetry in the planum has been confirmed and other asymmetries have been uncovered (LeMay, 1976). In particular, LeMay (1976) demonstrated a larger and wider left occipital and right frontal lobe that was related to handedness.

Neuropsychological testing methods have also been widely employed to study functional correlates of cerebral asymmetry in living humans. In some cases the left hemisphere has been characterized as the more analytic, rational, and propositional and the right as the more holistic, intuitive, and appositional (Bogen, 1969a,b; Bogen and Bogen, 1969). The most common characteristic of right hemisphere functions appears to be that they are all nonverbal (Milner, 1971; Corballis and Morgan, 1978a). Right hemisphere advantages have been reported for perception of musical chords (Gordon, 1970), melodies (Kimura, 1964), visuospatial configurations (Kimura, 1969), faces (Geffen et al., 1971), judgment of stereoscopic depth (Dunford and Kimura, 1971), color discrimination (Davidoff, 1976) and spatial processing (Witelson, 1976) as well as in the control of facial expression (Sackeim et al., 1978). Some investigators (Geschwind, 1976; Zangwill, 1960) have suggested that by developing a lateralized capacity for language and perhaps for fine motor control, the left hemisphere may have forfeited some of its capacity to carry out nonverbal perceptual functions. There has also been some interest in the notion that faulty cerebral asymmetry in humans is associated with behavioral dysfunction. Studies designed to evaluate this hypothesis will be reviewed below.

Until relatively recently, it had been assumed that cerebral lateralization of function is a distinctly human trait (e.g., Levy, 1977). Evidence that functional brain asymmetry occurs in species other than humans is of relatively recent origin, having largely accumulated only over the last 15 years. Because of this there has been considerable resistance on the part of many scientists to consider the possibility that variations in brain asymmetry among animals may be analogous to similar variations in brain asymmetry among humans. However, findings have suggested that cerebral lateralization is an evolutionary principle and not simply one indigenous to humans. Morphologic and functional brain asymmetries have been reported in species that range from songbirds (Arnold and Bottjer, 1985; Nottebohm, 1977) to nonhuman primates (LeMay, 1985; LeMay and Geschwind, 1975).

Research conducted in this laboratory and in others has established that normal rats have functional and neurochemical asymmetries in several brain regions.

Much of this work has focused on the dopamine-containing nigrostriatal pathways whereas other work has focused on the cortex, hippocampus, nucleus accumbens, and hypothalamus. Behaviorally, this asymmetry is most apparent as a rotational preference that appears to result from a persistent side or orienting bias; such a bias may serve as a proprioceptive reference, enabling discrimination between left and right as well as mediating lateralized and stereotypic motor reactions in times of exigency (Glick, 1983b; Glick and Ross, 1981a). It is also a primary behavioral index of cerebral asymmetry in rodent species and its direction and intensity have been shown to predict chemical and morphologic indices of asymmetry. Data have also been assembled to suggest that variation in rat brain asymmetry is associated with behavioral variation, which is analogous to behavioral dysfunction seen in humans. Examples of such data are presented below.

FUNCTIONAL, CHEMICAL, AND MORPHOLOGIC ASYMMETRIES IN ANIMAL AND HUMAN BRAINS

Functional asymmetry in the rat brain was first indicated by the observations that d-amphetamine would enhance side preferences (Glick, 1973; Glick and Jerussi, 1974) and induce rotation (or circling) in normal intact rats (Jerussi and Glick, 1974, 1976). As it was known (Ungerstedt and Arbuthnott, 1970; Ungerstedt, 1971a,b; Christie and Crow, 1971) at the time that rats with unilateral lesions of the nigrostriatal system would turn in circles at high rates (5–10 rotations/min), it was postulated that the less intense circling (0.5–4 rotations/min) in normal rats was due to an endogenous physiological asymmetry in the same dopaminergic system (Glick and Jerussi, 1974; Jerussi and Glick, 1974). Subsequently, normal asymmetries in striatal dopamine content (Zimmerberg et al., 1974), striatal dopamine metabolism, and dopamine-stimulated adenylate cyclase activity (Jerussi et al., 1977) were demonstrated directly and shown to be related to spontaneous side preferences (Zimmerberg et al., 1974) and to nocturnal (Jerussi et al., 1977) and d-amphetamine–induced (Jerussi and Glick, 1976; Glick et al., 1974) circling behavior. Numerous studies of asymmetry in the normal rat brain followed in this and many other laboratories. Thus, the striatal dopamine asymmetry was confirmed (Robinson et al., 1980; Robinson and Becker, 1983) and many other biochemical asymmetries were reported. The latter include side-to-side differences in norepinephrine content in thalamus (Oke et al., 1980) and striatum (Rosen et al., 1984); gamma-aminobutyric acid (GABA) levels in substantia nigra, nucleus accumbens, thalamus, and striatum (Rosen et al., 1984); serotonin levels in striatum (Rosen et al., 1984; Knapp and Mandell, 1980), nucleus accumbens (Rosen et al., 1984) and hippocampus (Knapp and Mandell, 1980); dopamine release (Robinson et al., 1982), dopamine uptake (Shapiro et al., 1986) and dopamine (D2) receptors (Schneider et al., 1982; Drew et al., 1986) in striatum; dopamine levels in nucleus accumbens (Rosen et al., 1984) and frontal cortex (Slopsema et al., 1982); and leutinizing hormone–releasing hormone (LHRH) content in hypothalamus (Gerendai, 1984). Other studies have described morphological asymmetries, for example in cortex (Diamond et al., 1981; Sherman et al., 1982; Fleming et al., 1986) and hippocampus (Diamond et al., 1982), or functional asymmetries, as evidenced by differential effects of lesions on either side of the brain, for example in striatum (Glick and Cox, 1978; Rothman and Glick, 1976; Glick and Cox, 1976; Mittleman et al., 1985), cortex (Denenberg et al., 1978; Pearlson and Robinson, 1981; Crowne et al., 1987), and hippocampus (Therrien et al., 1982). Many of these asymmetries are more complex than immediately apparent, there being both sex differences (e.g., Robinson et al., 1980; Becker et al., 1982; Brass and Glick, 1981) and strain differences (e.g., Glick and Ross, 1981b; Glick, 1985; Glick et al., 1986) in their magnitude and direction.

That humans and rats show many commonalities in brain asymmetry is manifest in many recent studies. Variation in asymmetry has been associated with behavioral dysfunction in both species. For example, a report that neuroleptic-induced dyskinesias are lateralized in humans (Waziri, 1980) suggested that an asymmetry in basal ganglia dopamine, similar to that observed in rats, would be identified in humans. Direct evidence of neurochemical asymmetries in post mortem human brain was subsequently obtained (Glick et al., 1982a). Dopamine was found to be significantly higher in the left than in the right globus pallidus, with the average asymmetry in basal ganglia dopamine levels approximately twice that observed in the rat (Zimmerberg et al., 1974). Asymmetries in GABA and choline acetyltransferase were observed in both cortical and subcortical structures of humans (Glick et al., 1982a; Amaducci et al., 1981) and similar asymmetries in GABA were observed in rats. Another similarity between species was the finding of left–right asymmetry in norepinephrine content in both rat (Oke et al., 1980) and human (Oke et al., 1978) thalamus. A particularly well controlled clinical study reported differences in neuropsychological impairments and cerebrospinal fluid (CSF) levels of dopamine metabolites in patients with predominantly left- or right-sided Parkinson's disease (Direnfeld et al., 1984): "asymmetries in the dopaminergic system, either anatomic or physiologic" were held responsible for the findings. Last, normal humans have recently been shown to have consistent left or right rotational biases (Bracha et al., 1987a) that are strikingly similar to those observed in rats and, similar to rats with unilateral nigrostriatal lesions, human patients with hemi-Parkinsonism have been found to rotate at higher than normal rates (Bracha et al., 1987b). It is thus becoming increasingly evident that animal models of brain asymmetry are both relevant and important for understanding related or common mechanisms of brain asymmetry in humans (cf. Geschwind, 1985). To the extent that this is true, as well as the notion that extreme variations in asymmetry may underlie some aspects of neuropsychiatric syndromes, it would be predicted that animal models of such syndromes could be developed and used to assess novel therapeutic interventions.

CEREBRAL LATERALITY AND BEHAVIORAL DYSFUNCTION AS A GENETICALLY DETERMINED TRAIT

Handedness and cerebral dominance (the greater proficiency of a single hemisphere in the performance of a particular function) have been long seen as intrinsic biological characteristics of humans and numerous genetic models have been presented (Levy, 1976; Corballis and Morgan, 1978b; Collins, 1977; Carter Saltzman, 1980; Annett, 1978). A current genetic theory (Annett, 1985) has argued for the existence of a single gene locus for the control of cerebral asymmetry. The right shift theory of cerebral laterality suggests a dominant right shift allele (RS+) that produces right handedness and left cerebral control of speech. Individuals that are homozygous (RS−) shown no disposition for cerebral dominance or handedness (although they may still exhibit it to a small degree) and tend to be susceptible to reading disabilities. Those homozygous for RS+ show extreme right-hand preference and cerebral dominance. Extremely right-handed individuals tend to suffer from deficiencies in mathematical ability (Annett and Kilshaw, 1982) and show poorer intelligence (Annett and Manning, 1989) and performance on motor skills tests (Annett and Kilshaw, 1984). Although the single-locus element of the theory is controversial, it has nonetheless stimulated much investigative effort. Data have been collected to suggest many significant correlations between degree of handedness and behavioral dysfunction, at least in dextrals. The exact genetic nature of these relationships remains speculative. However, indications of the contribution of genetic factors in determining cerebral dominance were seen in observations of familial clustering of functional asymmetries, especially sinistrality, as well as in studies of monozygotic and dizygotic twins (Annett, 1978).

An influential genetic approach to human cerebral laterality has been Geschwind's (Geschwind and Galaburda, 1987) theory of developmental learning pathology. Studies of families of left-handed individuals reported a greater frequency of immune diseases (Geschwind and Behan, 1982) and learning disabilities (Annett, 1978). It was suggested that genetic factors controlling handedness in humans might account for at least part of the differences occurring among these individuals. These divergences might also be associated with variation in genes controlling the major histocompatibility complex (MHC) in man. This hypothesis derives from data showing that genetic loci of the MHC control testosterone levels (Ivanyi, 1978), that testosterone slows the growth of the left hemisphere, and that variations in this influence, as they interact with genetic determinants, might account for individual differences in cerebral laterality and handedness.

Since handedness, the most obvious manifestation of asymmetry, occurs in humans and variations in asymmetry have been associated with behavioral variation, it is perhaps not too surprising that as far back as 1930, "handedness" or paw preference was investigated in the albino rat (Tsai and Maurer, 1930). Paw preference appeared to be a stable phenomenon that remained unchanged for the life of the animal (Peterson, 1934) but could be changed by surgical procedures and forced practice of the nonpreferred limb (Peterson and Barnett, 1961; Peterson and Devine, 1963). Unilateral lesions of the striatum contralateral to a preferred forepaw prevented bar-pressing with that paw, while performance with the ipsilateral paw remained unaffected (Hansing et al., 1968; Levine et al., 1971). When only the contralateral cortex was ablated, three out of four rats tested continued to respond with the preferred paw and the other rat was easily retrained to press as before (Hansing et al., 1968).

Stable paw preferences are also exhibited by mice and a major genetic model of of cerebral lateralization has also been undertaken using inbred strains of mice over the past 20 years (Collins, 1985). Mice exhibit a paw preference that is not task-specific and is consistent upon repeated measurement (Collins, 1968). Its strength may be assessed in a simple food-reaching task with animals reaching into a narrow tube. Studies in C57BL/6J mice indicated that although highly inbred, nearly maximum genetic variation in paw preference still occurred (Collins, 1970). When testing occurred in an "unbiased world" (tube located equidistant from chamber walls), a U-shaped distribution of paw preferences resulted. There was a sexual dimorphism in the strength of this preference, with females exhibiting a stronger bias than males. However, when the location of the tubes was "biased" (i.e., located adjacent to the left of right wall), the distribution was J-shaped, with 90% of mice exhibiting a bias consistent with the tube bias and males showing a stronger preference than females (Collins, 1975). From these studies it was concluded that highly inbred mice (possessing minimum genetic heterogeneity) exhibit an enduring, non–task-specific paw preference that is sexually dimorphic (stronger in females) and modifiable by environmental influences. Clearly, there was an indication that genetic factors might determine the strength, but not the direction, of lateralization. Attempts have been undertaken to derive, by selective breeding, two populations of mice that differ markedly in the degree of paw preference (Collins, 1985). Large potential genetic variability was insured by using an 8-way cross in parental strains. After 11 generations of selective breeding, mice from the two selected lines differed markedly in their degree of paw preference as well as from a random-bred control line (Collins, 1985). Morphologic studies have indicated that the brains of the two strains also differ. In four of five horizontal sections from mid-dorsoventral planes, strongly lateralized mice had greater asymmetries and the two lines could be best discriminated by asymmetries in the hippocampus and orbitofrontal cortex (Lipp et al., 1984). The brains of strongly lateralized mice were heavier and those of weakly lateralized mice lighter than those of control mice of the same body weight; in both selected lines the corpus callosum was found to be smaller than that in the random-bred control line (Ward and Collins, 1985). From these accumulated findings it has been suggested (Collins, 1985) that genetic elements control the degree or strength of brain asymmetry but that the direction of asymmetry is strongly influenced by environmental (nongenetic) processes that modulate the strength of directional bias. Other findings have shown that variations in paw or hand preferences also occur in other species such as cats (Warren et al., 1967; Forward et al., 1962) and monkeys (Ettlinger and Moffett, 1964; Warren, 1953), although there have been conflicting results concerning the relationship between paw preference and cerebral dominance in these species.

HERITABLE DIFFERENCES IN ROTATIONAL BEHAVIOR AND THEIR INTERACTION WITH HORMONAL INFLUENCES

The most studied indicator of cerebral asymmetry in the rat is that of circling or rotational behavior. Findings have indicated that there is a genetic influence on its development. The intensity of circling behavior, as well as the relative incidence of left- and right-directional preference, varies among rat strains (Brass and Glick, 1981; Glick and Ross, 1981b; Glick and Shapiro, 1985) as well as between sexes (Becker et al., 1982; Brass and Glick, 1981). These observations as well as the evidence that biochemical asymmetry and side preferences are present at birth (Ross et al., 1981) suggested that some aspects of this lateralized behavior in rats would be under heritable control. An earlier study from this laboratory indicated that whether a particular rat develops a left or right circling preference was probably determined by an interaction between heritable and sex-related factors. The offspring of rats having opposite or same-sided turning biases were tested for turning biases as adults and the degree of similarity to the parents' biases assessed. There were significant and equivalent tendencies for the male offspring to have the same bias as the male parent and the opposite bias as the female parent. Also reported was a significant tendency for female offspring to have biases that were different from those of the female parent in litters having more males than females (Glick, 1983a). Based on reports indicating that there is a relationship between the sex ratio of a litter and female levels of testosterone (e.g., Meisel and Ward, 1981; Vom Saal and Bronson, 1980), it was suggested that exposure to testosterone reverses the coding of a heritable female influence and induces a tendency for offspring to have biases opposite those of the female parent. Another indicator of this *in utero* hormonal modulation of heritable influences on cerebral laterality was seen during an attempt to breed strains of strongly and weakly rotating rats. The objective was to breed one strain that would exhibit intense circling behavior in a preferred direction and another strain that would exhibit little or no directional preference in circling behavior. A strain difference (approximately twofold) in rotation parameters was limited to females and developed gradually over eight generations before asymptoting (Glick and Shapiro, 1985), indicating strong genetic control at least in females. A serendipitous finding was that in both males and females of the weakly rotating strain there developed a left-sided population bias. A modulatory role of testosterone was postulated to account for the results with respect to sex dependence and the left bias in weak turners. Neonatal or perinatal exposure to testosterone is known to masculinize the central nervous system (e.g., McClusky and Naftolin, 1981). In this laboratory neonatal administration of testosterone propionate (1.25 mg, s.c.) reduced adult rotation in female rats. Thus, the absence of a breeding effect on the rates of male rotation might be attributable to a dampening effect of the obviously stronger and longer exposure to testosterone in males (females are exposed to some extent *in utero;* cf. Glick, 1983a). Testosterone has been hypothesized to delay preferentially the development of the left side of the brain (Geschwind and Behan, 1982) and to be responsible for the greater incidence of left-handedness in human males than females. The left bias in the weakly rotating strain may have developed because the parents were selected for having very weak or negligible left–right preferences and the effect of testosterone on brain maturation could be exerted independently of a parental directional influence.

EMOTION REACTIVITY AND DEPRESSION

In recent years human data from both normal and clinical populations have begun to accumulate that suggest that the two hemispheres of the human brain are differentially specialized for the regulation and processing of affect. Several lines of evidence have indicated that regions of the left hemisphere are specialized in processing positive affect whereas the right hemisphere is specialized for negative affect (Davidson, 1988; Flor-Henry, 1986b). A vast literature describes the differential effects of unilateral cerebral lesions on emotional processes (Robinson and Chait, 1985; Robinson, 1985). Emotionally expressive movements of the left or right side of the face have also been held to indicate lateral specialization of affect (Sackeim et al., 1978). Individual differences in emotional expression have also been reported. It has been shown (Davidson and Fox, 1989), for example, that a subset of infants, those showing profuse "stranger anxiety," show a unique distribution of asymmetrical brain activity. Infants who cried in response to maternal separation showed greater right frontocortical electroencephalogram (EEG) activation during the preceding baseline period compared with infants who did not cry. The findings suggest that a state-independent asymmetric difference in the brain can be used to predict the occurrence of a behavioral response in an emotion-arousing situation.

Data from this laboratory have indicated that the rat brain is also lateralized for affect. Based on the premise that differences in affect might result from differences in mechanisms mediating reward or pleasure, it was postulated and then demonstrated that the two sides of the brain are differentially sensitive to reinforcement from lateral hypothalamic stimulation and that these side differences are related to the direction of rotational behavior (Glick et al., 1980). Other work (Glick et al., 1981; Glick et al., 1982b; Glick and Badalamenti, 1986) has shown that drugs that are known to have "euphoric" or "depressant" effects preferentially affect different sides of the brain—and that these effects are related to the animal's preferred direction of rotational behavior. Similarly, differences in responding to a negative emotional stimulus (as measured by reactivity to a footshock stressor) have been seen. Left rotating rats are more reactive (display shorter latiencies) on the early trials of a simple shock escape task (Carlson and Glick, 1989). The accumulated results suggest that the rat brain is also lateralized in the regulation of emotional processes.

Assuming that the above is true, one might expect that variation in other more complex behavioral processes among rats would to some extent depend on differences in brain laterality. One preliminary approach to testing this hypothesis was through the assessment of behavioral differences between left- and right-sided rats. For several years, results in this laboratory had seemed to indicate that left and right biases in rotation and side preference occurred randomly in a rat population. A typical experiment generally found that approximately 50% of rats rotated to the right and 50% to the left. Nonetheless, several findings suggesting a left–right asymmetry in rat cortex had also appeared. Differences in cortical thickness (e.g., Diamond et al., 1975), differences in behavioral effects of left and right cortical lesions (e.g., Denenberg et al., 1978; Robinson, 1979) and, based on measurements of labeled deoxyglucose uptake, a difference in frontal cortical energy metabolism (Glick et al., 1979) had been seen. The latter deoxyglucose results suggested that the left–right asymmetry in frontal cortex modulated the nigrostriatal asymmetry: rats' side preferences were stronger if deoxyglucose uptake was greater in the contralateral than in the ipsilateral frontal cortex. Because left frontal cortex was usually more active in terms of deoxyglucose uptake, it was postulated that in a large population more rats should have right side preferences than left side preferences and right preferences should be stronger than left preferences. Data of 602 rats (all female Sprague–Dawley derived) tested for nocturnal or amphetamine-induced rotation were reviewed, and indeed there was a small (54.8%) but significant ($p < .025$) right population bias.

A large body of literature has indicated that the activity of the frontal cortex modulates mesolimbic and nigrostriatal dopamine function (Glick and Greenstein, 1973; Ross and Glick, 1981a; Carter and Pycock, 1980) through projections that use glutamate or aspartate as excitatory neurotransmitters (Christie et al., 1985; Fonnum, 1984). If frontal cortex is lateralized and modulates spatial bias in the proposed manner, it might be expected that bilateral lesions of frontal cortex, by removing this modulation, would decrease side preferences and activity in right-sided rats and have opposite effects in left-sided rats. This is exactly what occurred experimentally (Ross and Glick, 1981b). These findings were of significance to us because they suggested a brain mechanism for behavioral differences between left- and right-sided rats. They also indicated that for a given animal, the strength of a cortical influence (which is presumed to be crucial in the control of emotional behavior) could be predicted from rotational behavior. It was postulated that stimuli that activate the cortex would modulate the magnitude and direction of rotational behavior. The frontal cortex and especially mesocortical dopamine neurons are known to be selectively activated by inescapable footshock stressors (Thierry et al., 1976; Blanc et al., 1980; Herman et al., 1982). Sprague–Dawley rats were exposed to footshock stress between two tests of d-amphetamine–induced rotational behavior. Footshock induced a change in the direction and intensity of rotational behavior that was dependent on sex and pre-existing rotational bias. Right-rotating males and left-rotating females shifted their directional bias toward the opposite side, whereas left-rotating males and right-rotating females displayed increased rotation in their prestress direction (Carlson et al., 1987). The data indicated that activation of the mesocortical DA system by stressors could induce a change in the direction of rotational behavior. The directional difference between sexes was in agreement with the reported sexually dimorphic nature of cortical asymmetry in rats (Diamond et al., 1981).

In other experiments we saw behavioral differences between rats of differing rotational asymmetry. Right-sided rats were more active and had, as predicted, stronger side preferences than left-sided rats (Glick and Ross, 1981b). We also determined that right-sided rats, although initially less reactive to footshock stressors, learned to escape footshock better than their left-sided counterparts (Carlson and Glick, 1989). These differences in stressor reactivity suggested that animals might differ on other behaviors involving stressors.

In humans, diseases of affect have been strongly associated with brain asymmetry. For example, lateralized changes in brain function during affective disorders have been shown (Tucker, 1984; Schaffer et al., 1983; Robinson and Chait, 1985), and lateralized electroconvulsive therapy has been shown to have differential effects in depressive disorders (Sackeim, 1986; Flor-Henry, 1986a). Changes in information processing on the two sides of the brain as indicated by dichotic listening and tachistoscopic recognition tests has also been seen to vary with diagnostic subtype for depression in clinical patients (Lenhart and Katkin, 1986; Bruder et al., 1989; Otto et al., 1989). The general trend in these studies indicates that depression is associated with right-hemisphere dysfunction. In animals, the best characterized model of depression is the Learned Helplessness (LH) effect (Maier and Seligman, 1976), where exposure to uncontrollable stressors such as footshock provides a transient disruption of behaviors necessary for controlling stressors in subsequent behavioral tasks. The modeled behavior is analogous to depressive symptomatology in humans (Anisman and Zacharko, 1982). In order to assess the relevance of this phenomenon to cerebral laterality in the rat, we performed a systematic inquiry as to the relative susceptibilities to LH effects of left- and right-biased rats. Male Zivic–Miller Sprague–Dawley rats were subjected to inescapable shock. Twenty-four hr later the animals were tested in a shuttlebox footshock escape test. Left- and right-biased rats differed in their performance. Right-biased rats became helpless; that is, they failed to learn to escape footshock, whereas left-biased rats did not; they rapidly escaped shock on early trials and slightly improved their performance over trials. Since the time of these original observations we have examined the phenomenon in rats from two other suppliers. Previous findings had indicated to us that rats from seven different sources varied severalfold in their nocturnal rotational behavior (Glick et al., 1986). We were therefore interested in assessing the relative susceptibilities to LH among various strains of rats. The first of these was a Sprague–Dawley strain from Taconic Farms, Germantown, N.Y., U.S.A. Again we saw marked differences between left- and right-biased rats. Only right-biased rats exhibited the LH phenomenon whereas among left-biased rats neither LH-treated nor controls displayed short escape latencies. We have identified this strain as being predominantly right-biased. In another study we employed animals from the Long–Evans hooded strain that we have identified as left-biased. Again we saw that only the right-biased animal exhibited the LH phenomenon. The left-biased animals displayed extremely short shuttlebox escape latencies and the LH group actually performed better than controls.

A comprehensive study using rats from eight different stocks has shown that rats of differing genetic background vary widely in their susceptibility to LH (Wieland et al., 1986). When these findings are viewed in light of data on the rotational behavior of rats from different stocks, some interesting predictions emerge. We have already shown, for example, that male Sprague–Dawley rats from Zivic–Miller, (Pittsburgh, Penn.) that are strongly lateralized show large differences in both reactivity to stress and in sensitivity to uncontrollable stress depending on whether they are left- (insensitive) or right-biased (sensitive). These rats show maximal interindividual differences in sensitivity. Since at the population level there is a tendency for these rats to be left-biased, they are only moderately (i.e., <50%) sensitive to uncontrollable stress effects. We would hypothesize that a very sensitive line of rats (e.g., Charles River Holtzman) would contain many strongly lateralized right rotators. Conversely, in Fisher 344 rats, that are as a group insensitive to uncontrollable stressors (Wieland et al., 1986), strongly lateralized left rotators but weakly lateralized right rotators would be expected, and we have seen just this in preliminary observations (Glick and Greengard, 1980).

The findings outlined above suggest that intrinsic lateralized differences might exist in the brains of susceptible and nonsusceptible rats. We have thus conducted preliminary investigations to determine a lateralized neurochemical basis for the above behavioral effects. Since previous data have indicated that dopamine (DA) systems are lateralized in the rat cortex (Slopsema et al., 1982; Carlson et al., 1988) in a manner that varies among strains (see Rosen et al., 1984) and that DA projections to the medial prefrontal cortex (PFC) are activated upon exposure to mild stressors (Thierry et al., 1976; Roth et al., 1988), we focused on potential lateralized effects on the PFC (mesocortical) DA system (Fitzgerald et al., 1989). We exposed male Long–Evans rats to escapable footshock (ESC), identical inescapable footshock (YOK), or no shock at all (CTL) as in previous studies. Within 30 min of the end of this session, animals were removed from the stress chambers, their brains removed, and the PFC homogenized and analyzed for DA and its acid metabolite 3,4-dihydroxyphenylacetic acid (DOPAC). Figure 10.1 depicts bilateral levels of DA and in the PFC. Here, as seen in previous studies, a left greater than right asymmetry is seen for DA in the CTL group. Dopamine levels did not differ in the ESC group nor did DA asymmetry. A significant bilateral depletion of DA was seen, however, for the YOK group and the left greater than right asymmetry was abolished. When DA activation (as indexed by the ratio DOPAC to DA) was assessed among the groups, a significant elevation was seen in the the YOK right PFC; although both shocked groups showed increased PFC DA activation, this occurred to a greater extent on the RIGHT side in the YOK group (Fig. 10.2).

These findings suggest that lateralized DA activation occurs in projection areas on the right side of the mesocortical system in the LH animal model. This is of some significance in that depression and other affectual diseases in humans are associated with dysfunction of the right side of the brain.

DRUG SENSITIVITY AND ADDICTIVE BEHAVIOR

The influence of brain laterality on drug sensitivity has been relatively unexplored in humans. Some investigators have indicated that drugs such as neuroleptics and antidepressants can have asymmetrical influences across the brain hemispheres.

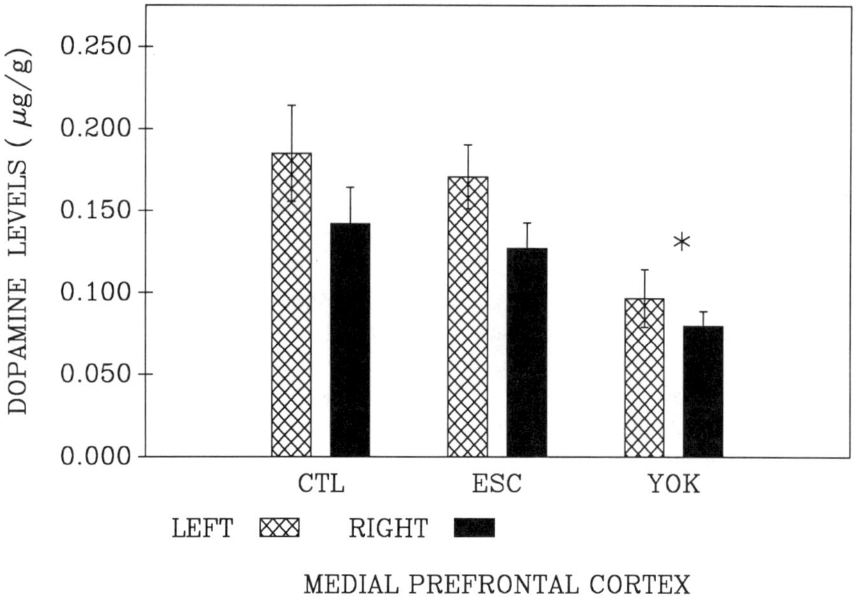

FIGURE 10.1. The effects of uncontrollable footshock stress (YOK) and identical controllable footshock stress (ESC) on dopamine levels on each side of the medial prefrontal cortex. Male Long—Evans rats were exposed to a footshock stressor and within 30 min of the end of the session were sacrificed, their brains were removed and the PFC homogenized and analyzed. A left greater than right asymmetry is seen for DA in the CTL group. DA levels did not differ in the ESC group nor did DA asymmetry. A significant bilateral depletion of DA was however seen for the YOK group and the left greater than right asymmetry was abolished. $^*p < 0.05$ significantly different (bilaterally) from control group.

The lateralized effects of these drugs are consistent with a dampening of the presumably overactive left side found in schizophrenia and of the right side in depression. Evidence can be found from tests of schizophrenics that left hemispheric dominance can be reduced by neuroleptics (Gruzelier, 1983). In normal subjects right hemispheric activation can be induced by the antidepressant drug imipramine (Von Knorring, 1983). There are data from human clinical studies to indicate that the antidepressant actions of tricyclic drugs, monoamine oxidase inhibitors (Fromm and Schopflocher, 1984), and electroconvulsive therapy (Flor-Henry, 1986a) involve changes in brain asymmetry. If psychoactive drugs can affect the two hemispheres asymmetrically, then it might be expected that among individuals, the magnitude of the drug response would vary with the degree and direction of cerebral asymmetry. There have been few studies to explore this idea in humans. In a study of hemispheric differences in drug response, Irwin (1985) found a greater effect of various psychoactive substances on electroencephalographic activation among left handers.

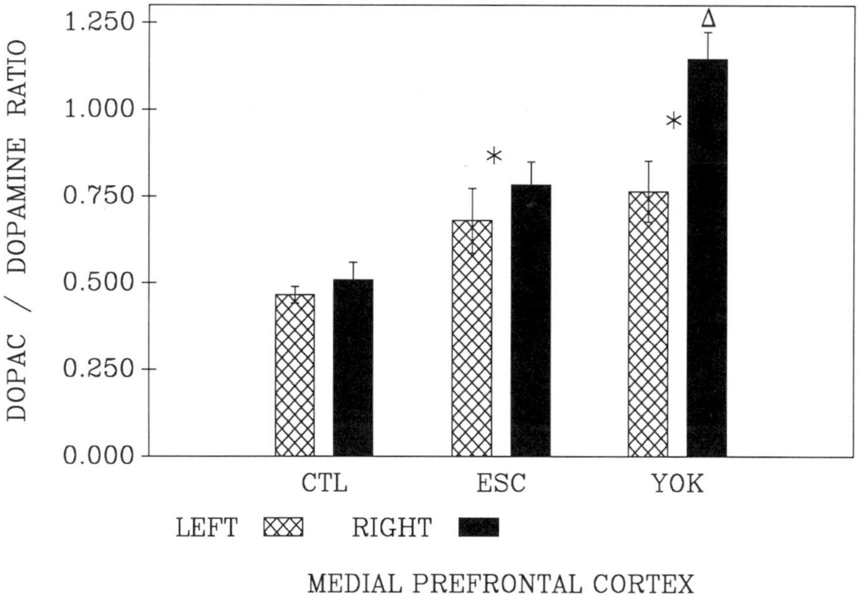

FIGURE 10.2. The effects of uncontrollable footshock stress (YOK) and identical controllable footshock stress (ESC) on the DOPAC/DA ratio on each side of the medial prefrontal cortex. Male Long–Evans rats were exposed to a footshock stressor and within 30 min of the end of the session were sacrificed, their brains were removed, and the PFC homogenized and analyzed. When DA activation (as indexed by DOPAC/DA) was assessed among the groups, a significant elevation was seen in the YOK right PFC; while both shocked (ESC and YOK) groups showed increased (DOPAC/DA), this effect occurred only to a significantly greater extent on the RIGHT side in the YOK group. $*p < 0.05$ significantly different (bilaterally) from control group (CTL). $\Delta P < 0.05$ significantly different from the other side.

Evidence that brain laterality is a key determinant of drug sensitivity in rats is more abundant. Both the degree and direction of asymmetry have been shown to be important in this regard. Lateralized rats have been shown to be more responsive to the locomotor stimulant effects of both d-amphetamine (d-A) and low doses of morphine (Glick and Hinds, 1985). Nonlateralized rats have been shown to be more sensitive to the depressant effects of high doses of morphine. The findings suggest that the magnitude of response to these drugs somehow depends on the asymmetric nature of the neuronal systems upon which they act. That "drugs of abuse" can have asymmetrical effects on behavior, as well as on brain chemistry and function, has been clear for several years. In addition to d-A, other drugs have been shown to induce rotation in naive rats. These include morphine (Morihisa and Glick, 1977), LSD (Fleisher and Glick, 1979), phencyclidine (Glick et al., 1980), and cocaine (Glick et al., 1983). Whereas the rotatory effects of most drugs appear to be directly (d-A, morphine, cocaine) or indirectly

(LSD) attributable to modulation of the striatal DA asymmetry, another mechanism involving the hippocampus appears to be largely responsible for the effect of phencyclidine (Glick et al., 1980); differences between d-A and cocaine may involve the participation of a mesocortical dopamine mechanism.

A relationship between brain asymmetry and the "euphoric" effects of drugs has been implicated by studies of intracranial self-stimulation (ICS). The two sides of the rat brain (medial forebrain bundle) have been shown to be differentially sensitive to reinforcing brain stimulation (Glick et al., 1980) and, in this context, the effects of different classes of abused drugs (e.g., opiates vs. stimulants) have been distinguishable on the basis of their predominant side of action (Glick et al., 1981; Fromm and Schopflocher, 1984; Glick and Badalamenti, 1986). The effects of d-A and morphine were tested on self-stimulation rates across the two sides of the brain. The effects of the two drugs with respect to lateralization were quite different. Although both drugs caused an increase in self-stimulation responding, d-A preferentially affected the low threshold (more rewarding) side of the brain whereas morphine preferentially affected the high threshold side. These differential drug effects were also dependent upon the individual animal's endogenous asymmetry. For d-A, the shift in reward threshold occurred more for the side contralateral to the direction of rotation whereas for morphine, the shift was more on the ipsilateral side. In addition, morphine had effects that depended on dose: low doses preferentially increased responding in the low threshold side and high doses decreased responding on the high threshold side. Clinical observations of addictive behavior have noted individual differences in the tendency to abuse certain classes of drugs selectively. There are also individual differences in introspective reports of response to these drugs. The above findings indicate that individual differences in the relative influence of lateralized reward systems may contribute in determining this behavior. In tests of drug self-administration, behaviorally lateralized and nonlateralized rats were found to be differentially sensitive to d-A and morphine. Lateralized rats self-administered more d-A than did nonlateralized rats, whereas nonlateralized rats self-administered more morphine (Glick and Hinds, 1985). Recent studies have suggested that differences exist between rodent strains in the extent to which they will work to obtain drugs that are abused by humans. Marked genetic differences in the self-administration of and preference for opioids (Carroll et al., 1986) cocaine (George and Goldberg, 1988), and ethanol (Ritz et al., 1986) have been reported. Clinical observations of humans have suggested that there is marked interindividual variability in the tendency to abuse drugs. The above data indicate that a source of these individual differences may be in part related to differences in lateralized brain systems.

STIMULANT SENSITIZATION, STRESSOR REACTIVITY, AND SCHIZOPHRENIA

With repeated administration of stimulants such as d-A or cocaine, rats display an enhanced behavioral response to the stimulant properties of the drug (Robinson and Becker, 1986). This is termed sensitization and it is accompanied by an enhancement of the activity of dopamine systems. Stressors that cannot be controlled (MacLennan and Maier, 1983) have also been shown to potentiate stimulant effects and the relationship between stimulant sensitization and stressor reactivity has been described (Antelman and Chiodo, 1983). Many of the behaviors that are augmented during drug sensitization are reminiscent of those seen in humans with psychotic disease. Overactive dopamine systems are also a hallmark of human pschosis. Stress, which also increases DA activity, substitutes for d-A in the animal model and precipitates psychotic symptoms in schizophrenics (Antelman and Chiodo, 1984; Dohrenwend and Egri, 1981). For these reasons amphetamine sensitization and stressor responsivity in rats have been shown to be useful models of human psychosis.

An important characteristic of schizophrenia is that it involves abnormal brain lateralization. Differences are seen between schizophrenics and normals on such measures as regional cerebral blood flow (Gur et al., 1985), electroencephalography (Flor-Henry and Koles, 1984; Shaw et al., 1983), evoked potentials (Connolly et al., 1985), and sensorimotor laterality (Gureje, 1988). The literature also suggests that schizophrenics exhibit reduced or reversed cerebral lateral dominance (i.e. left- or non–right-handness relative to normal control subjects) (Green et al., 1989; Taylor et al., 1982; Manoach et al., 1988). According to one account (Crow, 1987), the psychoses are viewed as a continuum extending from unipolar through bipolar and schizoaffective disorder to schizophrenia with increasing severity of defect occurring by genetic rearrangements. The locus at which these changes take place is envisaged as related to the genetic determinants of cerebral lateralization (the "cerebral dominance gene"). Lack of concordance for schizophrenia among monozygotic twins is often interpreted as evidence against a genetic basis for schizophrenia. However, in an influential study (Boklage, 1977), unusual brain function asymmetry was shown to be a factor in virtually every appropriately tested case of monozygotic discordant schizophrenia. Recently, other data have appeared to question this phenomenon (Lewis et al., 1989).

Experiments with rodents have focused on the potential contribution of differences in cerebral laterality to the process of stimulant sensitization. It was noted earlier that Sprague–Dawley derived rats from seven different sources vary substantially in their rotational behavior. This work was extended to include sensitivity to d-A and sensitization to d-A–induced rotational behavior upon repeated testing. Rats from three sources were tested twice (a week between tests) for rotation induced by d-A. Rats from two of these sources showed evidence of sensitization, there being significantly greater rotation in response to the second dose than in response to the first dose; the d-A–induced rotational behavior of rats from the third source did not significantly change from one week to the next. However, the latter rats had a greater initial response to the first dose of d-A than did rats from the other two breeders. Further analysis revealed that, among rats from all three breeders, rats rotating weakly in response to d-A on the first test tended to rotate more on the second test whereas rats rotating strongly in response to d-A on the first test tended to rotate less on the second test. Thus, Sprague–Dawley rats derived from three different sources displayed differences in d-A sensitization; to the extent that these differences are reflective of degrees of cerebral laterality, they indicate that it may be an important determinant of individual responsiveness within the model (Glick et al., 1986).

Subsequent to our characterization of these differences among Sprague–Dawley derived rats, we tested Long–Evans rats. Except for very weakly lateralized rats, there was no evidence of sensitization to d-A upon repeated testing. Further testing revealed that these Long–Evans derived (LEd) rats differed in another important respect from the Sprague–Dawley derived (SDd) rats we had used: population biases were sex-dependent for SDd rats but not for LEd rats. That is, the majority (55%) of female SDd rats had right-sided rotational (nocturnal and d-A–induced) preferences (Ross and Glick, 1981b) whereas males as a group tended (52%) to have left-sided biases (Robinson et al., 1985); in contrast, both female and male LEd rats had a preponderance of left-sided biases (62% and 56%, respectively, based on approximately 120 rats of each sex). Previous work had shown that cocaine elicited sex-dependent differences in rotation between left- and right-biased SDd rats (Glick et al., 1983): right-sided females rotated more than left-sided females whereas left-sided males rotated more than right-sided males. In view of the left–right population differences between SDd and LEd rats, we repeated the cocaine experiment in LEd rats: cocaine (20 mg/kg, i.p.) elicited more intense rotation in left- than in right-sided rats of both sexes (Carlson and Glick, 1987). Overall, in both LEd and SDd rats, females rotated more than males in response to d-A and cocaine as well as nocturnally. The differential interactions between sex and side variables in LEd and SDd rats should be very useful in assessing the relative contributions of these variables in determining individual differences in drug sensitivity in future studies. To the extent that stimulant reactivity is a useful animal model of schizophrenia, the findings indicate that genetically defined individual differences among rodents may provide a valuable parallel to human differences.

Schizophrenia in humans has been characterized as a disorder involving the left side of the brain (Gruzelier, 1984). Left brain hyperinvolvement for hypodensity (Reveley et al., 1987), various pathologic changes (Brown et al., 1986), amygdaloid DA (Reynolds, 1983) and glutamatergic dysfunction (Deakin et al., 1989) have been reported. Differences among rodents in stimulant sensitization and stressor reactivity may parallel these phenomena. Male left-sided SDd rats are much more reactive to footshock stressors (Carlson and Glick, 1989) as are left-sided LEd rats of both sexes to the stimulant cocaine (Carlson and Glick, 1987). It has also been shown that rats that show a left rotational bias develop a greater sensitization to d-A and show a greater plasma hormonal response (ACTH and norepinephrine) to stressors (LaHoste et al., 1988). In an animal analog of lateralized amygdaloid DA function, it has been shown that DA infusion into the left but not right amygdala caused hyperactivity that was blocked by neuroleptic drugs (Bradbury et al., 1985). In view of the left–right population differences between strains, as reviewed above, future work to establish strain-related differences in susceptibility within these models would be of interest.

IMMUNE SYSTEM DIFFERENCES AND BEHAVIORAL DYSFUNCTION

The idea that experiences associated with the generation of psychiatric disorders can also modulate the status of the human immune response has received much scientific and public attention of late. Immunocompetence is impaired during schizophrenia, depression, and Alzheimer's disease (Ader, 1981). Laboratory findings suggest numerous interrelationships between brain and immune function: brain lesions (Keller et al., 1980), circulating hypothalamic pituitary axis (HPA) hormones (Munck et al., 1984), classically conditioned stimuli (Cohen and Ader, 1988), and affective state (Schleifer et al., 1985a) all have been shown to have profound effects on immune system function.

A central tenet of the Geshwind and Behan (1982) hypothesis has been an association between left-handedness and immune disorders in man. However, strong evidence that the hypothesis is correct has not been forthcoming from human populations. It has been shown that a significantly higher incidence of left-handedness is observed in people with either Crohn's disease or ulcerative colitis and that males with Type I diabetes are more likely to be left-handed than males with Type II diabetes (Searleman and Fugagli, 1987). Left-handedness has also been associated with reduced longevity (Halpern and Coren, 1988). However, others (Satz and Soper, 1986) have raised concerns regarding the association between left-handedness and autoimmune disease and question its empirical support. Many negative findings have been reported (Satz and Soper, 1986; Schur, 1986; van Strien et al., 1987). Unfortunately, in human studies of this type, handedness has been the only measure of cerebral dominance available to investigators and this may not have been the most appropriate index to measure. For example, it has been pointed out that anomalous cerebral asymmetry may be represented either by an inverted left–right asymmetry or by simply a reduced strength of behavioral asymmetry (Bryden, 1982). Cerebral dominance as measured by means other than handedness has also been associated with immune system pathology. For example, anomalous cerebral asymmetries have been identified using computed tomographic scans in acquired immunodeficiency syndrome (AIDS) patients (Bear et al., 1988).

Animal data have accumulated to suggest that lateralized brain function does play a major role in the regulation of immune status. Brain immunomodulation has been shown to be lateralized by two different experimental approaches. The first is unilateral ablation of the brain cortex in rodents. Several lymphocyte and macrophage functions are depressed after lesion of the left cortex whereas the same functions are unchanged or enhanced after right cortical ablation. The second approach is a behavioral one. In rodents, a lateralized behavior appears to be related to differences in lymphocyte reactivity and the pathogenesis of autoimmunity.

Renoux et al. (1983) were the first to report the existence of a lateralized cortical influence on immune function using animals. In mice, lesions of the left frontoparietal cortex lead to a depression of T-lymphocyte responsiveness whereas lesions of the right frontoparietal cortex lead to an enhancement of the same measure. Since these seminal studies, brain neocortex immunomodulation has been also seen in rats (Barneoud et al., 1987). More recently, this pattern has been reproduced numerous times on immune system measures such as interleukin-2 production (Neveu et al., 1989c), mononuclear phagocyte activation (Neveu et al., 1989b), and spleen lymphocyte DNA synthesis (Barneoud et al., 1988). Functional laterality has also been associated with differences in immune system regulation. Neveu et al. (1989a) studied New Zealand black mice, a model of lupus erythematosus, using the Collins (1985) test of paw preference. It was found that left-sided mice produced pathogenic autoantibodies directed against red blood cells and DNA earlier than right-sided mice. These results provide evidence of a relationship between a functional brain asymmetry and autoantibody production and suggest the involvement of the central nervous system in the pathogenesis of autoimmune processes—precisely the prediction of the Geschwind hypothesis.

The immune system is a major target for stressors and has been shown to be susceptible to psychological factors on numerous occasions (Croiset et al., 1987; Dorian et al., 1982). In animal studies, it has been shown that the lack of control over stressors (as in the LH model reviewed above) results in profound alterations to the immune system: these include a reduced ability to reject tumors (Vinstainer et al., 1982) and a reduction in T-lymphocyte mitogen responsiveness (Schleifer et al., 1985b). Similar phenomena involving diminished immune responsiveness following a lack of stressor control have been shown in bereaved humans (Stein, 1981) and lymphocyte responsiveness has long been known to be impaired in endogenous depression (Leonard, 1988). Given the data reveiwed above suggesting that left- and right-sided rats differ in their susceptibility to stressor-evoked behavioral disturbances in the LH model, it follows that differences in stressor-associated immune suppression might also be seen. Since ablation of the right cortex in rodent species leads to an enhancement of immune system function, it would not be surprising to find that right-sided cortical activation is associated with immunosuppression. We could thus make predictions about individual differences in susceptibility to stressor-evoked immune disruptions based on predispositions involving cerebral laterality.

Evidence that right cortical anomalies may be associated with stressor-evoked immune dysfunction has been reported from clinical populations. Using standardized quantitative tomographic scans, Bear et al. (1988) examined cerebral asymmetries among homosexual men with confirmed AIDS. Significant anomalous frontocortical asymmetries associated with a reversal of the normally greater right frontcortical width were found in these patients. The inability to cope with stressors has been identified as a major predisposing factor in susceptibility to the AIDS virus (Glaser and Kiecolt-Glaser, 1988). It is tempting to speculate that variations associated with anomolous cortical asymmetry, stressor vulnerability, and right cortical activation also contribute to this disease process.

Acknowledgments. This work was supported in part by NIEHS Grant ES04032 (J.N.C.) and NIDA Grant DA03817 (S.D.G.).

REFERENCES

Ader R, ed (1981): *Psychoneuroimmunology*. New York: Academic Press

Amaducci L, Sorbi S, Albanese A, Gainotti G (1981): Choline acetyltransferase (ChAT) activity differs in right and left human temporal lobes. *Neurology* 31:799–805

Anisman H, Zacharko RM (1982): Depression: the predisposing influence of stress. *Behav Brain Sci* 5:89–137

Annett M (1978): Genetic and nongenetic influences on handedness. *Behav Genet* 8:227–249

Annett M (1985): *Left, Right, Hand and Brain: The Right Shift Theory*. Hillsdale, NJ: Erlbaum

Annett M, Kilshaw D (1982): Mathematical ability and lateral asymmetry. *Cortex* 18:547–568

Annett M, Kilshaw D (1984): Lateral preference and skill in dyslexics: implications of the right shift theory. *J Child Psychol Psychiatry* 25:357–377

Annett M, Manning M (1989): The disadvantages of dextrality for intelligence. *Br J Psychol* 80:213–226

Antelman SM, Chiodo LA (1983): Amphetamine as a stressor. In: *Stimulants: Neurochemical, Behavioral and Clinical Perspectives,* Creese I, ed. New York: Raven Press, pp 269–299

Antelman SM, and Chiodo LA (1984): Stress: its effect on interactions among biogenic amines and role in the induction and treatment of disease. In: *Handbook of Psychopharmacology, Vol. 18,* Iversen LL, Iversen SD, Snyder SH, eds. New York: Raven Press, pp 279–341

Arnold AP, Bottjer SW (1985): Cerebral lateralization in birds. In: *Cerebral Lateralization in Nonhuman Species,* Glick SD, ed. Orlando, FL: Academic Press, pp 11–39

Barneoud P, Neveu PJ, Vitiello S, Le Moal M (1988): Early effects of right or left cerebral cortex ablation on mitogen-induced spleen lymphocyte DNA synthesis. *Neurosci Lett* 90:302–307

Barneoud P, Neveu PJ, Vitiello S, Mormede P, Le Moal M (1987): Brain neocortex immunomodulation in rats. *Life Sci* 40:1127–1130

Bear DM, Agostini M, Saporta JA Jr (1988): Anomalous cerebral asymmetries in acquired immunodeficiency syndrome [letter]. *Arch Neurol* 45:248

Becker JB, Robinson TE Lorenz KA (1982): Sex differences and estrous cycle variations in amphetamine-elicited rotational behavior. *Eur J Pharmacol* 80:65–72

Blanc G, Herve D, Simon H, Lisoprawski A, Glowinski J, Tassin JP (1980): Response to stress of mesocortico-frontal dopaminergic neurones in rats after long-term isolation. *Nature* 284:265–267

Bogen JE (1969a): The other side of the brain I: Dysgraphia and dyscopia following cerebral commissurotomy. *Bull LA Neurol Soc* 34:73–105

Bogen JE (1969b): The other side of the brain II: an appositional mind. *Bull LA Neurol Soc* 34:135–162

Bogen JE, Bogen GM (1969): The other side of the brain III: the corpus collossum and creativity. *Bull LA Neurol Soc* 34:191–220

Boklage CE (1977): Schizophrenia, brain asymmetry development, and twinning: cellular relationship with etiological and possibly prognostic implications. *Biol Psychiatry* 12:19–35

Bracha HS, Seitz DJ, Otemaa J, Glick SD (1987a): Rotational movement (circling) in normal humans: sex difference and relationship to hand, foot and eye preference. *Brain Res* 411:231–235

Bracha HS, Shults C, Glick SD, Kleiman JE (1987b): Spontaneous asymmetric circling behavior in hemi-Parkinsonism: a human equivalent of the lesioned-circling rodent behavior. *Life Sci* 40:1127–1130

Bradbury AJ, Costall B, Domeney AM, Naylor RJ (1985): Laterality of dopamine function and neuroleptic action in the amygdala in the rat. *Neuropharmacol* 24:1163–1170

Brass CA, Glick SD (1981): Sex differences in drug-induced rotation in two strains of rats. *Brain Res* 223:229–234

Broca P (1861): Remarques sur le siege de la faculte du langage articule. *Bull Soc d'Anthropol Paris, 2nd series,* 6:398–407

Brown R, Colter N, Corsellis JA, Crow TJ, Frith CD, Jagoe R, Johnstone EC, Marsh L (1986): Postmortem evidence of structural brain changes in schizophrenia differences in brain weight, temporal horn area, and parahippocampal gyrus compared with affective disorder. *Arch Gen Psychiatry* 43:36–42

Bruder GE, Quitkin FM, Stewart JW, Martin C, Voglmaier MM, Harrison WM (1989): Cerebral laterality and depression: differences in perceptual asymmetry among diagnostic subtypes. *J Abnorm Psychol* 98:177–186

Bryden MH (1982): *Laterality, Functional Asymmetry in the Intact Brain.* New York: Academic Press

Carlson JN, Glick SD (1987): Stress- and cocaine-like effects on amphetamine induced rotational behavior cause by the anxiogenic benzodiazepine inverse agonist methyl B-carboline-3-carboxylate. *Soc Neurosci Abstr* 13:218

Carlson JN, Glick SD (1989): Cerebral lateralization as a source of interindividual differences in behavior. *Experientia* 45:788–798

Carlson JN, Glick SD, Hinds PA (1987): Changes in d-amphetamine elicited rotational behavior in rats exposed to uncontrollable footshock stress. *Pharmacol Biochem Behav* 26:17–21

Carlson JN, Glick SD, Hinds PA, Baird JL (1988): Food deprivation alters dopamine utilization in the rat prefrontal cortex and asymmetrically alters amphetamine-induced rotational behavior. *Brain Res* 454:373–377

Carroll ME, Pederson MC, Harrison RG (1986): Food deprivation reveals strain differences in opiate intake of Sprague–Dawley and Wistar rats. *Pharmacol Biochem Behav* 24:1095–1099

Carter CJ, Pycock CJ (1980): Behavioral and neurochemical effects of dopamine and noradrenaline depletion within the medial prefrontal cortex of the rat. *Brain Res* 192:163–176

Carter Saltzman L (1980): Biological and sociocultural effects on handedness: comparison between biological and adoptive families. *Science* 209:1263–1265

Christie JE, Crow TJ (1971): Turning behavior as an index of the action of amphetamine and ephedrines on central dopamine-containing neurons. *Br J Pharmacol* 43:658–667

Christie MJ, James LB, Beart PM (1985): An excitant amino acid projection from the medial prefrontal cortex to the anterior part of nucleus accumbens in the rat. *J Neurochem* 45:477–482

Cohen N, Ader R (1988): Immunomodulation by classical conditioning. *Adv Biochem Psychopharmacol* 44:199–202

Collins RL (1968): On the inheritance of handedness. I. Laterality in inbred mice. *J Hered* 59:9–12

Collins RL (1970): The sound of one paw clapping: an inquiry into the origin of left-handedness. In: *Contributions to Behavior-Genetic Analysis: The Mouse as a Prototype,* Lindzey G, Thiessen D, eds. New York: Appleton, pp 115–136

Collins RL (1975): When left-handed mice live in right-handed worlds. *Science* 187:181–184

Collins RL (1977): Origins of the sense of asymmetry: Mendelian and non-Mendelian models of inheritance. *Ann NY Acad Sci* 299:283–305

Collins RL (1985): On the inheritance of direction and degree of asymmetry. In: *Cerebral Lateralization in Nonhuman Species,* Glick SD, ed. Orlando, FL: Academic Press, pp 41–71

Connolly JF, Manchanda R, Gruzelier JH, Hirsch SR (1985): Pathway and hemispheric differences in the event-related potential (ERP) to monaural stimulation: a comparison of schizophrenic patients with normal controls. *Biol Psychiatry* 20:293–303

Corballis RL, Morgan M (1978a): On the biological basis of human laterality. *Behav Brain Sci* 2:261–336

Corballis RL, Morgan M (1978b): On the biological basis of human laterality. I. Evidence for a maturational left-right gradient. *Behav Brain Sci* 1:261–269

Croiset G, Heijnen CJ, Veldhuis HD, de Wied D, Ballieux RE (1987): Modulation of the immune response by emotional stress. *Life Sci* 40:775–782

Crow TJ (1987): Integrated viral genes as potential pathogens in the functional psychoses. *J Psychiatr Res* 21:479–485

Crowne DP, Richardson CM, Dawson KA (1987): Lateralization of emotionality in right parietal cortex of the rat. *Behav Neurosci* 101:134–138

Davidoff J (1976): Hemispheric sensitivity differences in the perception of colour. *Q J Exp Psychol* 28:387–394

Davidson RJ (1988): EEG measures of cerebral asymmetry: conceptual and methodological issues. *Int J Neurosci* 39:71–89

Davidson RJ, Fox NA (1989): Frontal brain asymmetry predicts infants' response to maternal separation. *J Abnorm Psychol* 98:127–131

Deakin JF, Slater P, Simpson MD, Gilchrist AC, Skan WJ, Royston MC, Reynolds GP, Cross AJ (1989): Frontal cortical and left temporal glutamatergic dysfunction in schizophrenia. *J Neurochem* 52:1781–1786

Denenberg VH, Garbanati J, Sherman DA, Yutzey DA, Kaplan R (1978): Infantile stimulation induces brain lateralization in rats. *Science* 201:1150–1152

Diamond MC, Dowling GA, Johnson RE (1981): Morphological cerebral cortical asymmetry in male and female rats. *Exp Neurol* 71:261–268

Diamond MC, Johnson RE, Ingham CA (1975): Morphological changes in the young, adult and aging rat cerebral cortex, hippocampus, and diencephalon. *Behav Biol* 14:163–174

Diamond MC, Murphy GM Jr, Akiyama K, Johnson RE (1982): Morphologic hippocampal asymmetry in male and female rats. *Exp Neurol* 76:553–565

Direnfeld LK, Albert ML, Volicer L, Langlais PJ, Marquis J, Kaplan E (1984): Parkinson's disease. The possible relationship of laterality to dementia and neurochemical findings. *Arch Neurol* 41:935–941

Dohrenwend BP, Egri G (1981): Recent stressful life events and episodes of schizophrenia. *Schizophr Bull* 7:12–23

Dorian B, Garfinkel P, Brown G, Shore A, Gladman D, Keystone E (1982): Aberrations in lymphocyte subpopulations and function during psychological stress. *Clin Exp Immunol* 50:132–138

Drew KL, Lyon RA, Titeler M, Glick SD (1986): Asymmetry in D-2 binding in female rat striata. *Brain Res* 363:192–195

Dunford M, Kimura D (1971): Right hemisphere specialization for depth perception reflected in visual field differences. *Nature* 231:394–395

Ettlinger G, Moffett A (1964): Lateral preferences in the monkey. *Nature* 204:606–607

Fitzgerald LW, Keller RW, Glick SD, Carlson JN (1989): The effects of stressor controllability on regional changes in mesocorticolimbic dopamine activity. *Soc Neurosci Abstr* 15:1316

Fleisher LN, Glick SD (1979): Hallucinogen-induced rotational behavior in rats. *Psychopharmacol* 62:193–200

Fleming DE, Anderson RH, Rhees RW, Kinghorn E, Bakaitis J (1986): Effects of prenatal stress on sexually dimorphic asymmetries in the cerebral cortex of the male rat. *Brain Res Bull* 16:395–398

Flor-Henry P (1986a): Electroconvulsive therapy and lateralized affective systems. *Ann NY Acad Sci* 462:389–397

Flor-Henry P (1986b): Observations, reflections and speculations on the cerebral determinants of mood and on the bilaterally asymmetrical distributions of the major neurotransmitter systems. *Acta Neurol Scand* 74 (suppl 109):75–89

Flor-Henry P, Koles ZJ (1984): Statistical quantitative EEG studies of depression, mania, schizophrenia and normals. *Biol Psychol* 19:257–279

Fonnum F (1984): Glutamate: a neurotransmitter in mammalian brain. *J. Neurochem* 42:1–11

Forward E, Warren JM, Hara K (1962): The effects of unilateral lesions in sensorimotor cortex on manipulation by cats. *J Comp Physiol Psychol* 55:1130–1135

Fromm D, Schopflocher D (1984): Neuropsychological test performance in depressed patients before and after drug therapy. *Biol Psychiatry* 19:55–72

Galaburda AM, LeMay M, Kemper TL, Geschwind N (1978): Right–left asymmetries in the brain. *Science* 199:852–856

Geffen G, Bradshaw JL, Wallace G (1971): Interhemispheric effects on reaction time to verbal and nonverbal stimuli. *J Exp Psychol* 87:415–422

George FR, Goldberg SR (1988): Genetic differences in responses to cocaine. *NIDA Res Monogr* 88:239–249

Gerendai I (1984): Lateralization of neuroendocrine control. In: *Cerebral Dominance: The Biological Foundations,* Geschwind N, Galaburda AM, eds. Cambridge, MA: Harvard University Press, pp 167–178

Geschwind N (1976): The apraxis: neural mechanisms of disorders of learned movement. *Am Scientist* 63:188–195

Geschwind N (1985): Implications for evoloution, genetics and clinical syndromes. In: *Cerebral Lateralization in Nonhuman Species,* Glick SD, ed. Orlando, FL: Academic Press, pp 247–278

Geschwind N, Behan P (1982): Left-handedness: association with immune disease, migrane, and developmental learning disorder. *Proc Natl Acad Sci USA* 79:5097–5100

Geschwind N, Galaburda AM (1987): *Cerebral Lateralization Biological Mechanisms, Associations, and Pathology*. Cambridge, MA: Bradford/MIT Press

Geschwind N, Levitsky W (1968): Human brain: left–right asymmetries in temporal speech region. *Science* 161:186–187

Glaser R, Kiecolt-Glaser JK (1988): Stress-associated immune suppression and acquired immune deficiency syndrome. In: *Psychological, Neuropsychiatric, and Substance*

Abuse Aspects of AIDS, Bridge PK, Mirsky AF, Goodwin FK, eds. New York: Raven Press, pp 203–215
Glick SD (1973): Enhancement of spatial preferences by (+) -amphetamine. *Neuropharmacology* 12:43–47
Glick SD (1983a): Heritable determinants of left-right bias in the rat. *Life Sci* 32:2215–2221
Glick SD (1983b): Cerebral lateralization in the rat and tentative extrapolations to man. In: *Hemisyndromes: Psychobiology, Neurology, Psychiatry,* Myslobodsky MS, ed. New York: Academic Press, pp 7–26
Glick SD (1985): Heritable differences in turning behavior of rats. *Life Sci* 36:499–503
Glick SD, Badalamenti JI (1986): Sex difference in reward asymmetry and effects of cocaine. *Neuropharmacology* 25:633–637
Glick SD, Cox RD (1976): Differential effects of unilateral and bilateral caudate lesions on side preferences and timing behavior in rats. *J Comp Physiol Psychol* 90:528–535
Glick SD, Cox RD (1978): Nocturnal rotation in normal rats: correlation with amphetamine-induced rotation and effects of nigrostriatal lesions. *Brain Res* 150:149–161
Glick SD, Greengard O (1980): Exaggerated cerebral lateralization in rats after early postnatal hyperphenylalaninemia. *Brain Res* 202:243–248
Glick SD, Greenstein S (1973): Possible modulating influence of frontal cortex on nigro-striatal function. *Br J Pharmacol* 49:316–321
Glick SD, Hinds PA (1985): Differences in amphetamine and morphine sensitivity in lateralized and non-lateralized rats: locomotor activity and drug self-administration. *Eur J Pharmacol* 118:239–244
Glick SD, Hinds PA, Shapiro RM (1983): Cocaine-induced rotation: Sex-dependent differences between left- and right-sided rats. *Science* 221:775–777
Glick SD, Jerussi TP (1974): Spatial and paw preferences in rats: their relationship to rate-dependent effects of d-amphetamine. *J Pharmacol Exp Ther* 188:714–725
Glick SD, Jerussi TP, Water DH, Green JP (1974): Amphetamine-induced changes in striatal dopamine and acetylcholine levels and relationship to rotation (circling behavior) in rats. *Biochem Pharmacol* 23:3223–3225
Glick SD, Meibach RC, Cox RD, Maayani S (1979): Multiple and interrelated functional asymmetries in the rat brain. *Life Sci* 25:395–400
Glick SD, Meibach RC, Cox RD, Maayani SS (1980): Phencyclidine-induced rotation and hippocampal modulation of nigrostriatal asymmetry. *Brain Res* 196:99–107
Glick SD, Ross DA (1981a): Lateralization of function in the rat brain: basic mechanisms may be operative in humans. *Trends Neurosci* 4:196–199
Glick SD, Ross DA (1981b): Right-sided population bias and lateralization of activity in normal rats. *Brain Res* 205:222–225
Glick SD, Ross DA, Hough LB (1982a): Lateral asymmetry of neurotransmitters in human brain. *Brain Res* 234:53–63
Glick SD, Shapiro RM (1985): Functional and neurochemical mechanisms of cerebral lateralization in rats. In: *Cerebral Lateralization in Nonhuman Species,* Glick SD, ed. Orlando, FL: Academic Press, pp 157–183
Glick SD, Shapiro RM, Drew KL, Hinds PA, Carlson JN (1986): Differences in spontaneous and amphetamine-induced rotational behavior, and in sensitization to amphetamine, among Sprague–Dawley derived rats from different sources. *Physiol Behav* 38:67–70
Glick SD, Weaver LM, Meibach RC (1980): Lateralization of reward in rats: differences in reinforcing thresholds. *Science* 207:1093–1095

Glick SD, Weaver LM, Meibach RC (1981): Amphetamine enhancement of reward asymmetry. *Psychopharmacology* 73:323–327

Glick SD, Weaver LM, Meibach RC (1982b): Asymmetrical effects of morphine and naloxone on reward mechanisms. *Psychopharmacology* 78:219–224

Gordon HW (1970): Hemispheric asymmetries in the perception of musical cords. *Cortex* 6:387–398

Green MF, Satz P, Smith C, Nelson L (1989): Is there atypical handedness in schizophrenia? *J Abnorm Psychol* 98:57–61

Gruzelier JH (1983): A critical assessment and integration of lateral asymmetries in schizophrenia. In: *Hemisyndromes, Psychobiology, Neurology, Psychiatry,* Myslobodsky MS, ed. London: Academic Press, pp 265–326

Gruzelier JH (1984): Hemispheric imbalances in schizophrenia. *Int J Psychophysiol* 1:227–240

Gur RE, Gur RC, Skolnick BE, Caroff S, Obrist WD, Resnick S, Reivich M (1985): Brain function in psychiatric disorders. III. Regional cerebral blood flow in unmedicated schizophrenics. *Arch Gen Psychiatry* 42:329–334

Gureje O (1988): Sensorimotor laterality in schizophrenia: which features transcend cultural influences? *Acta Psychiatr Scand* 77:188–193

Halpern DF, Coren S (1988): Do right-handers live longer? [letter]. *Nature* 333:213–213

Hansing RA, Schwartzbaum JS, Thompson JB (1968): Operant behavior following unilateral and bilateral caudate lesion in the rat. *J Comp Physiol Psychol* 66:378–388

Herman JP, Guillonneau D, Dantzer R, Scatton B, Semerdjian Rouquier L, Le Moal M (1982): Differential effects of inescapable footshocks and of stimuli previously paired with inescapable footshocks on dopamine turnover in cortical and limbic areas of the rat. *Life Sci* 30:2207–2214

Irwin P (1985): Greater brain response of left-handers to drugs. *Neuropsychologia* 23:61–67

Ivanyi P (1978): Some aspects of the H-2 system, the major histocompatibility system in the mouse. *Proc Roy Soc Lond [Biol]* 202:117–158

Jerussi TP, Glick SD (1974): Amphetamine-induced rotation in rats without lesions. *Neuropharmacology* 13:283–286

Jerussi TP, Glick SD (1976): Drug-induced rotation in rats without lesions: behavioral and neurochemical indices of a normal asymmetry in nigro-striatal function. *Psychopharmacology (Berlin)* 47:249–260

Jerussi TP, Glick SD, Johnson CL (1977): Reciprocity of pre- and postsynaptic mechanisms involved in rotation as revealed by dopamine metabolism and adenylate cyclase stimulation. *Brain Res* 129:385–388

Keller SE, Stein M, Camerino MS, Schleifer SJ, Sherman J (1980): Suppression of lymphocyte stimulation by anterior hypothalamic lesions in the guinea pig. *Cell Immunol* 52:334–340

Kimura D (1964): Left-right differences in the perception of melodies. *Q J Exp Psychol* 16:355–358

Kimura D (1969): Spatial localization in left and right visual fields. *Can J Psychol* 23:445–458

Knapp S, Mandell AJ (1980): Lithium and chlorimipramine differentially alter bilateral asymmetry in mesostriatal serotonin metabolites and kinetic confirmations of midbrain tryptophan hydroxylase with respect to tetrahydrobiopterin cofactor. *Neuropharmacology* 19:1–7

LaHoste GJ, Mormède P, Rivet J-M, Le Moal M (1988): Differential sensitization to amphetamine and stress responsivity as a function of inherent laterality. *Brain Res* 453:381–384

LeMay M (1976): Morphological cerebral asymmetries of modern man, fossil man, and non-human primate. *Ann NY Acad Sci* 280:349–366

LeMay M (1985): Asymmetries of the brains and skulls of nonhuman primates. In: *Cerebral Lateralization in Nonhuman Species,* Glick SD, ed. Orlando, FL: Academic Press, pp 233–245

LeMay M, Geschwind N (1975): Hemispheric differences in the brains of great apes. *Brain, Behav Evol* 11:48–52

Lenhart RE, Katkin ES (1986): Psychophysiological evidence for cerebral laterality effects in a high-risk sample of students with subsyndromal bipolar depressive disorder. *Am J Psychiatry* 143:602–607

Leonard BE (1988): Depression and the immune system. *Human Psychopharmacol* 3:157–158

Levine MS, Ferguson N, Kreinick CJ, Gustafson JW, Schwartzbaum JS (1971): Sensorimotor dysfunctions and aphagia and adipsia following pallidal lesions in rats. *J Comp Physiol Psychol* 77:282–293

Levy J (1976): A review of the evidence for a genetic component of handedness. *Behav Genetics* 6:429–449

Levy J (1977): The mamallian brain and the adaptive advantage of cerebral asymmetry. *Ann NY Acad Sci* 299:264–272

Lewis SW, Chitkara B, Reveley AM (1989): Hand preference in psychotic twins. *Biol Psychiatry* 25:215–221

Lipp HP, Collins RL, Nauta WJ (1984): Structural asymmetries in brains of mice selected for strong lateralization. *Brain Res* 310:393–396

MacLennan AJ, Maier SF (1983): Coping and the stress-induced potentiation of stimulant stereotype in the rat. *Science* 219:1091–1093

Maier SF, Seligman MEP (1976): Learned helplessness: theory and evidence. *J Exper Psychol Gen* 105:3–46

Manoach DS, Maher BA, Manschreck TC (1988): Left-handedness and though disorder in the schizophrenias. *J Abnorm Psychol* 97:97–99

McClusky NJ, Naftolin F (1981): Sexual differentiation of the central nervous system. *Science* 211:1294–1303

Meisel RL, Ward IL (1981): Fetal female rats are masculinized by male littermates located caudally in the uterus. *Science* 213:239–242

Milner B (1971): Interhemispheric differences in the the localization of psychological processes in man. *Br Med Bull* 27:272–277

Mittleman G, Fray PJ, Valenstein ES (1985): Asymmetry in the effects of unilateral 6-OHDA lesions on eating and drinking evoked by hypothalamic stimulation. *Behav Brain Res* 15:263–267

Morihisa JM, Glick SD (1977): Morphone-induced rotation (circling behavior) in rats and mice: species differences, persistence of withdrawal-induced rotation and antagonism by naloxone. *Brain Res* 123:180–187

Munck A, Guyre PM, Holbrook NJ (1984): Physiological functions of glucocorticoids in stress and their relation to pharmacological actions. *Endocr Rev* 5:25–44

Neveu PJ, Barneoud P, Georgiades O, Vitiello S, Vincendeau P, Le Moal M (1989b): Brain neocortex influence on the mononuclear phagocyte system. *J Neurosci Res* 22:188–193

Neveu PJ, Barneoud P, Vitiello S, Kelley KW, Le Moal MA (1989c): Brain neocortex modulation of mitogen-induced interleukin 2, but not interleukin 1, production. *Immunol Lett* 21:307–310

Neveu PJ, Betancur C, Barneoud P, PreudHomme JL, Aucouturier P, LeMoal M, Vitiello S (1989a): Functional brain asymmetry and murine systemic lupus erythematosus. *Brain Res* 498:159–162

Nottebohm F (1977): Asymmetries of neural control of vocalization in the canary. In: *Lateralization In the Nervous System,* Harnad S, Doty RW, Goldstein L, Jaynes J, Krauthamer G, eds. New York: Academic Press, pp 23–44

Oke A, Keller RW Jr, Mefford I, Adams RN (1978): Lateralization of norepinephrine in the human thalamus. *Science* 200:397–400

Oke A, Lewis R, Adams RN (1980): Hemispheric asymmetry of norepinephrine distribution in rat thalamus. *Brain Res* 188:269–272

Otto MW, Dougher MJ, Yeo RA (1989): Depression, pain, and hemispheric activation. *J Nerv Ment Dis* 177:210–218

Pearlson GD, Robinson RG (1981): Suction lesions of the frontal cerebral cortex in the rat induce asymmetrical behavioral and catecholaminergic responses. *Brain Res* 218:233–242

Peterson GM (1934): Mechanisms of handedness in the rat. *Comp Psychol Monogr* 9:1–67

Peterson GM, Barnett PE (1961): The cortical destruction necessary to produce a transfer of a forced-practice function. *J Comp Physiol Psychol* 54:382–385

Peterson GM, Devine JV (1963): Transfers in handedness in the rate resulting from small cortical lesions after limited forced practice. *J Comp Physiol Psychol* 56:752–756

Pfeifer RA (1936): Pathologie der Horstrahlung und der corticaler Horsphare. In: *Handbuch der Neurologie,* Vol. 6, Bumke O, Forster O, eds. Berlin: Springer–Verlag, pp 534–626

Renoux G, Biziere K, Renoux M, Guillaumin JM, Degenne D (1983): A balanced brain asymmetry modulates T cell-mediated events. *J Neuroimmunol* 5–227–238

Reveley MA, Reveley AM, Baldy R (1987): Left cerebral hemisphere hypodensity in discordant schizophrenic twins. A controlled study. *Arch Gen Psychiatry* 44:625–632

Reynolds GP (1983): Increased concentrations and lateral asymmetry of amygdala dopamine in schizophrenia. *Nature* 305:527–529

Ritz MC, George FR, deFiebre CM, Meisch RA (1986): Genetic differences in the establishment of ethanol as a reinforcer. *Pharmacol Biochem Behav* 24:1089–1094

Robinson RG (1979): Differential behavioral and biochemical effects of right and left hemispheric cerebral infarction in the rat. *Science* 205:707–710

Robinson RG (1985): Lateralized behavioral and neurochemical consequences of unilateral brain injury in rats. In: *Cerebral Lateralization in Nonhuman Species,* Glick SD, ed. Orlando, FL: Academic Press, pp 135–156

Robinson RG, Chait RM (1985): Emotional correlates of structural brain injury with particular emphasis on post-stroke mood disorders. *CRC Crit Rev Clin Neurobiol* 1:285–318

Robinson TE, Becker JB (1983): The rotational behavior model: asymmetry in the effects of unilateral 6-OHDA lesions of the substantia nigra in rats. *Brain Res* 264:127–131

Robinson TE, Becker JB (1986): Enduring changes in brain and behavior produced by chronic amphetamine administration: a review and evaluation of animal models of amphetamine psychosis. *Brain Res Rev* 11:157–198

Robinson TE, Becker JB, Camp DM, Mansour A (1985): Variation in the pattern of behavioral and brain asymmetries due to sex differences. In: *Cerebral Lateralization in Nonhuman Species,* Glick SD, ed. Orlando, FL: Academic Press, pp 185–231

Robinson TE, Becker JB, Presty SK (1982): Long-term facilitation of amphetamine-induced rotational behavior and striatal dopamine release produced by a single exposure to amphetamine: sex differences. *Brain Res* 253:231–241

Robinson TE, Becker JB, Ramirez VD (1980): Sex differences in amphetamine-elicited rotational behavior and the lateralization of striatal dopamine in rats. *Brain Res Bull* 5:539–545

Rosen GD, Finklestein S, Stoll AL, Yutzey DA, Denenberg VH (1984): Neurochemical asymmetries in the albino rat's cortex, striatum and nucleus accumbens. *Life Sci* 34:1143–1148

Ross DA, Glick SD (1981a): Lateralized effects of bilateral frontal cortex lesions in rats. *Brain Res* 210:379–382

Ross DA, Glick SD (1981b): Lateralized effects of bilateral frontal cortex lesions in rats. *Brain Res* 210:379–382

Ross DA, Glick SD, Meibach RC (1981): Sexually dimorphic brain and behavioral asymmetries in the neonatal rat. *Proc Natl Acad Sci USA* 78:1958–1961

Roth RH, Tam S-Y, Ida Y, Yang J-X, Deutch AY (1988): Stress and the mesocorticolimbic dopamine systems. *Ann NY Acad Sci* 537:138–147

Rothman AH, Glick SD (1976): Differential effects of unilateral and bilateral caudate lesions on side preference and passive avoidance behavior in rats. *Brain Res* 118:361–369

Sackeim HA (1986): The efficacy of electroconvulsive therapy. Discussion of Part I. *Ann NY Acad Sci* 462:70–75

Sackeim HA, Gur RC, Saucy MC (1978): Emotions are expressed more intensely on the left side of the face. *Science* 202:434–436

Satz P, Soper HV (1986): Left-handedness, dyslexia, and autoimmune disorder: a critique. *J Clin Exp Neuropsychol* 8:453–458

Schaffer CE, Davidson RJ, Saron C (1983): Frontal and parietal electroencephalogram asymmetry in depressed and nondepressed subjects. *Biol Psychiatry* 18:753–762

Schleifer SJ, Keller SE, Siris SG, Davis KL, Stein M (1985a): Depression and immunity. *Lymphocyte function in ambulatory depressed patients, hospitalized schizophrenic patients, and patients hospitalized for herniorrhaphy. Arch Gen Psychiatry* 42:129–133

Schleifer SJ, Keller SE, Stein M (1985b): Stress effects on immunity. *Psychiatr J Univ Ottawa* 10:125–131

Schneider LH, Murphy RB, Coons EE (1982): Lateralization of striatal dopamine (D2) receptors in normal rats. *Neurosci Lett* 33:281–284

Schur PH (1986): Handedness in systemic lupus erythematosus. *Arthritis Rheum* 29:419–420

Searleman A, Fugagli AK (1987): Suspected autoimmune disorders and left-handedness: evidence from individuals with diabetes. Crohn's disease and ulcerative colitis. *Neuropsychologia* 25:367–374

Shapiro RM, Glick SD, Hough LB (1986): Striatal dopamine uptake asymmetries and rotational behavior in unlesioned rats: revising the model. *Psychopharmacol* 89:25–30

Shaw JC, Colter N, Resek G (1983): EEG coherence, lateral preference and schizophrenia. *Psychol Med* 13:299–306

Sherman GF, Galaburda AM, Geschwind N (1982): Neuroanatomical asymmetries in nonhuman species. *Trends Neurosci* 5:429–431

Slopsema JS, Van der Gugten J, De Bruin JPC (1982): Regional concentrations of noradrenaline and dopamine in the frontal cortex of the rat: dopaminergic innervation of the prefrontal subareas and lateralization of prefrontal dopamine. *Brain Res* 250:197–200

Stein M (1981): A biopsychosocial approach to immune function and medical disorders. *Psychiatr Clin North Am* 4:203–221

Taylor P, Dalton R, Fleminger JJ (1982): Handedness and schizophrenic symptoms. *Br J Med Psychol* 55:287–291

Therrien BA, Camp DM, Robinson TE (1982): Sex differences in the effects of unilateral hippocampal lesions on spatial learning. *Soc Neurosci Abst* 8:312–312

Thierry AM, Tassin JP, Blanc G, Glowinski J (1976): Selective activation of the mesocortical DA system by stress. *Nature* 263:242–244

Tsai L, Maurer S (1930): Right handedness in white rats. *Science* 72:436–438

Tucker DM (1984): Lateral brain function in normal and disordered emotion: interpreting electroencephalographic evidence. *Biol Psychol* 19:219–235

Ungerstedt U (1971a): Postsynaptic supersensitivity after 6-hydroxydopamine induced degeneration of the nigro-striatal dopamine system. *Acta Physiol Scand* 367 (suppl):69–93

Ungerstedt U (1971b): Striatal dopamine release after amphetamine or nerve degeneration revealed by rotational behavior. *Acta Physiol Scand* 367 (suppl):49–68

Ungerstedt U, Arbuthnott G (1970): Quantitative recording of rotational behavior in rats after 6-hydroxydopamine lesions of the nigro-striatal dopamine system. *Brain Res* 24:485–493

Van Strien JW, Bouma A, Bakker DJ (1987): Birth stress, autoimmune disease and handedness. *J Clin Exp Neuropsychol* 9:775–780

Vinstainer MA, Volpicelli JR, Seligman MEP (1982): Tumor rejection in rats after inescapable or escapable shock. *Science* 216:437–440

Vom Saal FS, Bronson FH (1980): Sexual characteristics of adult female mice are correlated with their blood testosterone levels during prenatal development. *Science* 208:597–599

Von Knorring L (1983): Interhemispheric EEG differences in affective disorders. In: *Laterality and Psychopathology*, Flor-Henry P, Gruzelier J, eds. Amsterdam, Holland: Elsevier-North Holland, pp 315–338

Ward R, Collins RL (1985): Brain size and shape in strongly and weakly lateralized mice. *Brain Res* 328:243–249

Warren JM (1953): Handedness in the rhesus monkey. *Science* 118:622–623

Warren JM, Abplanalp JM, Warren HB (1967): The development of handedness in cats and rhesus monkeys. In: *Early Behavior: Comparative and Developmental Approaches*, Stevenson HW, Hess EH, Rheingold H, eds. New York: Wiley, pp 342–359

Waziri R (1980): Lateralization of neuroleptic-induced dyskinesia indicates pharmacological asymmetry in the brain. *Psychopharmacology* 68:51–53

Wieland S, Boren JL, Consroe PF, Martin A (1986): Stock differences in the susceptibility of rats to learned helplessness training. *Life Sci* 39:937–944

Witelson SF (1976): Sex and the single hemisphere: specialization of the right hemisphere for spatial processing. *Science* 193:425–427

Zangwill OL (1960): *Cerebral Dominance And Its Relation To Psychological Function*. Edinburgh: Oliver and Boyd

Zimmerberg S, Glick SD, Jerussi TP (1974): Neurochemical correlate of a spatial preference in rats. *Science* 185:623–625

Absence of the Corpus Callosum

HANS-PETER LIPP AND DOUGLAS WAHLSTEN

THE CORPUS CALLOSUM

Do we really need a corpus callosum (CC)? Historical answers to this problem have been rather diverse. The corpus callosum has been considered as the seat of the soul—because it decays last during post mortem brain autolysis (Colonnier, 1986). Followers of the idea that the soul resides in the ventricular fluid (Sömmering, 1796) were more inclined to believe that the CC was stringing together the hemispheres. The first modern view of structure and function of the CC was presented by the founder of both phrenology and modern neuropsychology, Franz Josef Gall (Gall and Spurzheim, 1810). He conceived of the CC as the main, but not the sole, fiber tract connecting subsystems located in the gray matter, and he proposed that the functions of commissural fiber tracts were the integration of processes taking place in various homotopic areas in order to perform behavioral and mental acts. Yet, he did not conceive of the hemispheres as independent brains. Rather, he assumed that homotopic locations were handling different aspects of a task requiring both hemispheres (Gall, 1979).

One-hundred-eighty years of brain research have not added much to this insight, as evidenced by a catalog of questions published by a leading expert in the field (Gazzaniga et al., 1989). Undeniably, there is a wealth of information describing the neuroanatomy of the CC and the behavioral consequences of surgical or natural lesions. For unknown reasons, the neurophysiology of callosal interactions has received comparatively less attention. Consequently, most of the functional information is inferred from the study of callosal malfunction. This source of information has its inherent limitations, however: behavioral sequelae of callosal damage reflect predominantly a disconnection syndrome of two hemispheres that have developed jointly and interactively. Thus, one must deduce normal function from patterns of malfunction, a heuristically undesirable situation.

Apart from neurophysiology, the two remaining possibilities for studying the relation between structure and function of the CC are (1) the investigation of normal variations of the CC and (2) studies in individuals showing congenital absence of the CC. In this chapter, we shall briefly summarize the current knowledge of the CC and will concentrate on the information provided by callosal agenesis in humans and mouse models. Our goal is to give neurologists and neuropsychologists enough information in order to evaluate the applicability of mouse models of callosal agenesis for studying normal and abnormal aspects of callosal function. In addition, we suggest some future lines of research in which mouse models of callosal agenesis might prove to be useful.

Normal Anatomy of the CC

The CC of adult, placental mammals connects homotopic or mirror-image locations in the two hemispheres, as well as heterotopic regions. Consequently, it is both a commissure and a decussation. Furthermore, it contains commissural fibers of the claustrum (Crescimanno et al., 1989) and many fibers of passage that connect the cerebral cortex with ipsilateral thalamus, caudate nucleus, septum, and hypothalamus (Hankin and Silver, 1988). Although synaptic terminations of CC axons are widely and diffusely distributed in the cerebral cortex in neonatal mice, rats, hamsters, and cats, many of these axons perish postnatally (Berbel and Innocenti, 1988; Hedin-Pereira et al., 1988), and eventually there are zones with very few CC connections interspersed among regions with dense callosal input, for example, the border between visual areas 17 and 18 (Innocenti, 1986). The adult human brain may contain at least 200 million callosal fibers (O'Kusky et al., 1988). Counts in other species have revealed about 56 million axons in the old world monkey *Macaca mulatta* (LaMantia and Rakic, 1990), 32 million in cats (Berbel and Innocenti, 1988), and about 11 million in rats (Gravel et al., 1990). It has been noted that the numbers of axons in the CC decline greatly a few weeks after birth in the rhesus monkey, for example (LaMantia and Rakic, 1990), but laborious axon counts in developing rats have revealed asymptotic growth rather than decline (Gravel et al., 1990). The patterns of intercortical connections of the CC are known to be modifiable by early postnatal experience, and this shaping is thought to contribute to adaptive plasticity in advanced mammals (Killackey, 1990). Until very recently, the neurotransmitter systems involved in intercortical communication were not known. Studies combining tract tracing with monoclonal antibodies to neurotransmitters have now demonstrated that some but not all callosal neurons stain for glutamate or somatostatin (Elberger and Honig, 1990). Thus, what is commonly termed the corpus callosum is, in fact, a collection of highly diverse and dynamic axon tracts.

Putative Functions of the Corpus Callosum

Today, the normal functions of the CC are virtually unknown. As reviewed by Cook (1986), the neurophysiological actions of callosal activation include a first burst of homotopic excitation, followed by a relatively long period of inhibition. The same author classifies functional theories into four groups. The *homotopic excitatory theory* (Berlucchi, 1983) holds that homotopic modules primarily symmetrize neuronal activity in the two hemispheres by conveying topographically bound information to the other side. In this case, functional lateralization must arise from genuine functional asymmetries in the two hemispheres. The *diffuse inhibitory theory* (Denenberg, 1981) assumes a relatively widespread contralateral suppression of neuronal activity. In this model, hemispheric specialization is thought to originate from rather subtle functional and morphological asymmetries that are shaped and reinforced by callosal activity, the hemisphere better suited for a given task paralyzing large regions of the other one in order to handle its proper task without disturbance. A third model (Guiard, 1980) has been classified by Cook as *nonspecific excitatory*. Here, the CC is thought to excite homotopic modules not for conveying specific information but rather to alert them, a form of transcallosal "arousal." Based on the topographical concepts of Doty et al. (1986), Gazzaniga (1989), and others, Cook (1986) proposes a *topographical inhibitory* model that assumes contralateral inhibition of cortical modules, resulting in topographically organized complementary specialization of homotopic regions. In this model, the phenomenon of cerebral lateralization is attributed predominantly to callosal fibers that shape small, genuine functional asymmetries of the hemispheres. For example, left modules may be concerned with the elaboration of phonemes, whereas cortical modules in approximate homotopic position handle the paralexical and contextual aspects of the left-hemispheric phonemes.

Normal Variations of the Corpus Callosum

Form and size of the human CC vary greatly and show substantial genetic influence, as evident from comparisons of monozygotic twins (Gazzaniga, 1989; Oppenheim et al., 1989) and from differences among inbred mouse strains (Rosen et al., 1990; Wahlsten, 1989a). There has been considerable dispute about possible sex differences in CC size and shape in humans. After initial reports that the most posterior fifth (splenium) of the CC is larger and more bulbous in women than men (de Lacoste-Utamsing and Holloway, 1982; de Lacoste et al., 1986; Holloway and de Lacoste, 1986), several attempts to replicate this finding with improved anatomical methods and larger samples were undertaken. As reviewed by Clarke et al. (1989) and Peters (1988), there has been little or no support for the original findings. Clarke et al. (1989) did find that females in their sample had a slightly more bulbous splenium of the CC, but the effect was closely related to brain size, which tends to be larger in males. Any sex difference in the CC must be exceedingly small, and links to possible sex differences in hemispheric specialization must be tenuous, especially in view of recent challenges to the interpretation of studies of lateralization of neural function (Efron, 1990). Small sex differences are sometimes reported in rats (e.g., Berrebi et al., 1988), but they do not always appear in mice (Bulman-Fleming and Wahlsten, 1988; Cassells et al., 1990; Lipp et al., submitted for publication).

Another line of research has tried to link size of the CC to handedness and speech. Witelson (1985, 1989) reported a comparatively larger CC (mostly in its caudal portions in front of the splenium) in mixed-handers, particularly males, the data being obtained in a study in which volunteer cancer patients were tested prior to death. Conflicting results, however, have been reported by others using magnetic resonance imaging (MRI) (Nasrallah et al., 1986a; O'Kusky et al., 1988; see also Witelson, 1989). At present, the dispute has not been resolved. The increasing facilities for using MRI, however, might clarify the issue in the near future. O'Kusky et al. (1988) reported a comparatively larger CC, mainly in its middle portion, in patients with right-hemisphere cerebral dominance. Thus far, this study has not been replicated.

AGENESIS OF THE CORPUS CALLOSUM

Lack of a CC does not necessarily mean abnormality. Among the vertebrates, only placental mammals typically possess a CC. In such distinguished animals as the platypus, the opossum, and the spiny anteater, direct communication between the cerebral hemispheres is possible only via axons that cross through the anterior commissure (AC). In one group of marsupials (diprotodonts) these axons enter the posterior branch of the AC via the internal capsule, whereas in another branch (polyprotodonts; also monotremes) they travel via the external capsule (Johnson, 1977). The anatomy of the cerebral commissures is thus a valuable indicator of phylogeny. Although there may be no CC in the opossum, for example, the topographical distribution of interhemispheric connections in the adult cortex resembles closely the patchy pattern seen in cats, rats, and primates (Cabana and Martin, 1985). The lack of a CC in marsupials, however, is associated with an extremely small neocortex (Fig. 11.1). Thus, one might speculate that animals with a relatively small neocortex can do well without a CC. But what about placental mammals such as humans?

Although 16 cases of complete absence of the human CC were documented in the 19th century (for a review see Probst, 1901), it was a relatively rare encounter for the neuropathologist (Cameron, 1917). More common has been a condition wherein the CC is abnormally small, which is often termed partial absence. There are also extreme cases lacking the hippocampal, or even anterior, commissure. Thus, there is evidently a continuous range of cerebral midplane defects. For the general population, the incidence is estimated to vary between 3 and 7 per 1000 (Serur et al., 1988). Comparing the various degrees of defect in the adult with the normal sequence of forebrain commissure development in the embryo and fetus, Loeser and Alvord (1968) concluded that partial CC absence results from arrested development at the telencephalic midline. Whenever a small fragment of the CC was present, it apparently occurred in the same location as seen in early stages of normal CC formation. This interpretation has since been challenged, however. Other investigators prefer to link the severity of callosal dysgenesis with developmental timing (Byrd and Naidich, 1988; Jeret et al., 1987). At gestational weeks 10 through 12, the formation of the CC begins with the growth of the commissural plate, a rostral neuroectodermal bridge between the two hemispheres that provides the substrate for crossing axons. Maldevelopments during this period are thought to result in complete agenesis, while later acting processes might damage parts of

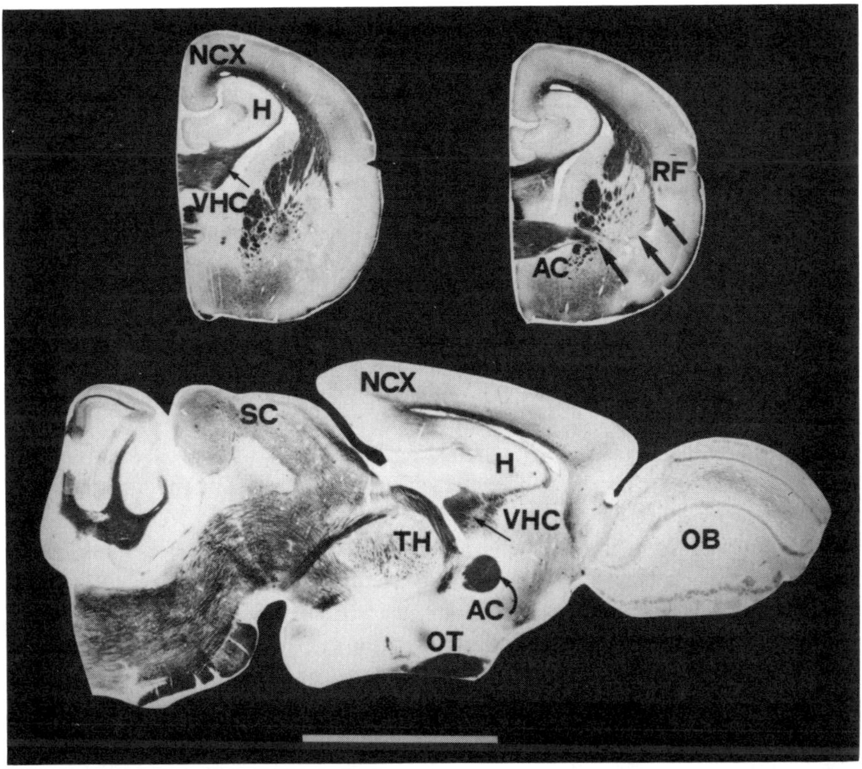

FIGURE 11.1. Fiber stain of an opossum brain *(Didelphis domestica)*. Parasagittal near-midline plane *(bottom)* and two coronal planes *(top)*, Sudan Black stain. Note the relatively small neocortex (extended between hippocampus and rhinal fissure), and the massive anterior and ventral hippocampal commissures. Abbreviations: AC, anterior commissure; H, hippocampus; NCX, neocortex; OB, olfactory bulb; OT, optic tract; RF, rhinal fissure (border between neocortex and paleocortex); SC, superior colliculus; TH, thalamus; VHC, ventral hippocampal commissure. Arrows in coronal section *(top right)* point at the course of the fibers of the anterior commissure, which travel via the capsula externa. Bar: 5 mm.

the developing CC (until week 20) or the established CC, for example, because of perinatal hypoxia, and result in partial absence or atrophy (Barkovich and Norman, 1988).

Developmental Syndromes

Absence of the CC occurs in a surprisingly large number of human neurological syndromes. In fact, its incidence among the developmentally disabled population ranges between 2% and 3% (Jeret et al., 1985; Serur et al., 1988). It is highly unlikely that these syndromes share a common etiology. Rather, it seems that early processes at the cerebral midplane are particularly prone to disruption by a variety of agents, including mutant genes, prenatal environment, and unusual timing of embryonic events with no external cause. There are four well established syndromes for which at least partial absence of the CC is typical.

1. The *Andermann syndrome* (Larbrisseau et al., 1984) is well documented in a large sample of cases deriving from Charlevoix County of Quebec in Canada. Almost all known cases can be traced through extensive pedigrees to a couple married in Quebec City in 1657, suggesting single-locus, recessive inheritance. The common ancestry and frequent marriages between relatives apparently yield the relatively high frequency of the syndrome in a small region. Affected children show moderate mental deficiency and progressive motor neuropathy, leading to an inability to walk by the age of 10 years. A small minority of cases shows a CC of normal size, and partial agenesis sometimes occurs. A recent study, however, reports a frequency of 32% normal CC and 9.4% partial agenesis of the CC in a sample of 64 patients investigated by means of computed tomography (CT) scanning (Mathieu et al., 1990). Detailed anatomical study of brains at autopsy has not yet been reported. The severe neuromuscular degeneration characteristic of the Andermann syndrome is not seen in any mouse strain known to suffer CC absence.

Other agenesis cases from Quebec have been the subjects of extensive neuropsychological testing by Lassonde and coworkers. Several of these children also come from the Saguenay–Lac St. Jean region of Quebec and show progressive myelopathy; hence they probably are instances of Andermann syndrome. Psychological problems are not restricted to tasks believed to depend on interhemispheric transmission; intrahemispheric processes may also be abnormal (Lassonde, 1986). Six children showed intelligence quotients (IQ) in the range 55 to 78, on the borderline between normal and mentally deficient, and their language functions appeared to be more strongly lateralized than in IQ-matched controls (Lassonde et al., 1990). Interpretation was rendered more difficult by an observation that low IQ is itself associated with accentuated asymmetry of function. Cook et al. (1990) present an entirely different conclusion derived from the same data set. Because so many aspects of the nervous and motor systems are impaired in Andermann syndrome, it is hazardous to attribute a psychological deficit specifically to the absence of the CC. Of special interest for future psychological research will be the cases of Andermann syndrome that do, in fact, have the CC present.

2. The *Aicardi syndrome* was first described by Aicardi et al. (1965). This syndrome includes infantile seizures, chorioretinal lacunar lesions, partial or complete agenesis of the CC, and mental retardation. It is commonly found in females only, except for one case with an XXY chromosome set (Donnenfeld et al., 1989). The genetic classification is disputed. Some consider it an X-linked dominant disorder with early embryonic lethality in hemizygous males (Aicardi et al., 1965; Donnenfeld et al., 1989), but other data appear incompatible with this notion (Jeret et al., 1987). Associated anomalies include costovertebral defects such as hemivertebrae or malformed ribs and cortical heterotopias (misplaced cortical gray matter due to neuronal migration failure). Thus far, no cases with normal intelligence have been reported, and the symptoms appear clinically

intractable. No comparable physical symptoms have been observed in mouse strains with callosal agenesis.

3. The *acrocallosal syndrome* was first described by Schinzel (1979) and includes agenesis of the CC, craniofacial dysmorphism, duplication of the phalanges in the hallux, postaxial polydactyly of the fingers and toes, and mental retardation. This syndrome was subsequently dubbed the acrocallosal syndrome (Schinzel and Schmid, 1980). Most cases found thus far lived in Appenzell, a restricted region of northeastern Switzerland. To date, about 20 cases have been noted. The mode of inheritance is assumed to be autosomal recessive (Schinzel, 1988). Because polydactyly (but not agenesis of the CC) also occurs in Greig's syndrome (Legius et al., 1985), it has been suggested that both syndromes might represent allelism of a common developmental gene (Hall, 1990).

Polydactyly has not been observed in mouse models of callosal agenesis. On the other hand, the mouse strain DDN (see below) occasionally shows a craniofacial anomaly (flat face), reportedly associated with agenesis of the CC (Ozaki et al., 1984). Also, it has been suggested that the mouse mutation dumpy, found in a BALB/c strain (Hollander, 1981), might provide a mouse model for Greig's syndrome (Hall, 1990).

4. *Shapiro's syndrome* represents a puzzling combination of callosal agenesis and recurrent episodes of hypothermia, the body temperature falling as low as 30°C. It is associated with physiological dysregulations manifested as polydipsia, polyuria, and hyponatremia (Mooradian et al., 1984; Shapiro et al., 1969), but apparently not with mental retardation (Edlund, 1989; Sanfield et al., 1989). An elevated plasma level of norepinephrine, possibly indicating overactivation of the sympathetic nervous system, has been found in a single case study (Sanfield et al., 1989). Since the clinical symptoms indicate hypothalamic malfunction, it has been proposed that septal or limbic anomalies associated with agenesis of the CC might play a causative role. Thus far, such associated anomalies have not been found, nor does callosotomy seem to influence thermoregulation (Berardi et al., 1988). On the other hand, agenesis of the CC has been found in association with metabolic disorders (see below). Thus, callosal agenesis in Shapiro's syndrome might perhaps follow, rather than cause, metabolic disorders. Among the mouse strains with callosal agenesis, the strain I/Ln (see below) would seem an interesting candidate for an animal model of human acallosy associated with metabolic problems.

Other Syndromes and Malformations Associated with Agenesis of the CC

As reviewed in depth by Jeret et al. (1987), clinical–developmental syndromes that are associated with agenesis of the CC beyond chance level are the fetal alcohol syndrome, the Arnold–Chiari II syndrome, West's syndrome, and the oral–facial–digital syndrome I. A possible new syndrome of multiple congenital anomalies (agenesis of the CC, facial anomalies, and Robin sequence) might also be added here (Toriello and Carey, 1988). Occasional cases with agenesis of the CC have been reported for dwarfism, Hirschsprung's disease, Hurler syndrome, oculocerebral syndrome (Jeret et al., 1987), and the Apert syndrome (De León et al., 1989; Jeret et al., 1989), whereas many other syndromes listed by Jeret et al. (1987) show sporadic CC agenesis.

Neurologically, frequent associations are mental retardation and/or retarded development, brain malformations such as migrogyria (Ferrer et al., 1986) heterotopia of gray matter, and seizures. Among 705 patients with diagnosed absence or partial absence of the CC, Jeret et al. (1987) found 85% with mental retardation. Although this number is high, it is not surprising in view of the large number of developmental syndromes involved. On the other hand, the lack of intellectual impairment in the 15% of acallosal patients that have been diagnosed because of other clinical symptoms clearly shows that agenesis per se does not necessarily cause severe mental retardation. Subtle intellectual impairments due to agenesis of the CC (Cook et al., 1990), however, cannot be ruled out by these findings. On the other hand, the high frequency of seizures (42%) may relate more directly to agenesis of the CC, suggesting that a normal CC inhibits the development of synchronized discharges in the healthy but not the epileptic brain. Somatically, agenesis of the CC is often associated with craniofacial, cardiovascular, and costovertebral malformations (Jeret et al., 1987). Genetically, chromosomal aberrations of chromosomes 8,11,13–15, and 18 were found most frequently (Jeret et al., 1987; Serur et al., 1988).

A new class of associations with deficient CC in humans has emerged recently. Dobyns (1989) reported an association between hyperglycinemia and agenesis of the CC (see also Kolodny, 1989) and reviewed other cases. A possible link between inherited metabolic disorders and callosal agenesis or dysgenesis has also been noted by others (Bamforth et al., 1988; Jeret et al., 1985; Johnson and Jones, 1985; Sadowsky and Reeves, 1975), and seems to present in Shapiro's syndrome as well.

Complete agenesis of the CC has been reported in one case of psychosis (Velek et al., 1988) and schizophrenia (Lewis et al., 1988), and partial agenesis in two other cases (Swayze et al., 1990). These observations are in line with a long series of reports claiming morphological changes of the CC in schizophrenic persons (Bigelow et al., 1983; Nasrallah et al., 1983; Rosenthal and Bigelow, 1972). However, morphometric data obtained post mortem and by means of MRI are contradictory, the former reporting a thickening of the CC, the latter a reduction (Rossi et al., 1988, 1989; Stratta et al., 1989) and some studies no differences (Kelsoe et al., 1988; Smith et al., 1987), except for a longer CC in schizophrenics. The most recent magnetic resonance imaging (MRI) data have failed to reveal substantial differences in the cross-sectional area between schizophrenic and control persons. The methodologically most convincing study thus far compared the CC of 12 pairs of monozygotic twins discordant for schizophrenia (Casanova et al., 1990). No area differences were found, but the affected twins appeared to have different shaped CC in the anterior and middle segments, and the shape of the splenium showed possible sex differences (see also Nasrallah et al., 1986b). Taken together, the available morphological data indicate that changes in area and form of the CC in schizophrenics are merely symptomatic of changes in ventricular

TABLE 11.1 Neurological problems in humans with absent CC and normal IQ.

Case	IQ	Originating problems	Reference
D.S.	106	Unconsciousness, staring	Dennis, 1981
B.F.	87	Epileptic seizures	Jeeves and Temple, 1987
K.C.	81	Headaches, fainting, cyst	Jeeves and Temple, 1987
S.B.	104	Nystagmus, hypotonia	Temple et al., 1989
K.W.	105	Severe myopia, memory loss	Temple et al., 1989

size and in limbic and temporal structures whose alterations are perhaps more directly associated with schizophrenia. Given the close anatomical relations, developmental abnormalities in the formation of the cingulate cortex and hippocampus might be reflected in ill-defined morphological alterations of the CC (Atlas et al., 1986; Swayze et al., 1990). This view is in agreement with the increasing number of observations linking developmental anomalies of the limbic cortices with schizophrenia (Conrad and Scheibel, 1987; Falkai and Bogerts, 1986; Jakob and Beckmann, 1989; Kovelman and Scheibel, 1984). Despite the failure to demonstrate a clear covariation of callosal cross-sectional area with schizophrenia, it should be kept in mind that schizophrenia appears to have a strong left hemispheric dominance in neurochemical studies (Reynolds et al., 1990). Also, callosal dysfunctions in schizophrenics have been noted, although inconsistently (for contrasting examples see David, 1987, and Merriam and Gardner, 1987). Finally, there are (speculative) theories postulating a more subtle involvement of callosal fibers in the pathogenesis of schizophrenia (Doty, 1989; Randall, 1983).

Asymptomatic Agenesis of the CC

A substantial number of "asymptomatic" cases of CC agenesis has been reported. The anatomist is sometimes genuinely surprised to behold an acallosal brain from what was thought to be a psychologically healthy person. Such cases studied after autopsy suffer from a lack of formal mental testing that might have revealed deficits not apparent to the family or physician. With the advent of x-ray and MRI techniques to view the living brain, many more people without a CC have been discovered, usually during examination for an apparently unrelated problem. Caution is also warranted here, because people with no neurological problems at all usually do not poke their heads into a CT scan or MRI machine. Indeed, among five recent cases with normal IQ scores who were examined for language functions, each came to the attention of researchers because of neurological difficulties (see Table 11.1).

The most striking feature of these cases is that language is quite normal in almost every respect, perhaps with a deficit in syntactic comprehension (Sanders, 1989). The only consistent finding in the four most recent cases is a deficit on tasks requiring rhyming (Temple et al., 1989). A mouse model of this uniquely human attribute is not expected. More promising is the observation that CC absence produces deficits in "simple tasks requiring bilateral motor coordination under speed stress" (Jeeves, 1986), probably because of longer interhemispheric transmission times. Also undisputed appears the observation that on average humans without a CC show a slightly subnormal IQ. Chiarello (1980) reports an average IQ of 89 for persons without overt signs of mental retardation. Although this may reflect associated, but undetected, brain anomalies, it should be kept in mind that no case of callosal agenesis with markedly superior intellect has been found thus far.

The other data about peculiarities in humans with agenesis of the CC are mostly conflicting: that is, potential sequelae of CC agenesis found in one of the samples were not replicated by other investigators or were inconsistent across individuals. There is agreement that acallosal humans rarely show the classical disconnection symptoms of callosotomy, most likely as a result of developmental plasticity (Lassonde et al., 1988). Given that the samples were extremely small, and given the extraordinary behavioral variability of humans, we refrain from an evaluation of the neuropsychological data, and refer the reader to the original literature (Cook et al., 1990; Ettlinger et al., 1972, 1974; Jeeves, 1986; Jeeves et al., 1988; Kinsbourne and Fisher, 1971; Lassonde, 1986; Lassonde et al., 1988, 1990; Milner and Jeeves, 1979; Milner, 1982; Milner, 1983; Milner et al., 1985; Reynolds and Jeeves, 1974, 1977; Sauerwein et al., 1981; Sauerwein and Lassonde, 1983).

ANIMAL MODELS OF CC AGENESIS

Absence of the CC is an obvious anatomical defect, even to the novice, but it has been reported only occasionally for large mammals such as horses (Cudd et al., 1989), cattle (Cho and Leipold, 1978), and monkeys (Tumbelaka, 1915). Perhaps surprisingly, it is quite common in laboratory mice, being found in four major strains and several of their substrains (Table 11.2). Because these animals have been inbred and reared under controlled conditions for many generations, they provide excellent material for investigating the genetics and development of the defect, as well as its functional consequences.

The following survey lists some noteworthy features of each major strain. The reader is referred to the original sources for further details. The main genetic, neuroanatomical, and behavioral data will be summarized across strains later.

BALB/c

The history and salient points about this common and extremely well documented strain have been reviewed recently (Wahlsten, 1989a). This albino strain frequently shows a partial agenesis of the CC, resulting in considerable variability of the cross-sectional area of the CC. Complete agenesis is relatively rare in most substrains, except for BALB/cWah1 (Table 11.2). The degree of callosal dysgenesis varies with strains and suppliers (Wahlstein, 1982b). There is no way, short of

TABLE 11.2 Mouse strains with defects of the corpus callosum (CC) and hippocampal commissure (HC).

Strain	Deficient or no CC (%)	Totally absent CC (%)	Defective HC?	Reference
Rodless x silver[a]	?	91	No	Keeler, 1933
BALB/cJ	39	16	No	Wahlsten, 1974
BALB/cByJ	18	3	No	Wahlsten and Smith, 1989
BALB/cCF	21	6	No	Schmidt et al., 1990
BALB/cWah1	55	21	No	Wahlsten, 1989a,b
BALB/cWah2	17	2	No	Wahlsten, 1989a,b
BALB/cCRBL	23	?	No	Cassells, 1988
CXBG/By	7	0	No	Wahlsten and Smith, 1989
129/J	71	30	No	Wahlsten, 1974
129/ReJ	77	2	No	Livy, 1990
I/LnJ	100	100	Yes	Lipp and Waanders, 1990; Livy, 1990; Livy and Wahlsten, 1991
DDN	8	?		Ozaki et al., 1984

[a] Now extinct.

direct examination of the brain, to detect which animal in a litter has no CC. The developmental origin of the callosal defect is the failure of commissural axons to grow across a glial sling present in the fetal commissural plate, presumably because of retarded growth of the axons. Most BALB/c fetuses show an obvious defect, but the adult phenotype is much more variable, because the growing axons often find their way to the opposite hemisphere via the hippocampal commissure. This process of developmental compensation is, in itself, dependent on environmental factors such as conditions *in utero* (Wahlsten, 1982a). The degree of compensation is unpredictable, but it provides a unique model for studying the effects of variable size of the CC in individuals that are otherwise isogenic. Thus, control animals with intact CC should differ from those with absent CC only in neuroanatomical and, perhaps, behavioral features associated with the developmental process leading to an altered CC. The human equivalent for this condition would be the paradigm of studying monozygotic twins discordant for schizophrenia (see above). However, the approach also has some subtle problems that will be discussed below.

The strain reproduces satisfactorily. The brain of BALB/c mice shows several other neuroanatomical anomalies, including a collision between the columns of the fornix and the anterior commissure (Cassells, 1988; Wahlsten, 1974), unusual foliation of the cerebellum (Wahlsten and Andison, 1989), and a minor heritable lamination defect in the hippocampal area CA3 (Nowakowski, 1984), which is irrelevant in a variety of behavioral tasks (unpublished data). Whether their visual system shows the same anomalously crossed optic axons as observed in other albino rodents (Jeffery, 1989) remains to be determined. Behaviorally, BALB/c mice have two features that may pose problems. They are albino and thus highly sensitive to variable levels of illumination, probably being handicapped in tasks requiring visual acuity (Upchurch and Wehner, 1988). In this regard, a pigmented BALB/c strain produced by more than 25 generations of backcrossing may prove useful in future research (Bulman-Fleming and Wahlsten, 1988). Also, many BALB/c substrains show behavioral immobility ("freezing") when exposed to stressful environments. For example, in Zürich we have been unable to conduct studies in swimming navigation with these mice, because many animals showed motionless floating (with high-frequency defecation) during many trials. Similar swimming behavior has been observed elsewhere (Schwegler and Buselmaier, 1981). Others have had some success in swimming tasks with BALB/c, although the animals tend to swim rather poorly compared with C57BL/6 mice (Lassalle et al., 1979).

129/J

For information on this strain, the reader is referred to Green (1966, 1981) and Heiniger and Dorey (1980). In our hands, these mice performed extremely poorly in swimming navigation and complex Y-maze learning. Findings on handedness are given below. It reproduces satisfactorily. In our samples, the incidence of complete agenesis of the CC was about 25% (see also Wahlsten, 1982c). The apparent dullness of this strain in more demanding tasks makes behavioral comparisons difficult. This judgment, however, must await further confirmation.

DDN

This relatively unknown Japanese strain has been used for extensive experiments on cortical connections using horseradish peroxidase (HRP; see below). Within the strain, certain mice show an unusual flat face, and these all suffer nearly total absence of the CC (Ozaki et al., 1984). The CC defect is unrelated to rate of two-way avoidance learning. Electroencephalogram analysis has revealed a lower correlation between the EEGs of the hemispheres (Okada et al., 1989). Given the incidence of craniofacial malformation associated with agenesis of the CC, the strain might prove to be a useful animal model for studying the developmental mechanisms of the acrocallosal syndrome and related genetic defects (see above).

I/Ln

Agenesis of the CC in this mouse strain was first reported by Wimer, who discovered this condition (Fig. 11.2) by screening the brains of mice from strains of the Jackson Laboratory (personal communication). Thus far, all animals investigated have shown the deficit (Table 11.2). The strain reproduces poorly (Hoover-Plow et al., 1988). We have found that about 1 out of 10 breeding pairs will produce offspring, and only a few mothers will eventually raise their pups.

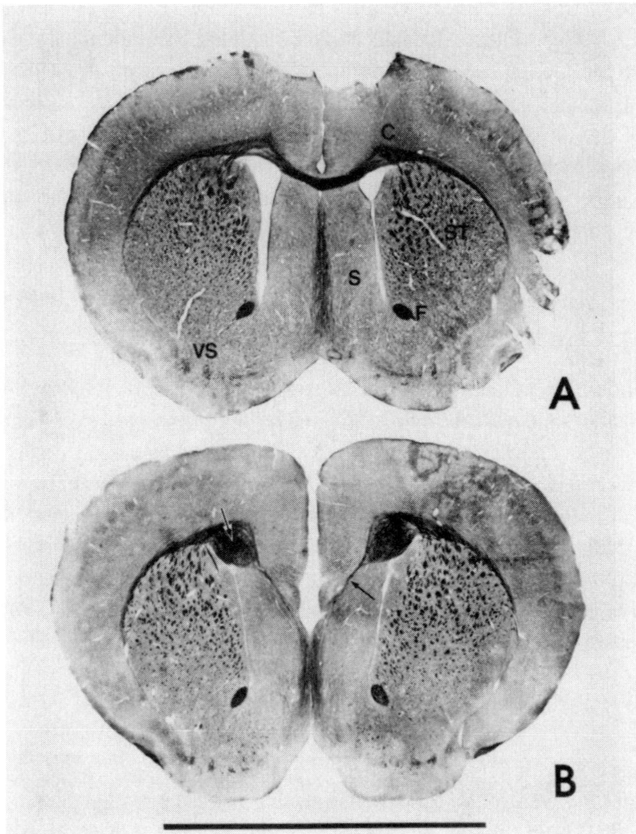

FIGURE 11.2. Coronal section of normal mouse brain **(A)** and brain with congenital agenesis of the CC **(B).** Liesegang silver stain. Left arrow in **B** points at a Probst bundle, that is, commissural fibers forced to grow ipsilaterally, right arrow at fibers leaving Probst's bundle to reach the septal area. Abbreviations: C, corpus callosum; F, frontal (anterior) branch of the anterior commissure; S, septum; ST, striatum; VS, ventral striatum. Bar: 5 mm.

Perhaps because of this, I/Ln mice cannot be purchased in large numbers. Recently, we have had more success in breeding, perhaps because of a devoted caretaker.

The developmental action of the genes causing callosal agenesis in the I strain is unknown and its biology is poorly documented. According to the compilations of the Jackson Laboratory (Green, 1966, 1981; Heiniger and Dorey, 1980) it carries the alleles pink-eye dilute and piebald. A high incidence of stomach ulcers has been reported as well (Grüneberg, 1952). The I strain shows a deficiency in muscular phosphorylase-b kinase activity (Chamberlain et al., 1987; Gross et al., 1978), and has been used in studies of metabolism because of its resistance to fat accumulation and sensitivity to vitamin B_6 deficiency (Bell et al., 1971; Hoover-Plow et al., 1987). These findings may be related to evidence linking human agenesis of the CC to metabolic deficiencies (Bamforth et al., 1988; Dobyns, 1989; Johnson and Jones, 1985; Kolodny, 1989). Another possible link to the agenesis of the CC is the observation of occasional vertebral fusions in the tail (kinky tails), which suggests mutations either in the *t* locus complex or in the fused locus (P. Roubertoux, personal communication), both located on chromosome 17. These gene complexes are known to control developmental steps in embryonic differentiation at various age levels (Bennett, 1981; Greenspan and O'Brien, 1986), which might explain the low fertility and high mortality of pups in the I/Ln strain. One may note here again that malformations associated with human agenesis of the CC often include costovertebral malformations.

The spontaneous behavior of the I/Ln strain is characterized by high locomotor activity, a rather unusual finding in a strain with nonpigmented eyes. Thus far, few neuroanatomical anomalies apart from CC absence have been discovered, but systematic investigation will be required. The ratio of brain weight to body weight is comparable to that of other mouse inbred strains (unpublished data). The infrapyramidal mossy fiber projection in the hippocampus, known to be of behavioral relevance in several learning tasks (Lipp et al., 1989b; Schöpke et al., 1991; Schwegler et al., 1990), is extremely small. In many I/LnJ mice the hippocampal commissure is unusually small (Livy and Wahlsten, 1990, 1991).

The Genetics of Callosal Agenesis in Mice

If the defect observed in the four mouse strains were the result of a single mutant gene, this would aid biochemical studies of the processes of axon guidance and allow attempts at genetic engineering. The original reports of absent CC in mice (Keeler, 1933; King and Keeler, 1932) claimed single-locus, recessive inheritance, and the symbol "ac" was then proposed. However, a later report by King (1936) stated explicitly that further data indicated a need for further study of the mode of inheritance. This became impossible when the stock of mice became extinct. After being displayed in standard listings of mouse genes for nearly 40 years, this ineradicable "ac" was tagged with a disclaimer in the February, 1990, issue of *Mouse Genome,* on the basis of doubts raised by Wahlsten (1982c).

Until a distinct gene is identified, research on absent CC will continue to use the inbred strains listed in Table 11.2. Several studies have shown beyond doubt that the relevant difference between a normal strain (e.g., C57BL/6J) and one having absent CC is caused by more than one locus. Mendelian backcrosses and F_2 hybrid crosses have documented multiple locus inheritance for BALB/cJ (Wahlsten, 1974), BALB/cWah (Wahlsten, 1987), BALB/cCRBL (Cassells, 1988), 129/J (Wahlsten, 1979), and I/LnJ (R.S. Nowakowski, personal communication). Recombinant inbred strains made the same point for BALB/cByJ (Wahlsten and Smith, 1989), and successive backcrossing confirmed this for I/LnJ (Lipp and Waanders, 1990).

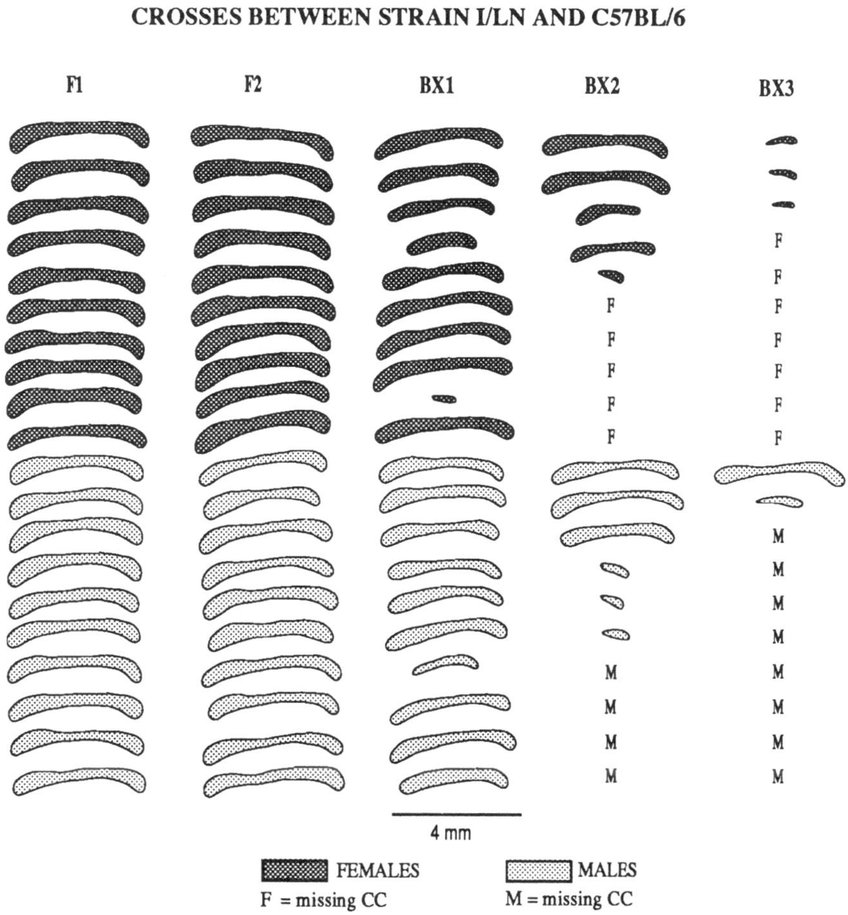

FIGURE 11.3. Morphological variability of the CC *(sagittal view)* in Mendelian crosses between normal mice (C57BL/6) and mice with agenesis of the CC (I/Ln). F1, first hybrid generation; F2, second generation (F1 × F1); BX1, first backcross generation (F1 × I/Ln); BX2, second backcrosses (BX1 × I/Ln); BX3, third backcross (BX2 × I/Ln). Note the normal size and shape of the CC in generations F1 and F2, and the increasing partial and complete agenesis in the backcross generations.

Two recent observations have led to optimism about further genetic analysis. Wahlsten and Smith (1989) presented evidence that the difference between C57BL/6 and BALB/c involves only two autosomal loci. Lipp and Waanders (1990) found a rapid reappearance of total CC absence in successive backcrosses to I/LnJ (Fig. 11.3). Because the F_1 hybrid between BALB/c and I/LnJ also shows CC defects (Livy and Wahlsten, 1991), the two strains must have essentially the same genetic defect, varying only in severity, and relatively few loci are probably involved.

In comparison with the bewildering variety of syndromes associated with callosal agenesis in humans, defects in mice offer one great advantage because the genetic effects are probably more specific. The reason is different developmental timing. In humans, the developmental window of vulnerability (the "critical period") is much more extended. Hence, more and diverse genetic and environmental factors can have an impact on formation of the CC. In mice, in which the critical phase of CC formation is extremely short, one might be able to identify a restricted set of genes responsible for the formation of a phylogenetically rather new structure, perhaps a new homeobox gene (Balling et al., 1989). Since mice are the genetically best investigated mammals, this speculation would seem not overly unrealistic.

Neuroanatomical Correlates of Callosal Agenesis in Mice

Most studies have either used the strain BALB/c or DDN. In BALB/c, Olavarria et al. (1988) used the tracer molecule horseradish peroxidase (HRP) to demonstrate that the pattern of intercortical connections in mice with a very small CC is normal but the number of connections is reduced. This leads to a suspicion that behavioral effects of CC defects will be most apparent in mice with total CC absence, which is rare in certain substrains. They also found no CC axons rerouted through the anterior commissure (AC), confirming an earlier claim by Wahlsten and Jones (1983).

Horseradish peroxidase tract tracing in DDN mice indicated that in complete absence, the CC axons form a typical ipsilateral Probst bundle and do not reroute through the AC (Ozaki and Shimada, 1988), and that the ipsilateral corticocortical projections are quite normal when there is a Probst bundle (Ozaki et al., 1989) (see also Fig. 11.2).

Today, the most promising approach for learning more about the morphological processes underlying the formation of normal and abnormal CC is given by the opportunity of post mortem tract tracing by means of the fluorescent carbocyanine dyes DiI and DiO (Elberger, 1989; Hogan and Berman, 1990). Because the tracer migrates very slowly, the small brains of mouse fetuses and pups are ideally suited for this type of investigation. The availability of differential dyes will enable investigators to discriminate axonal paths originating from different commissures, and to study the precise timing of growth. Finally, the availability of various mouse models with high incidence of callosal agenesis eliminates the problem of treating many brains in order to find a few with defective CC.

Behavioral Correlates of Callosal Agenesis in Mice

The ample availability of mice with partial and total agenesis of the CC provides an opportunity to study the behavioral consequences of such a morphological deficit. Before proceeding further, however, some points need clarification. Acallosal mice do not represent a model of a physiological split-brain preparation. Rather, the behavioral sequelae of CC agenesis will reflect the capacities of a brain in which the developmental and functional interactions between the hemispheres have become established without the presence of one commissure primarily connecting the neocortical areas. Because an absent CC in mice is inevitably associated with the development of ipsilateral fronto-occipital fiber tracts (the Probst bundles), agenesis of the CC automatically means connective reorganization, albeit in an orderly manner (Ozaki et al., 1989). The one exception to this rule is the defect produced by high doses of gamma radiation prenatally, which produce absent CC without a Probst bundle, presumably because the neurons of CC origin are badly damaged and cannot recover (Lent and Schmidt, 1986). At present, there is no convenient way of disentangling the effects of CC absence and Probst bundle formation; these two phenomena must be taken as one entity. Such connective reorganization also occurs in humans with agenesis of the CC. There is, however, one fundamental neuroanatomical difference between mice and men in this regard. Rodents have an extremely well developed commissural system between the hippocampal formations and their associated archicortices, which occupy as much as 50% of the total cortical area in rodents (Swanson et al., 1987). In primates and in humans, the commissural connections between the hippocampi are extremely reduced (Amaral et al., 1984). Although it is not clear whether the main commissural communication between the primate hippocampi occurs by way of the CC or the anterior commissure, one must assume that rodents with agenesis of the CC might not show many behavioral sequelae in tasks depending on interhippocampal communication, with the possible exception of the strain I/LnJ. In simplified terms, hippocampal functions are probably much more lateralized in humans. Thus, the study of mice with agenesis of the CC is likely to prove rewarding in tasks that depend heavily on neocortical commissural interaction, but perhaps less in what is commonly labeled cognitive processing. With these precautions in mind, however, the analysis of behavior of acallosal mice can be informative.

Strategies for Investigation

The study of callosal variation and behavioral covariation has three options that should be employed jointly. The first is the comparison between strains with normal CC and full CC agenesis in order to note strong behavioral differences possibly due to agenesis of the CC. There is a large behavioral data base for common mouse strains such as C57BL/6 or DBA/2 that facilitates the search for behavioral anomalies. Clearly, such strain comparisons may suggest a role for the CC but cannot prove it. Inbreeding fixes alleles randomly at many loci. Thus, the CC defect may be confounded with numerous other strain-specific characteristics. A suspected behavioral role of the CC can be tested subsequently by using two paradigms: studying developmental variation of the CC within one of the inbred strains BALB/c, 129/J, and DDN, or by producing variation of the CC by means of cross-breeding.

As explained above, studying the ontogenetic variation of the CC in an inbred strain with incomplete penetrance of the trait has the advantage of comparing virtually identical animals except for those nongenetic anatomical and behavioral factors associated with the defect. Problems arise, however, if the behavior one wishes to study is dominated by another strain-specific peculiarity. Many BALB/c mice, for example, show uniformly poor performance in swimming navigation (see Fig. 11.3). Thus, there is no way to detect whether variations of the CC might affect this type of learning.

I/Ln mice can be crossed with mice having a normal CC, and the backcross generations 2 and 3 show a high incidence of missing or variably reduced CC (Fig. 11.3). In this case, strain-specific genetic factors influencing a given behavior are mostly segregating (except for cases with linkage disequilibrium), and the backcross mice are active and healthy due to hybrid vigor. Statistically, the size of the CC is then considered as one variable randomly distributed among other genetic variables. Morphobehavioral covariations that can be observed in such samples are said to be robust, because they express themselves against the masking effects of other genetic factors (Lipp et al., 1989b). Conversely, a behaviorally dominant factor may no longer mask a potential correlation between variation of the CC and behavior. For examples, see below.

Handedness

The most investigated behavior in relation to agenesis of the CC in mice is handedness, measured as paw preference in a food-reaching task. Using BALB/c mice, Ward et al. (1987) found no correlation between CC size and the degree of paw preference, but their sample of only 35 mice included none with total CC absence. This negative result was confirmed in a thorough study of paw preference in 85 BALB/c mice (Schmidt et al., 1990), although their mice with abnormal CC were more likely to use the left paw. It is to be hoped that future work on BALB/c mice will be done with a substrain such as BALB/cWah1, in which the bimodal distribution of CC size (Wahlsten, 1989a) will yield greater statistical power. As things stand now, BALB/c mice can be regarded as behaviorally asymptomatic.

The situation appears rather different in the mouse strains 129/J and I/Ln. Both Ward et al. (1987) and Fride et al. (1989) found that 129/J mice with a small or absent CC showed a lower degree of paw preference when reaching for food. As compared to other mouse strains, the completely acallosal strain I/Ln showed an extremely low paw preference score (Fig. 11.4) that was equal to, or even lower than, the scores shown by a strain selectively bred for low paw preference scores (Gruber et al., 1991; Lipp and Waanders, 1990; Waanders and Lipp, 1989). This result was obtained by using two independent samples of 81 and 93 I/Ln mice. The data do not imply that all I/Ln mice are ambidextrous. However, the usual frequency distribution of paw preference in mice is U-shaped, the majority of animals being found with relatively high preference scores for either the left or the right paw. This is not the case in the I/Ln strain, which has many ambidextrous animals with no clear paw preference, particularly among the younger mice (Gruber et al., 1991). In one of the two samples, we also found a clear sex difference, acallosal males being significantly more ambidextrous than females.

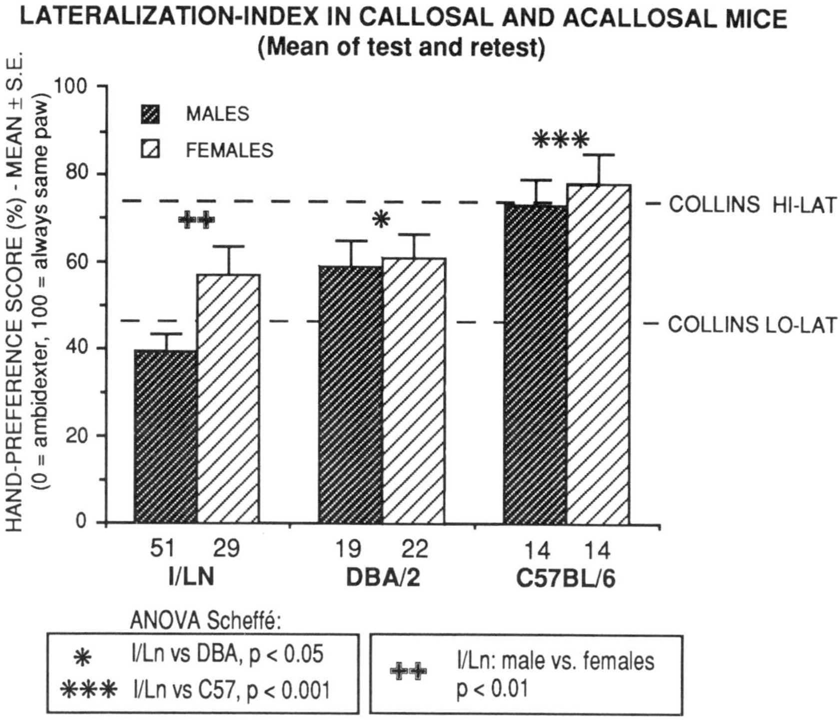

FIGURE 11.4. Paw lateralization scores in mouse strains with CC (DBA/2 and C57BL/6) and with agenesis of the CC (I/Ln). The score corresponds to the percentage of preferred paw entries. One hundred percent means a mouse using exclusively either the right or the left paw for reaching food in a small tube; 0% means complete ambidexterity. The procedure for testing paw preference is given by Collins (1979). The dashed lines correspond to the lateralization scores of a small sample of mice selectively bred for differential degree of paw lateralization (Collins, 1979), the mice having been provided by R.L. Collins.

On the other hand, a sample of 24 mice from the second backcross generation (I/Ln × C57BL/6) did not show a significant correlation between paw preference scores and differential size of the CC.

Thus, the actual situation is contradictory. Taken together, however, the data indicate that agenesis of the CC, or having a small CC only, is negatively associated with the development of paw preference, and there may be a subtle bias toward a left paw preference associated with agenesis of the CC. The lack of a clear correlation in BALB/c mice may represent a sampling problem (too few mice with small or no CCs) or a strain-dependent behavioral peculiarity dominating the expression of paw preference. However, the missing correlation in backcross mice and the phenomenon of many well lateralized I/Ln mice suggests that agenesis of the CC is only one of many conditions that might influence the expression of paw preference. For example, it has been shown that genetic asymmetries in arrange-

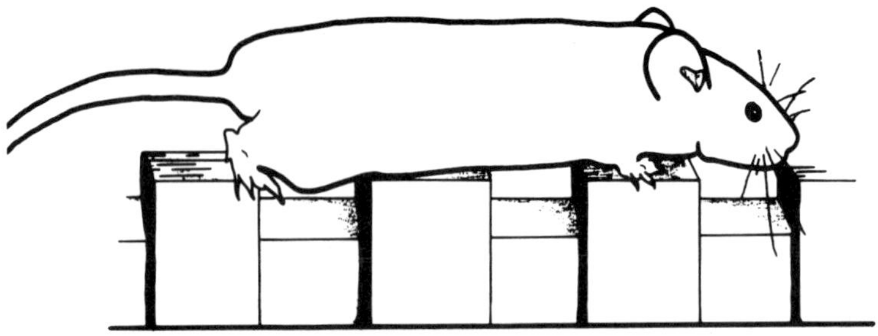

FIGURE 11.5. Mouse on a carved bar. Slips of the hindpaws are taken as a measure of hindlimb coordination. Drawing made by Rob Waanders.

ment of the mystacial vibrissae correlate with the side of paw preference (Barnéoud and Van der Loos, 1989, personal communication). Also, cerebral asymmetries are much more pronounced in mice selected for strong paw preference (Collins, 1979; Lipp et al., 1984). In line with this observation, Schmidt and Caparelli-Dáquer (1989) reported that mice with small or absent CC had more symmetrical cerebral hemispheres. Perhaps the bias toward reduced paw preference in acallosal mice is associated with deficits in the immune system, as reported by Fride et al. (1989). They found that 129/J mice with deficient CC showed weak paw lateralization and suffered from deficits of several immune system functions. Since the mouse strain I/Ln appears to be much more prone to husbandry problems and infections, this line of research certainly warrants future attention.

Hindpaw Coordination

One of the more frequently observed associations with agenesis of the CC in humans is bimanual coordination problems under speed stress (Jeeves, 1986). Since mice cannot be trained conveniently for manipulations with the forepaws, they were tested on a bar stepping test. The mice were placed first on a bar with a smooth surface from where they could escape to platforms located at both ends of the bar. The behavioral score was the number of slips with the hindpaws, which increased with faster locomotion. (Mice investigate the stepping area in front of them with their vibrissae; thus, they rarely misplace the forepaws.) The second day, the mouse was placed on a bar with slots that forced the mouse to step along little platforms (Fig. 11.5). The difference in slips made on the smooth versus the carved bar was then taken as a measure of deficient coordination between forepaws and hindpaws.

Overall, the results showed a clear genetic correlation between the size of the CC and the number of errors due to increasing the difficulties of the stepping task. The strain with an absent CC showed the highest number of errors, whereas the mean error score of strains and backcrosses with an increasing size of the CC showed a reduction of errors (Fig. 11.6). However, analysis of the within-group and within-strain correlations did not reveal significant correlations. The main picture was that animals with a fully developed CC rarely made errors, but that

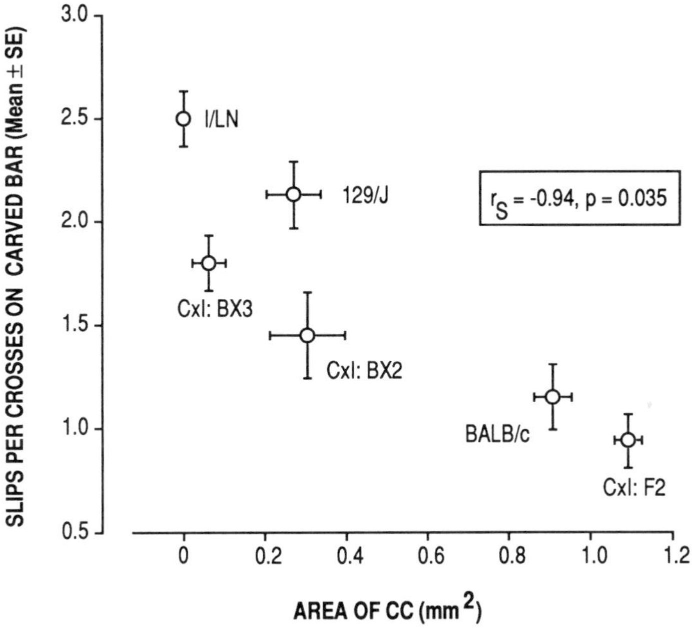

FIGURE 11.6. Genetic correlation between the mean values of the area of the CC and the mean number of hindpaw slips per crossing. I/Ln, 129/J, BALB/c: mouse strains. C × I: backcross generations BX1, BX2, and BX3; see legend to Figure 11.3.

mice with no CC would either show very poor or quite excellent scores. Thus, lack of the CC does not necessarily entail motor coordination problems, but the smaller the CC, the more likely problems will occur.

Spontaneous Behavior

I/Ln mice have been compared with the strain C57BL/6 in the open field. Since the I/Ln strain is highly active, the extreme strain difference was not surprising. However, three noteworthy features emerged: (1) a few I/Ln mice suddenly shifted into a circling running pattern, predominantly to the same side, (2) about half of the animals did not show the commonly observed overnight habituation, but rather increased their activity, and (3) there appeared to be a behavioral dichotomy among the mice, observed in locomotor activity as well as in the (noncorrelated) habituation scores.

Another test of spontaneous behavior is tube dominance. Two mice are placed at each end of a tube. Upon meeting each other, one of the animals will be pushed back. The behavior has a strong genetic component that permitted selective breeding for winners and losers over three generations. Perhaps surprisingly, the acallosal mice won clearly against members of the mouse strains C57BL/6, BALB/c, and DBA/2 (Lipp and Waanders, 1990). It seems unlikely that agenesis

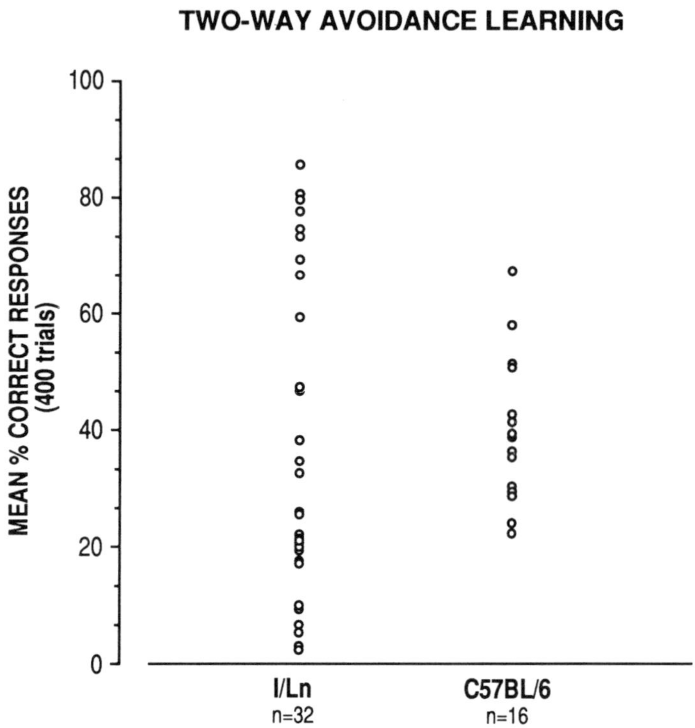

FIGURE 11.7. Two-way avoidance scores in C57BL/6 and DBA/2 mice. Note the extreme variability of the shuttlebox performance in the acallosal strain I/Ln.

of the CC can be responsible for this superiority, because variations in size of the CC in backcross mice did not correlate with the number of wins (unpublished observations), but acallosal mice are certainly not handicapped in socially competitive situations.

Conditioning Studies

Thus far, acallosal mice (mostly I/Ln) have been tested in two-way avoidance (Lipp et al., 1989a,c), swimming navigation (Lipp et al., 1989c), and aversively motivated Y-maze discrimination learning (unpublished data). Comparing two-way avoidance in the strains BALB/c, C57BL/6, DBA/2, and I/Ln revealed a clear superiority of the DBA/2 strain, a superiority known to correlate with hippocampal mossy fiber pattern (Schwegler and Lipp, 1983), whereas the other strains showed a comparable performance to one another. Yet, detailed analysis of the I/Ln mice again showed an unusually wide range of performance for an inbred strain: some animals showed excellent performance, whereas others behaved extremely poorly (Fig. 11.7). These dichotomous performance levels were clearly linked to a dichotomous presession activity in the shuttlebox. The highly active animals became excellent performers and the less active ones developed into poor performers. Note, however, that the well performing DBA/2 mice showed low presession activity. No details have been provided about the apparently normal two-way avoidance learning of DDN mice reported by Ozaki et al. (1984).

In swimming navigation, the animals are required to find an invisible target platform in a swimming pool. They must first recognize their release site from surrounding spatial cues, and must then navigate to coordinates stored mentally—a classical spatial task (Morris, 1984). Strain comparisons, within-strain correlations, and correlations in backcross mice have uniformly indicated that performance in swimming navigation is neither dependent on absence nor on variations of the CC. For example, acallosal I/Ln mice compare favorably with the inbred strains C57BL/6 and DBA/2, whereas backcross mice (I/Ln × C57) of the second generation, with quite small or absent CC, show excellent performance (Fig. 11.8). Swimming navigation is known to depend heavily on hippocampal functions. In mice, correlations with the extent of the infrapyramidal mossy fiber projections have been observed in four experiments using the same set-up as for studies of callosal variation (Schöpke et al., 1991). Other important factors influencing swimming navigation of mice appear to be associated with eye color (Upchurch and Wehner, 1988) and hybrid vigor (Upchurch and Wehner, 1989). The only handicap of acallosal I/LnJ mice in swimming navigation observed thus far was the strong tendency of a few mice to swim in narrow circles (for an example, see Fig. 11.9).

Discriminative Y-maze learning is a nonspatial, but fairly complex, task in which the mice must discriminate between two arms of an automatized Y-maze that present visual or vibrissotactile cues (smooth vs. slotted brass plates) in random order. Wrong choices are punished by means of air blasts. Learning the correct solution requires the suppression of spatially motivated strategies ("win–stay, lose–shift") and the ability to discriminate between left and right. Mouse strains with intact CC such as C57BL/6 and DBNA/2 learn the task with differential success, the performance difference being associated with superior learning in the strain showing a larger intrapyramidal mossy fiber projection. The strain 129/J showed a relatively poor performance in this set-up, while practically all I/Ln mice tested thus far were unable to learn correct left/right discrimination based on visual or vibrisso-tactile cues located in the arms (Fig. 11.10).

Preliminary Conclusions

It is clear that much remains to be learned about the behavioral abilities of mice with agenesis of the CC. Behavioral studies have, however, revealed two important aspects. First, as expected, tasks involving sensorimotor integration suggest parallels between functional consequences of agenesis of the CC in humans and mice; for example, tests for handedness and limb coordination. Perhaps the uncontrollable circling is also due to callosal deficiency. So far, it has never been observed in a mouse with a fully developed CC. The competency of acallosal mice in more "cognition-dependent" tasks may reflect unimpaired

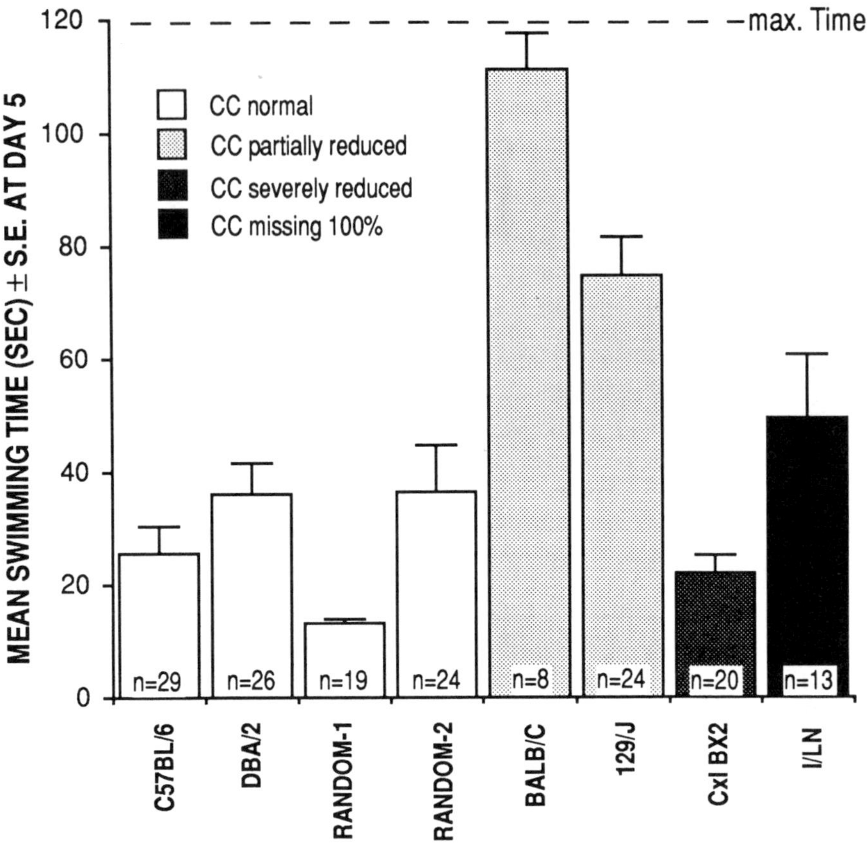

FIGURE 11.8. Performance of various mouse strains in Morris swimming navigation during the second day of reversal learning. RANDOM-1 and RANDOM-2 denote two random-bred populations with different progenitor strains. Note that the average performance of the second backcross generation (BX2) compares well with the performance of mouse samples with normal CC (this sample of BX2 mice had more acallosal individuals as compared to the sample shown in Fig. 11.3). BALB/c mice were completely inept because they spent most of their time floating passively. The strain 129/J also showed much passive floating. The comparatively poorer average performance of I/Ln mice reflects a tendency for excessive circling as observed in a few animals (see Fig. 11.9).

communication of the hippocampi in acallosal rodents or another unknown feature of brain organization. Yet, the data clearly show that acallosal mice are not mentally retarded and behave asymptomatically in some of the most complex tasks devised for mice, whereas they appear to have difficulties in others. If the brains of these mice show malformations and associated anomalies, as are typically observed in human developmental syndromes, one would expect uniformly poor performance across all tasks. This does not imply that the brains of these mice

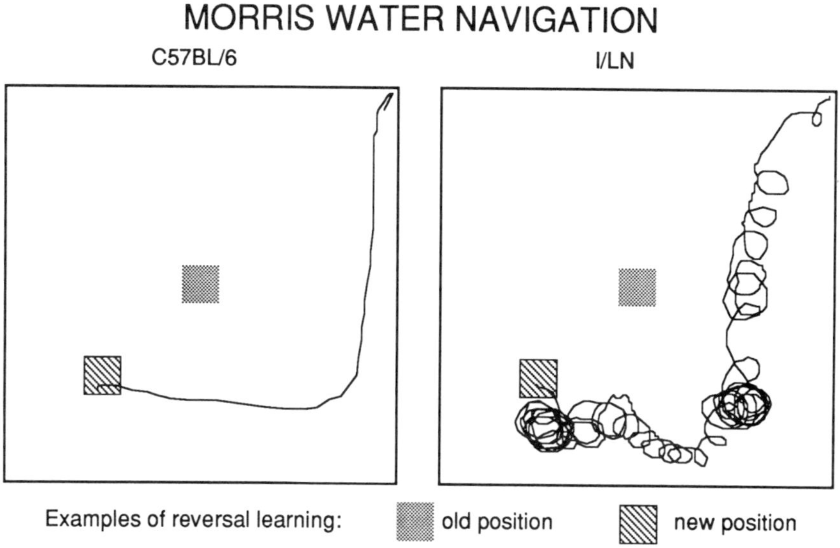

FIGURE 11.9. Circling swimming patterns in an I/Ln mouse during the last phases of reversal learning. Note that both the I/Ln and the C57BL/6 control mouse appeared to have learned the new platform location well. However, the I/Ln mouse was handicapped because of its uncontrollable circling.

possess no hidden anomalies, but it corroborates the impression of normality gained by routine neuroanatomical scanning.

Usually, behavioral scores of inbred strains show an approximatively normal distribution around a strain mean. With the exception of handedness data, distributions of behavioral scores approaching bimodality are very rare. An unexpected finding, therefore, was the behavioral dichotomies observed in the acallosal I/Ln strain. The significance of this observation is not yet clear. Two interesting speculations come to mind, though. One is behavioral lateralization. A given hemisphere may be dominant for a given task. In mice with normal CC, interactions between the neocortices ensure that the proper hemisphere is handling the task-relevant processes. In mice with agenesis of the CC, each hemisphere may have the same chance of taking over the processing required for the task but, once specialized, will then dominate the activity of the other hemisphere. If, by chance, the appropriate hemisphere becomes the dominant one, performance in the task will be excellent. If the other dominates the behavioral processing, poor performance is likely because of a lack of callosal feedback. The other hypothesis is that absence of the CC simply increases a general unpredictability of behavior. The normal role of the CC would then be a smoothing one: reducing the occurrence of chaotic states in symmetrical neural networks (Hoppensteadt, 1989). In that case, the CC might be considered as the physical substrate of an "attractor," creating mental orderliness by means of tiny little strings extended between the hemispheres. Will there even be a return to the most ancient views of callosal functions?

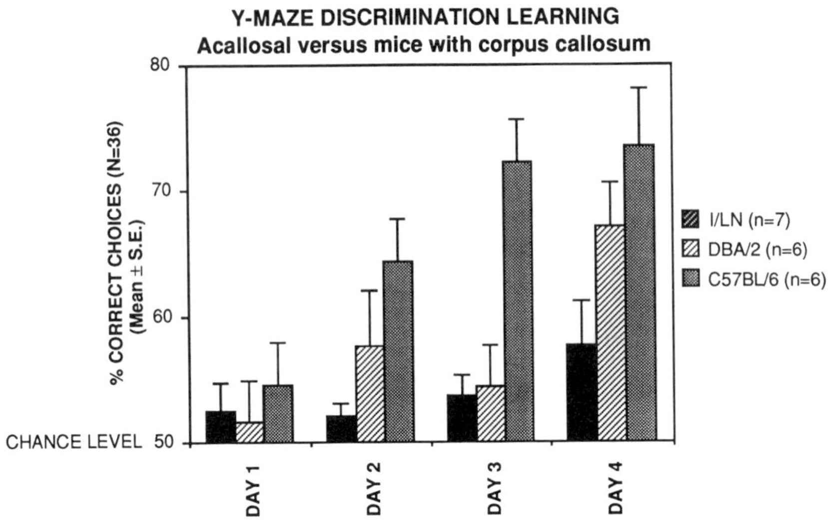

FIGURE 11.10. Y-maze discrimination learning in mouse strains with CC (DBA/2 and C57BL/6) and without CC (I/Ln). There was one I/Ln mouse showing sporadical bouts of discrimination. In another sample of 10 I/Ln mice (not shown here), none of them learned the task.

WHY STUDY MICE AT ALL

1. Studies of correlates of agenesis of the CC in humans are faced with the problem of multifarious associated malformations. By contrast, all four mouse models presented here lack gross anomalies. Contrary to a widely held belief, inbred strains are not merely some form of sick mice. The process of inbreeding itself automatically eliminates severe developmental malformations (unless they are specifically selected by the breeder). Common inbred strains show many minor peculiarities, some perhaps reducing biological fitness, but none of them important for survival. Thus, studies in mice offer the chance to find functional correlates of CC agenesis that are much less contaminated by associated malformations.

2. Agenesis of the CC can be studied in large numbers of individuals with different genetic backgrounds, or even in coisogenic mice. This permits classical and molecular genetic studies that are impossible to undertake in humans. The mouse models might, perhaps, not arise from the same genetic defects causing agenesis of the CC in humans, but they certainly allow a detailed analysis of the gene-to-phenotype pathway. For example, it was initially believed that agenesis of the CC was a consequence of arrested development at the telencephalic midline (Loeser and Alvord, 1968). Recent evidence from the mouse has cast doubt on this interpretation, and it illustrates the hazards of inferring the sequence of events in abnormal fetuses by an analogy between normal fetuses and abnormal adults. In

adult BALB/c mice, CC sizes range from totally absent to very large; indeed, less than half are abnormally small (Wahlsten, 1982b). However, in BALB/c fetuses at about 0.5 to 0.7 g body weight, when other strains all show a distinct CC, almost every BALB/c animal shows total CC absence, and the CC axons from the two hemispheres are separated by a gap, or bulge, in the longitudinal cerebral fissure (Wahlsten, 1987). Nevertheless, in almost all fetuses the CC axons eventually do find an unusual path across the hippocampal commissure, and the CC then grows rapidly to reach normal adult size. Plasticity of development leads to compensation for a severe defect in most fetuses.

3. Studies aimed at the elucidation of normal function of the CC can profit considerably from mouse models. Investigating completely acallosal animals with relatively small brains and limited intellectual capacities might not reveal the functions mediated by the CC in the normal human brain, but they have the potential of unraveling the kernel functions of the great neocortical commissure. Reliable behavioral deficits of acallosal mice with differential genetic background must correspond to functional processes that cannot be compensated by developmental and behavioral plasticity. Given that the effects of structural variation in hippocampal connectivity can be made evident in mice, there is justifiable hope that the remodeled connections in the neocortex of acallosal mice will eventually reveal those behaviors for which an intact CC is absolutely essential, that is, to show the *raison d'être* of the CC.

4. Studies on malformations associated with agenesis of the CC can be performed more easily in mice in which the agenesis is attributable to the action of a limited number of genes. For example, if variable agenesis of the CC in an inbred strain can be correlated reliably with a biochemical, metabolic, or growth defect, unconditional associations can be discriminated from occasional ones. In comparison with (potentially possible) studies in humans, the chief advantage of the mouse models is economy.

5. Experimental studies designed to verify theories based on observations in humans can only be performed in animals. Given that the population structure of mice is based on the production of large surplus numbers of individuals, the use of mice is also admissible from an ethical point of view. Apart from the experimental perspectives offered in the field of neuroanatomy, mouse models of callosal agenesis are likely to be of high interest for neurophysiology, epilepsy research, and neural network modeling (Anninos and Cook, 1988). For example, the model of Cook (1986) predicts that high-level intelligence is a function of the number of callosal fibers. Testing this hypothesis appears to be impossible except for mice in which individuals with variable size of the CC can be produced reliably and in large numbers.

Acknowledgments. This contribution was supported by the Swiss National Foundation for Scientific Research (Grant 31-9470 to HPL), by the Julius-Klaus-Stiftung at the University of Zürich (HPL), and by the Natural Sciences and Engineering Research Council of Canada and the Alberta Heritage Foundation for Medical Research (DW). We appreciate the expert photographic artwork of Helga Weber.

REFERENCES

Aicardi J, Lefebvre J, Lerique-Koechlin A (1965): A new syndrome: spasm in flexion, callosal agenesis, ocular abnormalities. *Electroencephalogr Clin Neurophysiol* 19:609–610

Amaral DG, Insausti R, Cowan WM (1984): The commissural connections of the monkey hippocampal formation. *J Comp Neurol* 224:307–336

Anninos PA, Cook ND (1988): Neural net simulation of the corpus callosum. *Int J Neurosci* 38:381–391

Atlas SW, Zimmerman RA, Bilaniuk LT, Rorke L, Hackney DB, Goldberg HI, Grossman RI (1986): Corpus callosum and limbic system: neuroanatomic MR evaluation of developmental anomalies. *Radiology* 160:355–362

Balling R, Mutter G, Gruss P, Kessel M (1989): Craniofacial abnormalities induced by ectopic expression of the homeobox gene *Hox-1.1* in transgenic mice. *Cell* 58:337–347

Bamforth F, Bamforth S, Poskitt K, Applegarth D, Hall J (1988): Abnormalities of corpus callosum in patients with inherited metabolic diseases. *Lancet* 2:451

Barkovich AJ, Norman D (1988): Anomalies of the corpus callosum: correlation with further anomalies of the brain. *Am J Radiol* 151:171–179

Barnéoud P, Van der Loos, H. (1989): Association between a genetically-expressed neocortical asymmetry and the direction of handedness in mice. *Eur J Neurosci* Suppl 2:176.

Bell RR, Blanchard CA, Haskell BE (1971): Metabolism of vitamin B 6 in the I-strain mouse. II. Oxidation of pyridoxal. *Arch Biochem Biophys* 147:602–611

Bennett D (1981): The T/t-complex in the mouse: mutations that impair differentiation. *Prog Clin Biol Res* 66:375–379

Berardi N, Bisti S, Fiorentini A, Maffei L (1988): The transfer of visual information across the corpus callosum in cats, monkeys and humans: spatial and temporal properties. *Prog Brain Res* 75:181–185

Berbel P, Innocenti GM (1988): The development of the corpus callosum in cats: a light- and electron-microscopic study. *J Comp Neurol* 276:132–156

Berlucchi G (1983): Two hemispheres but one brain. *Behav Brain Sci* 6:171–173

Berrebi AS, Fitch RH, Ralphe DL, Denenberg JO, Friedrich VL Jr, Denenberg VH (1988): Corpus callosum: region-specific effects of sex, early experience and age. *Brain Res* 438:216–224

Bigelow LB, Nasrallah HA, Rauscher FP (1983): Corpus callosum thickness in chronic schizophrenia. *Br J Psychiatry* 142:284–287

Bulman-Fleming B, Wahlsten D (1988): Effects of a hybrid maternal environment on brain growth and corpus callosum defects of inbred BALB/c mice: a study using ovarian grafting. *Exp Neurol* 99:636–646

Byrd SE, Naidich TP (1988): Common congenital brain anomalies. *Radiol Clin North Am* 26:755–772

Cabana T, Martin GF (1985): The development of commissural connections of somatic motor-sensory areas of neocortex in the North American opossum. *Anat Embryol (Berl)* 171:121–128

Cameron J (1917): The corpus callosum—a morphological and clinical study. *Can Med Assoc J* 7:609–616

Casanova MF, Sanders RD, Goldberg TE, Bigelow LB, Christison G, Torrey EF, Weinberger DR (1990): Morphometry of the corpus callosum in monozygotic twins discordant for schizophrenia: a magnetic resonance imaging study. *J Neurol Neurosurg Psychiatry* 53:416–421

Cassells S (1988): *Hereditary Influences on the Morphology of the Anterior Commissure and Columns of Fornix in* Mus musculus. Unpublished Ph.D. Dissertation, Waterloo, Canada: University of Waterloo

Cassells S, Collins RL, Wahlsten D (1990): Path analysis of sex difference, forebrain commissure area and brain size in relation to degree of laterality in selectively bred mice. *Brain Res,* in press

Chamberlain JS, VanTuinen P, Reeves AA, Philip BA, Caskey CT (1987): Isolation of cDNA clones for the catalytic gamma subunit of mouse muscle phosphorylase kinase: expression of mRNA in normal and mutant Phk mice. *Proc Natl Acad Sci USA* 84:2886–2890

Chiarello C (1980): A house divided? Cognitive functioning with callosal agenesis. *Brain Language* 11:128–158

Cho DY, Leipold HW (1978): Agenesis of corpus callosum in calves. *Cornell Vet* 68:99–107

Clarke S, Kraftsik R, Van der Loos H, Innocenti GM (1989): Forms and measures of adult and developing human corpus callosum: is there sexual dimorphism? *J Comp Neurol* 280:213–230

Collins RL (1979): Selective breeding for the degree of functional lateralization in mice. *Behav Genet* 9:443–444

Colonnier M (1986): Notes on the early history of the corpus callosum with an introduction to the morphological papers published in this festschrift. In: *Two hemisperes—One brain: Functions of the Corpus Callosum,* Lepore F, Ptito M, Jasper HH, eds. New York: Alan R. Liss, 37–45

Conrad AJ, Scheibel AB (1987): Schizophrenia and the hippocampus: the embryological hypothesis extended. *Schizophr Bull* 13:577–587

Cook ND (1986): *The Brain Code. Mechanisms of Information Transfer and the Role of the Corpus Callosum.* London/New York: Methuen

Cook ND, Brugger P, Regard M, Landis T (1990): On the role of the corpus callosum in cerebral laterality: a comment on Lassonde, Bryden and Demers. *Brain Lang,* 39:471–474

Crescimanno G, Salerno MT, Cortimiglia R, Amato G (1989): Claustral influence of ipsi- and contralateral motor cortical areas, in the cat. *Brain Res Bull* 22:839–843

Cudd TA, Mayhew IG, Cottrill CM (1989): Agenesis of the corpus callosum with cerebellar vermian hypoplasia in a foal resembling the Dandy-Walker syndrome: pre-mortem diagnosis by clinical evaluation and CT scanning. *Equine Vet J* 21:378–381

David AS (1987): Tachistoscopic tests of colour naming and matching in schizophrenia: evidence for posterior callosum dysfunction. *Psychol Med* 17:621–630

de Lacoste-Utamsing C, Holloway RL (1982): Sexual dimorphism in the human corpus callosum. *Science* 216(4553):1431–1432

de Lacoste M-C, Holloway RL, Woodward DJ (1986): Sex differences in the fetal human corpus callosum. *Hum Neurobiol* 5:93–96

De León GA, De León G, Grover WD, Zaeri N, Alburger PD (1989): Agenesis of the corpus callosum in Apert syndrome. *Arch Neurol* 46:479

Denenberg VH (1981): Hemispheric laterality in animals and the effects of early experience. *Behav Brain Sci* 4:1–49

Dennis M (1981): Language in congenitally acallosal brain. *Brain Language* 12:33–53

Dobyns WB (1989): Agenesis of the corpus callosum and gyral malformations are frequent manifestations of nonketotic hyperglycinemia. *Neurology* 39:817–820

Donnenfeld AE, Packer RJ, Zackai EH, Chee CM, Sellinger B, Emanuel BS (1989): Clinical, cytogenetic, and pedigree findings in 18 cases of Aicardi syndrome. *Am J Med Genet* 32:461–467

Doty RW (1989): Schizophrenia: a disease of interhemispheric processes at forebrain and brainstem levels. *Behav Brain Res* 34:1–33

Doty RW, Ringo JL, Lewine JD (1986): Interhemispheric mnemonic transfer in macaques. In: *Two hemisperes—One Brain: Functions of the Corpus Callosum,* Leporé F, Ptito M, Jasper HH, eds. New York: Alan R. Liss, 269–279

Edlund MJ (1989): Social phobia secondary to pathological sweating. *Br J Psychiatry* 155:403–405

Efron R (1990): *The Decline and Fall of Hemispheric Specialization.* Hillsdale, NJ: Erlbaum

Elberger AJ (1989): Corpus callosum connections in visual cortex of neonatal cat examined by double-labeling with a carbocyanine dye and flourescence immunoreactivity. *Soc Neurosci Abstr* 15:527.21

Elberger AJ, Honig MG (1990): Double-labeling of tissue containing the carbocyanine dye DiI for immunocytochemistry. *J Histochem Cytochem* 38:735–739

Ettlinger G, Blakemore CB, Milner AD, Wilson J (1972): Agenesis of the corpus callosum: a behavioural investigation. *Brain* 95:327–346

Ettlinger G, Blakemore CB, Milner AD, Wilson J (1974): Agenesis of the corpus callosum: a further behavioural investigation. *Brain* 97:225–234

Falkai P, Bogerts B (1986): Cell loss in the hippocampus of schizophrenics. *Eur Arch Psychiatry Neurol Sci* 236:154–161

Ferrer I, Cusi MV, Liarte A, Campistol J (1986): A Golgi study of the polymicrogyric cortex in Aicardi syndrome. *Brain Dev* 8:518–525

Fride E, Skolnick P, Arora PK (1989): Influence of the corpus callosum on immune function. *Soc Neurosci Abstr* 15:291.9 (Abstr)

Gall FJ (1979): *Franz Josef Gall, Naturforscher und Anthropologe. Ausgewählte Texte,* Lesky E, ed. Bern: Hans Huber

Gall FJ, Spurzheim JC (1810): *Anatomie und Physiologie des Nervensystems im allgemeinen, und des Gehirns insbesondere*. Paris.

Gazzaniga MS (1989): Organization of the human brain. *Science* 245:947–952

Gazzaniga MS, Kutas M, Van Petten C, Fendrich R (1989): Human callosal function: MRI-verified neuropsychological functions. *Neurology* 39:942–946

Gravel C, Sasseville R, Hawkes R (1990): Maturation of the corpus callosum of the rat: II. Influence of thyroid hormones on the number and maturation of axons. *J Comp Neurol* 291:147–161

Green CM (1981): *Genetic Variants and Strains of the Laboratory Mouse*. Stuttgart/New York: Gustav Fischer

Green EL (1966): *Biology of the Laboratory Mouse*. New York: Dover Publications

Greenspan RJ, O'Brien MC (1986): Genetic analysis of mutations at the fused locus in the mouse. *Proc Natl Acad Sci USA* 83:4413–4417

Gross SR, Bromwell K, Baanante IV (1978): Comparison of the mechanism is isoproterenol-stimulated glycogenolysis in skeletal muscle of normal and phosphorylase kinase-deficient mice (I strain). *J Pharmacol Exp Ther* 205:732–742

Gruber D, Waanders R, Collins RL, Wolfer DP, Lipp H-P (1991): Congenital absence of the corpus callosum in mice is associated with weak cerebral lateralization. *Behav. Brain Res* 46:9–16.

Grüneberg H (1952): *The Genetics of the Mouse*. Den Haag: Martinus Nijhoff

Guiard Y (1980): Cerebral hemispheres and selective attention. *Acta Psychol* 46:41–61

Hall JG (1990): Could acrocallosal syndrome and Greig syndrome affect the same developmental gene: (letter). *Am J Med Genet* 36:368

Hankin MH, Silver J (1988): Development of intersecting CNS fiber tracts: the corpus callosum and its perforating fiber pathway. *J Comp Neurol* 272:177–190

Hedin-Pereira C, Lent R, Jhaveri S (1988): The development of callosal and corticocortical innervation in the neocortex of the hamster. *Braz J Med Biol Res* 21:499–502

Heiniger H-J, Dorey JJ (1980): *Handbook of Genetically Standardized JAX Mice*. Bar Harbor, ME: The Jackson Laboratory

Hogan D, Berman NEJ (1990): Growth cone morphology, axon trajectory and branching patterns in the neonatal rat corpus callosum. *Dev Brain Res* 53:283–287

Hollander WF (1981): Linkage relations of dumpy, a recessive mutant on chromosome 13 of the mouse. *J Hered* 72:358–359

Holloway RL, de Lacoste MC (1986): Sexual dimorphism in the human corpus callosum: an extension and replication study. *Hum Neurobiol* 5:87–91

Hoover-Plow J, Elliott P, Moynier B (1988): Reproductive performance in C57BL and I strain mice. *Lab Anim Sci* 38:595–602

Hoover-Plow J, Guerre Millo M, Lavau M (1987): Increased insulin mediated glucose metabolism in fat cells from I versus C57BL mice. *Proc Soc Exp Biol Med* 184:410–415

Hoppensteadt FC (1989): Intermittent chaos, self-organization, and learning from synchronous synaptic activity in model neuron networks. *Proc Natl Acad Sci USA* 86:2991–2995

Innocenti GM (1986): General organization of the callosal connections in the cerebral cortex. In: *Cerebral Cortex,* Vol. 5, Jones EG, Peters A, eds. New York: Plenum Press, 291–351

Jakob H, Beckmann H (1989): Gross and histological criteria for developmental disorders in brains of schizophrenics. *J R Soc Med* 82:466–469

Jeeves M (1986): Callosal agenesis: neuronal and developmental adaptations. In: *Two Hemisperes—One Brain: Functions of the Corpus Callosum,* Leporé F, Ptito M, Jasper HH, eds. New York: Alan R. Liss, 403–421

Jeeves MA, Silver PH, Jacobson I (1988): Bimanual co-ordination in callosal agenesis and partial commissurotomy. *Neuropsychologia* 26:833–850

Jeeves MA, Temple CM (1987): A further study of language function in callosal agenesis. *Brain Lang* 32:325–335

Jeffery G (1989): Distribution and trajectory of uncrossed axons in the optic nerves of pigmented and albino rats. *J Comp Neurol* 289:462–466

Jeret JS, Serur D, Wisniewski K (1989): Agenesis of the corpus callosum and limbic malformation in Apert's syndrome (letter). *Arch Neurol* 46:10.

Jeret JS, Serur D, Wisniewski K, Fisch C (1985): Frequency of agenesis of the corpus callosum in the developmentally disabled population as determined by computerized tomography. *Pediatr Neurosci* 12:101–103

Jeret JS, Serur D, Wisniewski KE, Lubin RA (1987): Clinicopathological findings associated with agenesis of the corpus callosum. *Brain Dev* 9:255–264

Johnson J Jr (1977): Central nervous system of marsupials. In: *The Biology of Marsupials,* Hunsaker DII, ed. New York: Academic Press, pp 157–278

Johnson MH, Jones SN (1985): Status epilepticus, hypothermia and metabolic chaos in a man with agenesis of the corpus callosum. *J Neurol Neurosurg Psychiatr* 48:480–483

Keeler CE (1933): Absence of the corpus callosum as a mendelizing character in the house mouse. *Proc Natl Acad Sci USA* 19:609–611

Kelsoe JR Jr, Cadet JL, Pickar D, Weinberger DR (1988): Quantitative neuroanatomy in schizophrenia. A controlled magnetic resonance imaging study. *Arch Gen Psychiatry* 45:533–541

Killackey H (1990): Neocortical expansion: an attempt toward relating phylogeny and ontogeny. *J Cogn Neurosci* 2:1–17

King LS (1936): Hereditary defects of the corpus callosum in the mouse, Mus musculus. *J Comp Neurol* 64:337–363

King LS, Keeler CE (1932): Absence of corpus callosum, a hereditary brain anomaly of the house mouse. Preliminary report. *Proc Natl Acad Sci USA* 18:525–528

Kinsbourne M, Fisher M (1971): Latency of uncrossed and of crossed reaction in callosal agenesis. *Neuropsychologia* 9:471–473

Kolodny EH (1989): Agenesis of the corpus callosum: a marker for inherited metabolic disease. *Neurology* 39:847–848

Kovelman JA, Scheibel AB (1984): A neurohistological correlate of schizophrenia. *Biol Psychiatry* 19:1601–1621

LaMantia A-S, Rakic P (1990): Axon overproduction and elimination in the corpus callosum of the developing rhesus monkey. *J Neurosci* 10:2156–2175

Larbrisseau A, Vanasse M, Brochu P, Jasmin G (1984): The Andermann syndrome: agenesis of the corpus callosum associated with mental retardation and progressive sensorimotor neuronopathy. *Can J Neurol Sci* 11:257–261

Lassalle JM, Medioni J, LePaper G (1979): A case of behavioral heterosis in mice: quantitative and qualitative aspects of performance in a water-escape task. *J Comp Physiol Psychol* 93:116–123

Lassonde MC (1986): The facilitatory influence of the corpus callosum on intrahemispheric processing. In: *Two Hemisperes—One Brain: Functions of the Corpus Callosum,* Leporé F, Ptito M, Jasper HH, eds. New York: Alan R. Liss, pp 385–401

Lassonde MC, Bryden M and Demers P (1990) The corpus callosum and cerebral speech lateralization. *Brain Language* 38:195–206

Lassonde MC, Sauerwein H, McCabe N, Laurencelle L, Geoffroy G (1988): Extent and limits of cerebral adjustment to early section or congenital absence of the corpus callosum. *Behav Brain Res* 30:165–181

Legius E, Fryns JP, Casaer P, Boel M, Eggermont E (1985): Schinzel acrocallosal syndrome: a variant example of the Greig syndrome. *Ann Genet* 28:239–240

Lent R, Schmidt SL (1986): Dose-dependent occurrence of the aberrant longitudinal bundle in the brains of mice born acallosal after prenatal gamma irradiation. *Brain Res* 390:127–132

Lewis SW, Reveley MA, David AS, Ron MA (1988): Agenesis of the corpus callosum and schizophrenia: a case report. *Psychol Med* 18:341–347

Lipp H-P, Berther R, Glanzmann P, Bolla C, Waanders R (1989a): Two-way avoidance learning of acallosal mice. *Experientia* 45:A12

Lipp H-P, Collins RL, Nauta WJH (1984): Structural asymmetries in brains of mice selected for strong lateralization. *Brain Res* 310:393–396

Lipp H-P, Schwegler H, Crusio WE, Wolfer DP, Heimrich B, Driscoll P, Leisinger-Trigona M-C (1989b): Using genetically-defined rodent strains for the identification of hippocampal traits relevant for two-way avoidance learning: a non-invasive approach. *Experientia* 45:845–859

Lipp H-P, Waanders R (1990): The acallosal mouse strain I/Ln: behavioral comparisons and effects of cross-breeding. *Behav Genet* 20:728–729

Lipp H-P, Waanders R, Berther R, Glanzmann P, Bolla C (1989c) Two-way avoidance learning and swimming navigation in acallosal mice. *Eur J Neurosci* 2(suppl): 29

Livy D (1990): *Tests of Genetic Allelism between four Inbred Mouse Strains with Absent Corpus Callosum.* Unpublished M.D. Thesis, Waterloo, Canada: University of Waterloo

Livy D, Wahlsten D (1991): Tests of genetic allelism between four inbred mouse strains with absent corpus callosum. *Heredity,* in press

Loeser JD, Alvord EC Jr (1968): Agenesis of the corpus callosum. *Brain* 91:553–570

Mathieu J, Bedard F, Prevost C, Langevin P (1990): Motor and sensory neuropathies with or without agenesis of the corpus callosum: a radiological study of 64 cases. *Can J Neurol Sci* 17:103–108

Merriam AE, Gardner EB (1987): Corpus callosum function in schizophrenia: a neuropsychological assessment of interhemispheric information processing. *Neuropsychologia* 25:185–193

Milner AD (1982): Simple reaction times to lateralized visual stimuli in a case of callosal agenesis. *Neuropsychologia* 20:411–419

Milner AD (1983): Neuropsychological studies of callosal agenesis. *Psychol Med* 13:721–725

Milner AD, Jeeves MA (1979): A review of behavioural studies of agenesis of the corpus callosum: In: *Structure and Function of Cerebral Commissures,* Russell LS, Van Hof MW, Berlucchi G, eds. Baltimore: University Park Press, pp 428–448

Milner AD, Jeeves MA, Silver PH, Lines CR, Wilson J (1985): Reaction times to lateralized visual stimuli in callosal agenesis: stimulus and response factors. *Neuropsychologia* 23:323–331

Mooradian AD, Morley GK, McGeachie R, Lundgren S, Morley JE (1984): Spontaneous periodic hypothermia. *Neurology* 34:79–82

Morris RGM (1984): Development of a water-maze procedure for studying spatial learning in the rat. *J Neurosci Meth* 11:47–60

Nasrallah HA, Andreasen NC, Coffman JA, Olson SC, Dunn V, Ehrhardt JC (1986a): The corpus callosum is not larger in left-handers. *Soc Neurosci Abstr* 12:194.9

Nasrallah HA, Andreasen NC, Coffman JA, Olson SC, Dunn VD, Ehrhardt JC, Chapman SM (1986b): A controlled magnetic resonance imaging study of corpus callosum thickness in schizophrenia. *Biol Psychiatry* 21:274–282

Nasrallah HA, McCalley-Whitters M, Bigelow LB, Rauscher FP (1983): A histological study of the corpus callosum in chronic schizophrenia. *Psychiatry Res* 8:251–260

Nowakowski RS (1984): The mode of inheritance of a defect in lamination in the hippocampus of BALB/c mice. *J Neurogenet* 1:249–258

Okada Y, Negi T, Hori Y, Murakami TH (1989): The cortical EEG in the acallosal mouse. *Nippon Seirigaku Zasshi* 51:165–171

O'Kusky J, Strauss E, Kosaka B, Wada J, Li D, Druhan M, Petrie J (1988): The corpus callosum is larger with right-hemisphere cerebral speech dominance. *ann Neurol* 24:379–383

Olavarria J, Serra-Oller MM, Yee KT, Van Sluyters RC (1988): Topography of interhemispheric connections in neocortex of mice with congenital deficiencies of the callosal commissure. *J Comp Neurol* 270:575–590

Oppenheim JS, Skerry JE, Tramo MJ, Gazzaniga MS (1989): Magnetic resonance imaging morphology of the corpus callosum in monozygotic twins. *Ann Neurol* 26:100–104

Ozaki HS, Iwahashi K, Shimada M (1989): Ipsilateral corticocortical projections of fibers which course within Probst's longitudinal bundle seen in the brains of mice with congenital absence of the corpus callosum: a study with the horseradish peroxidase technique. *Brain Res* 493:66–73

Ozaki HS, Murakami TH, Toyoshima T, Shimada M (1984): Agenesis of the corpus callosum in ddN strain mouse associated with unusual facial appearance (flat-face). *Neurosci Res* 1:81–87

Ozaki HS, Shimada M (1988): The fibers which course within the Probst's longitudinal bundle seen in the brain of a congenitally acallosal mouse: a study with the horseradish peroxidase technique. *Brain Res* 441:5–14

Peters M (1988): The size of the corpus callosum in males and females: implications of a lack of allometry. *Can J Psychol* 42:313–324

Probst M (1901): Ueber den Bau des vollständig balkenlosen Grosshirnes sowie über Mikrogyrie und Heterotopie der grauen Substanz. *Arch Psychiatrie* 34:709–786

Randall PL (1983): Schizophrenia, abnormal connection, and brain evolution. *Med Hypotheses* 10:247–280

Reynolds DM, Jeeves MA (1974): Further studies of crossed and uncrossed pathway responding in callosal agenesis—reply to Kinsbourne and Fisher. *Neuropsychologia* 12:287–290

Reynolds DM, Jeeves MA (1977): Further studies of tactile perception and motor coordination in agenesis of the corpus callosum. *Cortex* 13:257–272

Reynolds GP, Czudek C, Andrews HB (1990): Deficit and hemispheric asymmetry of GABA uptake sites in the hippocampus in schizophrenia. *Biol Psychiatry* 27:1038–1044

Rosen GD, Sherman GF, Emsbo K, Mehler C, Galaburda AM (1990): The midsagittal area of the corpus callosum and total neocortical volume differ in three inbred strains of mice. *Exp Neurol* 107:271–276

Rosenthal R, Bigelow LB (1972): Quantitative brain measurements in chronic schizophrenia. *Br J Psychiatry* 121:259–264

Rossi A, Stratta P, Gallucci M, Passariello R, Casacchia M (1988): Brain morphology in schizophrenia by magnetic resonance imaging (MRI). *Acta Psychiatr Scand* 77:741–745

Rossi A, Stratta P, Gallucci M, Passariello R, Casacchia M (1989): Quantification of corpus callosum and ventricles in schizophrenia with nuclear magnetic resonance imaging: a pilot study. *Am J Psychiatry* 146:99–101

Sadowsky C, Reeves AG (1975): Agenesis of the corpus callosum with hypothermia. *Arch Neurol* 32:774–776

Sanders RJ (1989): Sentence comprehension following agenesis of the corpus callosum. *Brain Language* 37:59–72

Sanfield JA, Linares OA, Cahalan DD, Forrester JM, Halter JB, Rosen SG (1989): Altered norepinephrine metabolism in Shapiro's syndrome. *Arch Neurol* 46:53–57

Sauerwein H, Lassonde MC (1983): Intra- and interhemispheric processing of visual information in callosal agenesis. *Neuropsychologia* 21:167–171

Sauerwein HC, Lassonde MC, Cardu B, Geoffroy G (1981): Interhemispheric integration of sensory and motor functions in agenesis of the corpus callosum. *Neuropsychologia* 19:445–454

Schinzel A (1979): Postaxial polydactyly, hallux duplication, absence of the corpus callosum, macrencephaly and severe mental retardation: a new syndrome? *Helv Paediatr Acta* 34:141–146

Schinzel A (1988): The acrocallosal syndrome in first cousins: widening of the spectrum of clinical features and further support for autosomal recessive inheritance. *J Med Genet* 25:332–336

Schinzel A, Schmid W (1980): Hallux duplication, postaxial polydactyly, absence of the corpus callosum, severe mental retardation, and additional anomalies in two unrelated patients: a new syndrome. *Am J Med Genet* 6:241–249

Schmidt S, Manhaes A, DeMoraes V (1991): The effects of total and partial callosal agenesis on the development of paw preference performance in the BALB/cCF mouse, submitted Brain Res. 545:123–130

Schmidt SL, Caparelli-Dquer EM (1989): The effects of total and partial callosal agenesis on the development of morphological brain asymmetries in the BALB/cCF mouse. *Exp Neurol* 104:172–180

Schöpke R, Lipp H-P, Wolfer DP (1991): Swimming navigation and structural variations of the infrapyramidal mossy fibers in the hippocampus of the mouse. *Hippocampus* 1:345–328

Schwegler H, Buselmaier W (1981): Behavior genetic analysis of water-T-maze learning in inbred strains of mice, their hybrids, and selected second generation crosses. *Psychol Res* 43:335–345

Schwegler H, Crusio WE, Brust I (1990): Hippocampal mossy fibers and radial-maze learning in the mouse: A correlation with spatial working memory but not with non-spatial reference memory. *Neuroscience* 34:293–298

Schwegler H, Lipp H-P (1983): Hereditary covariations of neuronal circuitry and behavior: correlations between the proportions of hippocampal synaptic fields in the regio inferior and two-way avoidance in mice and rats. *Behav Brain Res* 7:1–39

Serur D, Jeret JS, Wisniewski K (1988): Agenesis of the corpus callosum: clinical, neuroradiological and cytogenetic studies. *Neuropediatrics* 19:87–91

Shapiro WR, Williams GH, Plum F (1969): Spontaneous recurrent hypothermia accompanying agenesis of the corpus callosum. *Brain* 92:423–436

Smith RC, Baumgartner R, Calderon M (1987): Magnetic resonance imaging studies of the brains of schizophrenic patients. *Psychiatry Res* 20:33–46

Sömmering ST (1796): *Ueber das Organ der Seele*. Königsberg: Friedrich Nicolovius

Stratta P, Rossi A, Gallucci M, Amicarelli I, Passariello R, Casacchia M (1989): Hemispheric asymmetries and schizophrenia: a preliminary magnetic resonance imaging study. *Biol Psychiatry* 25:275–284

Swanson LW, Köhler C, Björklund A (1987): The limbic region. I: the septohippocampal system. In: *Handbook of Chemical Neuroanatomy Vol. 5: Integrated Systems of the CNS, Part I. Hypothalamus, Hippocampus, Amygdala, Retina,* Björklund A, Hökfelt T, Swanson LW, eds. Amsterdam: Elsevier, pp 125–278

Swayze VW, Andreasen NC, Ehrhardt JC, Yuh WT, Alliger RJ (1990): Developmental abnormalities of the corpus callosum in schizophrenia. *Arch Neurol* 47:805–808

Temple CM, Jeeves MA, Villaroya O (1989): Ten pen men: rhymimg skills in two children with callosal agenesis. *Brain Lang* 37:548–564

Toriello HV, Carey JC (1988): Corpus callosum agenesis, facial anomalies. Robin sequence, and other anomalies: a new autosomal recessive syndrome. *AmJ Med Genet* 31:17–23

Tumbelaka R (1915): Das Gehirn eines Affen, worin die interhemisphäriale Balkenverbindung fehlt. *Folia Neurobiol* 9:1–64

Upchurch M, Wehner JM (1988): Differences between inbred strains of mice in Morris water maze performance. *Behav Genet* 18:55–68

Upchurch M, Wehner JM (1989): Inheritance of spatial learning ability in inbred mice: a classical genetic analysis. *Behav Neurosci* 103:1251–1258

Velek M, White LE Jr, Williams JP, Stafford RL, Marco LA (1988): Psychosis in a case of corpus callosum agenesis. *Alabama J Med Sci* 58:27–29

Waanders R, Lipp H-P (1989): Handedness and hind-limb coordination in mice with and without a corpus callosum. *Eur J Neurosci* 2(suppl):29

Wahlsten D (1974): Heritable aspects of anomalous myelinated fibre tracts in the forebrain of the laboratory mouse. *Brain Res* 68:1–18

Wahlsten D (1979): Mode of inheritance of deficient corpus callosum in the brains of laboratory mice. *Behav Genet* 9:484

Wahlsten D (1982a): Mice in utero while their mother is lactating suffer higher frequency of deficient corpus callosum. *Dev Brain Res* 5:354–357

Wahlsten D (1982b): Deficiency of the corpus callosum varies with strain and supplier of the mice. *Brain Res* 239:329–347

Wahlsten D (1982c): Mode of inheritance of deficient corpus callosum in mice. *J Hered* 73:281–285

Wahlsten D (1987): Defects of the fetal forebrain in mice with hereditary agenesis of the corpus callosum. *J Comp Neurol* 262:227–241

Wahlsten D (1989a): Genetic and developmental defects of the mouse corpus callosum. *Experientia* 45:828–838

Wahlsten D (1989b): Deficiency of the corpus callosum: incomplete penetrance and substrain differentiation in Balb/c mice. *J Neurogenet* 5:61–76

Wahlsten D, Jones G (1983): Structural changes in brains of mice with agenesis of the corpus callosum. *Soc Neurosci Abstr* 9:494

Wahlsten D, Andison M (1989): Patterns of cerebellar foliation in recombinant inbred mice. *Soc Neurosci Abstr* 15:102

Wahlsten D, Smith G (1989): Inheritance of retarded forebrain commissure development in fetal mice: results from classical crosses and recombinant inbred strains. *J Hered* 80:11–16

Ward R, Tremblay L, Lassonde MC (1987): The relationship between callosal variation and lateralization in mice is genotype-dependent. *Brain Res* 424:84–88

Witelson SF (1985): The brain connection: the corpus callosum is larger in left-handers. Science 229:665–668

Witelson SF (1989): Hand and sex differences in the isthmus and genu of the human corpus callosum. A postmortem morphological study. *Brain* 112:799–835

Genetic Animal Models of Endogenous Depression

DAVID H. OVERSTREET

There have been an extensive number of animal models developed to mimic some aspects of human affective disorders (Katz, 1981; Maier, 1984; McKinney, 1984; Porsolt, 1981; Willner, 1984). However, virtually all of these reviews have neglected the genetic component even though it has long been recognized that genetic factors contribute to the pathogenesis of affective disorders (Akiskal and McKinney, 1973; Gershon, 1982). Recently, interest has turned to the increased cholinergic sensitivity that has been induced by selective breeding for anti-cholinesterase sensitivity (Overstreet et al., 1979b, 1986a) as a potential model for the cholinergic supersensitivity reported to exist in at least some human depressives. In this chapter we will examine these selectively bred rats in detail as well as inbred mouse strains and psychogenetically selected rat lines; in so doing, we will suggest that the findings are consistent with the hypothesis of supersensitive cholinergic mechanisms in affective disorders.

Following the original proposal of Janowsky and colleagues (1972) that depression in humans may be associated with either an underactive noradrenergic system or an overactive cholinergic system, the progress in understanding cholinergic mechanisms in affective disorders has been gradual. Recently, there has been a renewed interest in this area because of the following observations: increased sensitivity of affective disorder patients to cholinergic agonists (Janowsky et al., 1980; Risch et al., 1981a,b), increased muscarinic receptor binding in affective disorder patients (Meyerson et al., 1982), cholinergically mediated rapid eye movement (REM) sleep abnormalities in depressed patients (Sitaram et al., 1980, 1982), and increased cholinergic sensitivity in antidepressant withdrawal phenomena (Dilsaver and Greden, 1984a,b). Several recent reviews have highlighted the evidence from human studies (Dilsaver, 1986; Janowsky and Risch, 1984; Janowsky et al., 1985); our purpose here will be to summarize these results as a background to the animal studies that will be discussed in greater detail.

HUMAN STUDIES

A number of studies have confirmed earlier observations (Janowsky et al., 1972) regarding the behavioral suppressing effects of physostigmine (Janowsky et al.,

1974, 1980, 1985). Similar observations have been reported following the administration of the muscarinic agonist arecoline (Nurnberger et al., 1982; Risch et al., 1981a, 1983). Other data suggest that patients with a history of depression may be more sensitive to the REM sleep-inducing (Jones et al., 1985; Sitaram et al., 1980, 1982, 1987), neuroendocrine (Akiskal and McKinney, 1973; Risch et al., 1983) and behavioral (Risch et al., 1981a,b) effects of muscarinic agonists. However, there are some reports that failed to find a significantly greater behavioral response to arecoline in depressed patients (Berger et al., 1983; Nurnberger et al., 1983).

There is some evidence that the increased cholinergic sensitivity may be related to a common neuroendocrine abnormality in affective disorders—the failure to suppress adrenocorticotropic hormone (ACTH) and cortisol following administration of dexamethasone. Carroll et al. (1979) demonstrated that physostigmine administration reversed dexamethasone-induced cortisol suppression and Janowsky et al. (1985) have reported partial confirmation of these observations. It is also well known that physostigmine administration leads to increased serum concentrations of ACTH, beta endorphine, and cortisol (Davis and Davis, 1980; Janowsky and Risch, 1984; Risch et al., 1983). Since dexamethasone-induced suppression of cortisol recurs with remission of depression or after discontinuation of physostigmine treatment, these data support the hypothesis that cholinergic supersensitivity may exist only during the state of depression (i.e., a state marker). However, some of the agonist-induced REM sleep studies suggest that the cholinergic supersensitivity may be a trait marker (Sitaram et al., 1987). Further studies are necessary to resolve the issue of whether the cholinergic supersensitivity model is more suitable for the trait and/or state of affective disorders (see Nurnberger et al., 1989; Berger et al., 1989).

There have been only a few studies of muscarinic acetylcholine receptors (mAChR) in patients with affective disorders. An early report suggested that mAChRs were elevated in individuals who had committed suicide (Meyerson et al., 1982). However, a more recent report failed to demonstrate any difference in mAChR binding between normals and suicides (Kaufman et al., 1984). There are several problems with these studies that preclude any definite statements at this stage. First, there were very little clinical data, so it was not certain how many of the suicides were in fact suffering from affective disorders. Second, the cerebral cortex was selected for receptor binding analyses and data about mAChR concentrations in other brain regions are not available. It is possible that the lack of mAChR changes in the cortex of suicides may simply reflect the lack of cognitive differences between suicides and controls. Binding data on mAChR in limbic or other brain regions in suicides and/or affective disorder patients could be illuminating.

A report suggesting that elevated mAChR on cultured fibroblasts may be a biological marker in bipolar affective disorder patients (Nadi et al., 1984) has engendered a considerable amount of interest and controversy. There is now considerable doubt about whether mAChR on fibroblasts are functionally relevant (Kelsoe et al., 1985). Certainly, more work on mAChR binding in affective disorder patients is necessary before we can determine whether the cholinergic model is supported.

In sum, several reports are consistent with a cholinergic supersensitivity model of depression in humans, although some negative reports also exist. Because of the likelihood of heterogeneous populations of human depressives, the human literature may be confounded. By studying animals with more uniform genetic backgrounds, additional evidence for the involvement of cholinergic mechanisms in depressive behavioral states has been collected. We now devote our attention to these animal studies.

Genetic Animal Models

Because of the evidence presented above that cholinergic supersensitivity may be a trait/state marker for depressive disorders in humans, this section will focus on genetic animal models in which changes in cholinergic functions either have been demonstrated or are likely to be involved because of the pattern of behavioral differences. The Roman High-Avoidance (RHA) and the Roman Low-Avoidance (RLA) lines of rats were selectively bred to differ in their ability to acquire an active avoidance response (Driscoll and Battig, 1982), as were the Syracuse High-Avoidance and Syracuse Low-Avoidance rats (Brush et al., 1979). On the other hand, the Maudsley Reactive and Maudsley Nonreactive rats were selectively bred for their high and low rates of defecation in an open field (Broadhurst, 1975), whereas the Tsukuba low-emotional and Tsukuba high-emotional rats were selectively bred for emotionality on a runway (Fujita et al., 1976). Finally, the Flinders Sensitive Line (FSL) and Flinders Resistant Line (FRL) of rats selectively bred to differ in their sensitivity to the anticholinesterase, diisopropyl fluorophosphate (DFP) (Overstreet et al., 1979b; Russell et al., 1982). Before reviewing the literature on these rats, however, we will devote some attention to studies on inbred mouse strains that have explored the relationship between cholinergic mechanisms and behavioral and physiological variables.

Inbred Mouse Strains

Although none of the studies to be considered in this section specifically discusses how the findings may be related to the cholinergic hypothesis of affective disorders, the reports do illustrate the complexity of the relationships between neurochemical and behavioral variables. The reader is referred to the excellent review by Ingram and Corfman (1980) for a discussion of other neurobiological differences among inbred mouse strains.

In one of the earliest reports, van Abeelen and colleagues noted a differential response to cholinergic drugs in inbred mouse strains (van Abeelen and Strijbosch, 1969; van Abeelen et al., 1971, 1972). The C57BL/6 strain, which had high initial levels of exploration, exhibited decreases after scopolamine, whereas the DBA/2 strain, which had low initial levels of exploration, exhibited increases (van

Abeelen and Strijbosch, 1969). The D2B6F1 strain, a genetic cross between the two inbred strains, had initial levels of exploration similar to the C57BL/6 strain and exhibited smaller decreases after scopolamine. Other experiments indicated that the strain differences were centrally mediated (van Abeelen et al., 1971) and that the hippocampal cholinergic neurons were involved (van Abeelen et al., 1972). Other workers also noted strain-dependent effects of scopolamine in mice (Anisman et al., 1975; Oliverio et al., 1973).

More recent studies have concentrated on the strain-dependent effects of direct and indirect cholinergic agonists. Thus, there have been reports of differences among inbred mouse strains for the muscarinic agonist oxotremorine (Marks et al., 1981), for the nicotinic agonist nicotine (Marks et al., 1983, 1984; Miner et al., 1984; see chapter by Collins et al.), and for the anticholinesterase agent DFP (Smolen et al., 1985). In general, the C57BL mice were more affected by the hypothermia-inducing effect of both oxotremorine and DFP than were DBA mice, whereas the C3H mice were least sensitive to oxotremorine and DFP (Marks et al., 1981; Smolen et al., 1985). On the other hand, C3H mice were most sensitive to the stimulant effects of nicotine whereas C57BL and DBA mice were most sensitive to the depressant effects (Marks et al., 1983; Miner et al., 1984). A valuable feature of these studies has been the measurement of cholinergic receptor binding and acetylcholinesterase (AChE) activity in addition to the pharmacological responses to the drugs. These data permitted the investigators to conclude that the genotype-dependent drug effects could not be easily explained by strain differences in either AChE activity or receptor binding (Marks et al., 1981; Smolen et al., 1985). They suggested "that the genetic influences on response to oxotremorine are due to some other factor such as genetic influence on rèceptor coupling processes" (Smolen et al., 1985, p. 624).

On the basis of these studies, the C57BL mice exhibit the greatest sensitivity to oxotremorine, a muscarinic agonist; therefore, they are similar to depressed humans and further work oriented toward the cholinergic hypothesis of depression could be illuminating. An example of the type of work that could be done is the recent preliminary report from Nurnberger et al. (1988): they studied six recombinant inbred strains from a DBA × C57 cross for open-field activity, responses to scopolamine and arecoline, and neurochemical indices of cholinergic function. Their results suggested that cholinergic sensitivity and open-field activity might be genetically correlated; they are continuing this work with a larger number of recombinant inbred strains (Nurnberger, personal communication).

Another series of reports has focused on presynaptic cholinergic processes and behavior. An early report suggested that the capacity for long-term memory is greater in the BALB/C strain than in the C57BL/6 or C57BR strains (Jaffard et al., 1977). These behavioral differences appeared to be correlated with differences in choline acetyltransferase (CAT) activity; the BALB/C mice have greater values. This study also reported that the improvement in memory following hippocampal stimulation was strain-dependent. However, in a more recent study, radial maze performance was examined in these same three strains (Reinstein et al., 1983). The C57BR performed most efficiently, followed by the BALB/C and C56BL. It was also reported that there were no significant differences in CAT activity between the strains, although the trends were similar to the earlier report (Jaffard et al., 1977). At any rate, there was no correlation betweeen behavioral performance and CAT activity in these strains. Thus, there is some discrepancy between the studies. A factor that may have been overlooked in these studies is the possibility that some of the behavioral differences among the mouse strains may be related to differences in postsynaptic cholinergic mechanisms (e.g., mAChR binding). Also, within-strain differences in mice between laboratories are known to exist, as Frischknecht et al. (1988) have pointed out for the C57BL/6 and DBA/2 strains.

Still another possibility that needs to be considered is that the behavioral differences among the mouse strains may be related to differences in other neurotransmitter systems (Ingram and Corfman, 1980). For example, because it is well known that the cholinergic system interacts with the dopaminergic system in the modulation of locomotor activity, it is possible that the differential behavioral effects of scopolamine on exploratory activity in mouse strains (van Abeelen et al., 1971) may be related to differences in dopaminergic function (Durkin et al., 1983). The possibility of opiate–cholinergic interactions could also be examined (Ebel et al., 1980).

This brief and selective summary has not been exhaustive but should be sufficient to indicate that there definitely are genotype-dependent cholinergic responses in mice. There are also genotype-dependent behavioral differences in mice. However, these behavioral and pharmacological differences cannot be easily accounted for by either presynaptic (CAT activity) or postsynaptic (mAChR binding) cholinergic mechanisms. Further behavioral and biochemical comparisons of the C57BL (cholinergic-sensitive) and DBA (cholinergic-resistant) strains of mice could be fruitful.

Roman Lines

Since their original development by Giorgio Bignami (1965) in Rome, the RHA and RLA lines of rats have been maintained in separate colonies in Switzerland, England, and Canada, among other places. In this brief section we wish to highlight some of the findings that are relevant to cholinergic supersensitivity and affective disorders. Driscoll and Battig (1982) have reviewed much of the data available on the two lines and concluded that the main behavioral differences between them were related to emotional behavior and reactions to stress, variables that are known to be influenced by the cholinergic system (Janowsky and Risch, 1984).

A number of studies have reported differences in sensitivity to cholinergic agonists in the two lines. The RLA line, which is regarded to be the more emotional of the two lines (Driscoll and Battig, 1982), was more sensitive to the behavioral effects of oxotremorine and physostigmine (Brimblecombe et al., 1975; Martin et al., 1981, 1984). However, these differences in drug sensitivity could not be attributed to differences in mAChR, there being no differences in receptor concentrations in the several brain regions studied (Overstreet et al., 1981). It is possible that the differences in sensitivity to cholinergic agonists between the RLA and RHA lines may be related to the reported differences in benzodiazepine receptors (Gentsch et al., 1981) and/or to differences in serotonin turnover (Driscoll, 1988; Driscoll et al., 1980), as both of these systems may interact with the cholinergic system (Janowsky and Risch, 1984; Janowsky et al., 1985; Pradhan and Bose, 1978). Alternatively, they may be related to genetic influences on receptor coupling, as suggested for the inbred mice earlier.

However, a recent study reported that the RLA rats had higher rates of spontaneous alternation and more rapid acquisition of a delayed reinforced alternation test than the RHA rats (Guenaire and Delacour, 1985). This study is consistent with other reports that the RLA rats perform better on some learning tasks than the RHA rats (Driscoll and Battig, 1982) and that their poor performance in the active avoidance task is a consequence of motor performance factors, not of learning ability. It is pertinent to point out here that the behavioral differences between the RLA and RHA rats can be related to postulated differences in the cholinergic system: an overactive cholinergic system in the RLA rats would be expected to lead to poorer active avoidance, better passive avoidance performance, and higher rates of spontaneous alternation (Driscoll and Battig, 1982; Guenaire and Delacour, 1985). A recent, interesting finding in this regard has been that RLA rats show higher levels of CAT activity in striatum and hippocampus than do RHA rats (Driscoll et al., 1987). Thus, the RLA strain of rats exhibits behaviors and sensitivity to cholinergic drugs that indicate that its cholinergic system is overactive. Consequently, these rats may provide a useful animal model for human affective disorder patients who are also more sensitive to muscarinic agonists, as indicated above. However, the concept of anxiety (emotionality) must also be considered in this context. It could be argued, for example, that the RLA rats are genetically more anxious, which predisposes them to becoming behaviorally depressed or immobile. A potential direction for further research is interbreeding studies between the lines to assess genetic relatedness of the various behavioral and neurochemical differences.

Two other recent developments with these Roman rats require comment. Lipp, Schwegler, and colleagues have reported that the avoidance differences between RHA and RLA rats may be related to differences in hippocampal morphology; specifically, the better performing rat (RHA) will have small infrapyramidal mossy fiber projections (see Lipp et al., 1988, for references). Interestingly, they have also reported that the DBA strain of mouse, concluded to be cholinergically resistant, also exhibits a high rate of avoidance and a low extent of mossy fiber projection. Thus, DBA mice are similar to RHA rats. In their most recent study the developmental profile of animals was altered by injecting thyroid hormone (Lipp et al., 1988). They found that the high negative correlation between avoidance performance and extent of mossy fiber projection was maintained. What is unknown as yet is how cholinergic sensitivity fits into this scheme. Cholinergic challenges in these developmentally altered rats could be informative.

Because of earlier studies of differences in dopaminergic sensitivity in RLA and RHA rats (Driscoll et al., 1985; Durcan et al., 1983), more recent studies have focused on the dopamine system. These studies have shown that dopamine turnover in the prefrontal cortex is increased during exposure to various stressful conditions (Scatton et al., 1988) and that the hypoemotional RHA rats showed elevation of dopamine turnover whereas the hyperemotional RLA rats did not (D'Angio et al., 1988; Scatton et al., 1988). In contrast, the RLA rats exhibited behavioral signs of emotional response such as freezing and defecation. They suggested that the lack of increased dopamine turnover in the RLA rats may be a sign of their inability to cope with the stressful environment. It would be intriguing to study the turnover of acetylcholine under identical experimental conditions, as immobility during stress has also been related to cholinergic mechanisms (Janowsky and Risch, 1984).

Syracuse Lines

The Syracuse Lines of rats were also psychogenetically selected to be high (SHA) or low (SLA) avoiders on a two-way active avoidance task (Brush et al., 1979). A fairly thorough behavioral analysis of these two lines, derived from Long–Evans rats, has been conducted in recent years, but there is relatively little information about the neurochemical mechanisms underlying these behavioral differences. Interestingly, however, Brush et al. (1988) have concluded that the behavioral evidence is "consistent with the hypothesis that SLA animals are more emotionally reactive than SHA animals," virtually identical to the conclusion that Driscoll and Battig (1982) made about the RLA and RHA rats. In this section the main behavioral differences between the SLA and SHA lines will be described in some detail; this will be followed by a briefer discussion on the likely mechanisms mediating these behavioral differences.

At present the SHA rats average more than 40 avoidances in a 60-trial session (67%), whereas the performance of the SLA rats is near 0. (Recent data indicate that the FRL rats, selectively bred for low cholinergic sensitivity, avoid on 60% of the trials, while the FSL rats, selectively bred for high cholinergic sensitivity, avoid on only 30% of the trials; see below) Therefore, a clear separation in their active avoidance behavior is seen. The SLA rats also showed greater suppression of intertrial behavior, but this and the learning deficit were not associated with differences in shock sensitivity or in speed of escape or avoidance responding (Brush et al., 1985).

The SHA and SLA rats also differ on a number of other behavioral measures. In an open field the levels of ambulation are similar in the two lines, but the SLA defecates at a higher rate (Brush et al., 1985). This finding is also consistent with the hypothesis that SLA rats are more emotionally reactive. Several studies have shown that the SLA rats learn more readily tasks that involve inhibitory responding, such as passive avoidance learning and conditioned emotional responses (Iso et al., 1988; Brush et al., 1988). Note that the RLA and FSL rats also perform better on inhibitory tasks.

There are also differences between the SLA and SHA rat lines for pain sensitivity, as assessed by the hot plate, and for stress-induced analgesia (Nagase and Brush, 1986; Nagase et al., 1985). However, there were no differences in pain sensitivity measured by the tail-flick test and in basal or stress levels of cardiovascular activity. Neither the pain sensitivity nor the stress-induced analgesia could be blocked by narcotic antagonists, so it appears that they are not mediated by endogenous opioids.

No other information is available as yet on which neurochemical mechanisms mediate these differential effects. However, the pattern of behavioral differences is virtually identical to those seen in the RLA and RHA rats on the one hand, and the FSL and FRL rats on the other. Since these rats differ in their sensitivity to cholinergic agonists, it is tempting to propose that differences in cholinergic sensitivity may mediate the differences in SLA and SHA rats. It is well known that cholinergic mechanisms may be involved in nonopiate-mediated stress analgesia (Chance, 1980) as well as in passive avoidance learning (see later). The cholinergic system is also known to regulate the pituitary–adrenal axis, which is thought to be involved in the SLA and SHA differences (Brush et al., 1980; Janowsky and Risch, 1984). Certainly, there is a need for psychopharmacological and biochemical studies of SLA and SHA rats; there are good reasons why the cholinergic system should be the first to be studied.

Other Psychogenetically Selected Lines

Two other groups of rats that have been psychogenetically selected will be briefly mentioned here. The Maudsley Reactive (MR) and Maudsley Nonreactive (MNR) lines of rats were originally developed by Broadhurst (1975) to be either high or low defecators in the open field. Some work suggests that the MR line is more reactive than the MNR line and is therefore comparable to the RLA line (Driscoll and Battig, 1982). However, these rats have been regarded more as animal models of anxiety rather than depression because the MR rats appear to be more "anxious" on some tests reflecting anxiety (Commissaris et al., 1986), but not all (Beardslee et al., 1989). Again, there is a paucity of information about the neurochemical mechanisms underlying these behavioral differences (Blizard, 1981) and no conclusions are possible at this stage.

The Tsukuba low-emotional (TLE) and Tsukuba high-emotional (THE) rats were selectively bred for low or high "emotionality," as indexed by a runway test (Fujita et al., 1976). The TLE rats enter more sections on a brightly lit runway than the THE rats. The two lines also differ on a number of other behavioral tasks, including open-field activity and various learning tasks. Iso et al. (1988) recently compared the THE and TLE lines with the SHA and SLA lines on a number of tasks. They found a high degree of correspondence, with the THE and SLA lines closely resembling each other—mutually reinforcing the views that each is more emotional than its counterpart. Even so, there is still very little information about the neurochemical mechanisms underlying these behavioral differences. A strong candidate is the cholinergic system because pharmacological selection for differences in cholinergic sensitivity has resulted in two lines of rats that vary in similar behaviors. We now turn our attention to these rats.

Flinders Lines

In the mid-1970s Overstreet and Russell initiated a selective breeding project in Australia in order to study the mechanisms underlying sensitivity and resistance to anticholinesterase compounds (Russell and Overstreet, 1987). It was known at the time that insects frequently developed resistance to organophosphate anticholinesterase compounds as a consequence of changes in the enzyme AChE. On the other hand, it was also well known that tolerance development to anticholinesterase compounds was readily observed and occurred in the absence of marked changes in AChE, which remained chronically low. In order to pursue this question, selective breeding of rats that were either highly sensitive or relatively resistant to the behavioral and physiological effects of DFP were conducted (Overstreet et al., 1979b). It soon became apparent that only unidirectional selection was occurring: the Flinders sensitive line (FSL) of rats was more sensitive to DFP than randomly bred Sprague–Dawley rats but the Flinders Resistant Line (FRL) was not more resistant (Overstreet et al., 1979b; Russell et al., 1982).

In the course of developing the FSL and FRL rats, a sex difference in sensitivity to DFP was discovered, with females being less affected than males at the standard dosing regimen (Overstreet et al., 1979a). This sex difference complicated the breeding program and necessitated the testing of both males and females at each generation. Nevertheless, it became evident that both males and females of the FSL rats were more sensitive to DFP than their like-sexed counterparts of the FRL (Overstreet et al., 1979b; Russell et al., 1982). Some time ago, we carried out an interbreeding study in which we compared the F1 and F2 progeny with the parental lines. This study suggested that the sensitivity to DFP was related to one or more recessive genes in the males, but the genetic regulation was more complex in females (Overstreet and Russell, 1984).

Because several experiments indicated that lack of involvement of AChE in the sensitivity of FSL rats (Overstreet et al., 1979b; Sihotang and Overstreet, 1983), we turned our attention to other possible mechanisms. Early psychopharmacological studies indicated that the FSL rats were more sensitive to the behavioral suppressing effects of muscarinic agonists and, under certain conditions, were less sensitive to the locomotor stimulatory effects of muscarinic antagonists (Overstreet and Russell, 1982). These results suggested that the FSL rats had increased mAChR sensitivity. Subsequent studies have confirmed the increased sensitivity of the FSL rats to muscarinic agonists, using a range of behavioral and physiological indices (Crocker and Overstreet, 1984; Netherton and Overstreet, 1983; Overstreet et al., 1986a,b; Pepe et al., 1988; Wallis et al., 1988). Most recently, the sensitivity of FSL and FRL rats to a number of selective muscarinic agonists

TABLE 12.1 Altered drug sensitivity in FSL and FRL rats

Agonist drug	Receptor	Measure	Line with greater effect
Arecoline	mACh	Operant Response	FSL
Arecoline	mACh	Corticosterone	FSL
Oxotremorine	mACh	Temperature	FSL
Pilocarpine	mACh	General activity	FSL
McN-343	m1ACh	Operant Response	FSL
Oxotremorine-M	m2ACh	Operant Response	FSL
Apomorphine	DA	Temperature	FSL
Apomorphine	DA	Stereotypy	FRL
LY157555	DA2	Temperature	FSL
LY157555	DA2	Sterotypy	FRL
mCPP	5-HT1	General activity	FSL
mCPP	5-HT1	Operant response	FSL
mCPP	5-HT1	Temperature	FSL
Diazepam	BZD	General activity	FSL
Diazepam	BZD	Operant response	FSL
Muscimol	GABA	General activity	FSL
Muscimol	GABA	Operant response	FSL

mACh, muscarinic acetylcholine; 1 and 2, subtypes of muscarinic receptors; DA, dopamine; 2, D2 dopamine receptor; 5-HT1, serotonin S1 receptor; BZD, benzodiazepine; GABA, gamma-aminobutyric acid.

was examined. The FSL rats exhibited supersensitive behavioral and/or physiological responses to McN-A-344, an m1 agonist; oxotremorine-M, an m2 agonist; pirenzepine, an ml antagonist; and AFDX-116, an m2 antagonist (Schiller et al., 1988). These findings are summarized in Table 12.1.It will be remembered that human depressed patients also exhibited increased behavioral and hormonal responses to muscarinic agonists (Janowsky et al., 1980; Risch et al., 1981a,b). Thus, there is a parallel between the FSL rats and human depressives.

Recent studies indicate that the muscarinic sensitivity in the FSL rats is influenced by both additive and dominance genetic variance. Cross-mating studies were carried out between the FRL and FSL rats to obtain F1 and F2 progeny and F1 backcrosses to each line. Then operant responding for water reward, open-field activity, and core body temperature were recorded after administration of the muscarinic agonists oxotremorine and arecoline, and the anticholinesterase physostigmine. Biometrical genetic analysis of the data indicated that genetic variance was due exclusively to additive and dominance factors, with the former being involved in 70% of the responses. When dominance genetic variance was involved, the F1 and F2 progeny more closely resembled the FRL parents, suggesting that muscarinic sensitivity may be influenced by recessive genes (Overstreet et al., in press). This finding may be relevant to studies of inheritance of depressive disorders, which conclude that genetic factors are definitely involved but the exact mode of inheritance has been difficult to discern (Gershon et al., 1987).

A study of sleep parameters in FSL and FRL rats under baseline conditions has demonstrated that FSL rats exhibit REM sleep abnormalities similar to those seen in human depressives (e.g., Sitaram et al., 1982; Shiromani et al., 1988). These include an increased amount of REM sleep and a reduced latency to the onset of REM sleep. There is good reason to believe that these abnormalities are related to the cholinergic supersensitivity observed in the depressed humans and FSL rats (Shiromani et al., 1987). Recently, using a sleep deprivation paradigm, we have confirmed that the FSL rats exhibit increased REM sleep (Shiromani et al., 1991).

It is abundantly clear, therefore, that FSL rats exhibit cholinergic (muscarinic) supersensitivity. However, as shown in Table 12.1, there are also some differences between the FSL and FRL rats in their responses to drugs acting on other neurotransmitter systems. The stereotypy-inducing effects of apomorphine, a dopamine agonist, were reduced in the FSL rats (Crocker and Overstreet, 1984, 1991)., whereas the locomotor depressive effects of m-chlorophenylpiperazine, a serotonin agonist, diazepam, a benzodiazpine agonist, and muscimol, a gamma-amino butyric acid agonist, were increased (Pepe et al., 1988; Wallis et al., 1988). In contrast, the hypothermia induced by all agents was significantly greater in the FSL rats (Crocker and Overstreet, 1984, 1991; Wallis et al., 1988). At first glance, these findings may seem confusing; however, it is possible to relate them to the way in which the four systems interact in regulating the two functions. The serotonergic, GABAergic, and cholinergic systems are parallel systems in the regulation of both locomotor activity and temperature; that is, stimulation of either leads to behavioral depression and hypothermia. By contrast, the dopaminergic system is an opposing system in the regulation of locomotor activity; that is, stimulation leads to behavioral activation and a parallel system in the regulation of temperature. It is noteworthy that the RLA line of rats, which are more sensitive to muscarinic agonists, are also more sensitive to the hypothermic effects of apomorphine (Driscoll et al., 1985) and less sensitive to its stereotypy-inducing effects (Durcan et al., 1983).

The fact that FSL rats with genetically influenced cholinergic supersensitivity are also more sensitive to agonists of other neurotransmitters raises questions about the value of single-system challenge studies in humans. Although there is an increasing trend for investigators to give selective agonists to depressives and controls and to measure blood hormone levels, there are virtually no studies that examine more than one neurotransmitter system. Such studies are very much needed, particularly since it is generally agreed that the depressive disorders are heterogeneous conditions.

A potentially important observation about the FSL rats is that the cholinergic supersensitivity in these animals may be mediated by different mechanisms in different brain regions. In rats that were shipped to the UCLA Department of Pharmacology, a tail vein injection of deuterium-labeled choline was administered 1 min before sacrifice by microwave fixation (Overstreet et al., 1984). This procedure permitted the simultaneous assessment of both endogenous and labeled ACh, as well as its precursor, choline. There were no differences in endogenous choline or ACh, or labeled choline. The main finding was a regionally specific increase in ACh synthesis in the cerebral cortex in the FSL rats. In parallel studies of mAChR binding in the lines of rats, significant increases in the concentrations of mAChRs in the striatum and hippocampus of FSL rats were detected, but there were no differences in the cerebral cortex (Overstreet et al., 1984). These data provide evidence to a presynaptic cholinergic overactivity in the cerebral cortex (increased ACh synthesis) and a postsynaptic cholinergic supersensitivity in the striatum and hippocampus (increased mAChRs). The increased concentrations of mAChRs in the striatum and hippocampus, but not the cortex, have been recently replicated; interestingly, binding sites for benzodiazepines were also elevated in the same brain regions (Pepe et al., 1988). Preliminary studies suggest that the mAChR elevation is not confined to one subtype: both m1 and m2 subtypes appear to be increased (Schiller et al., 1989), as might be expected from the functional studies discussed above). These preliminary studies have also indicated that the FSL rats have significantly higher binding sites in the hypothalamus, the site of temperature regulation, and midbrain/pons, the site of generation of REM sleep (Shiromani et al., 1987, 1988).

The regional specificity of the mAChR elevations may be relevant to the lack of mAChR differences between suicides and normals (Kaufman et al., 1984); that is, the cerebral cortex may have been an inappropriate region to examine. As indicated previously, data on mAChR concentrations in limbic and other regions likely to participate in the behavioral and physiological dysfunctions observed in depressed individuals would be useful.

Given the strong suggestion that the muscarinic receptor supersensitivity in the FSL rats is influenced by additive and dominance genetic factors and is correlated with mAChR elevations in specific brain regions, it was decided to conduct a developmental study of muscarinic supersensitivity. It was found that the FSL rats were more sensitive to the hypothermic and activity-reducing effects of oxotremorine than were the FRL rats at all ages tested, including 21-day-old rats (Daws et al., 1991). However, elevated mAChR concentrations were found in only 61-day-old rats, suggesting that muscarinic supersensitivity is not necessarily correlated with increases in mAChRs. Studies are currently in progress to determine whether second messengers of muscarinic stimulation may be involved in the supersensitivity exhibited by the FSL rats.

These findings indicate that the muscarinic supersensitivity is present at a very early age and suggest that a similar early appearance may occur in humans who later become depressed. Surprisingly, although many investigators have compared the muscarinic supersensitivity of human depressives with normals (see above), no one has yet studied young children. Such a study is likely to be quite informative because behavioral studies of children from depressed parents have indicated that some children might exhibit depressive-like states (e.g., Weissman et al., 1987).

Because the cholinergic system is involved in virtually all behavioral and physiological variables, the established differences in cholinergic functions between the lines (Overstreet et al., 1984) may result in some behavioral and physiological differences. Indeed, a number of baseline differences between the FSL and FRL rats has been observed, including decreased body weight and locomotor activity (Netherton and Overstreet, 1983; Overstreet et al., 1979b; Russell et al., 1982). Both of the latter changes are well documented to occur in humans who become depressed, but less information is known about these variables in euthymic individuals with a history of affective disorders. Interestingly, a recent report has established that euthymic patients do indeed show a reduced activity compared to control subjects (Wolff et al., 1985). This reduction was most apparent during the daylight hours, the active portion of the circadian cycle, which parallels our recent finding that the FSL rats were less active than the FRL rats during the dark, active portion of the circadian cycle (Overstreet, unpublished observations). The parallels between FSL rats and depressives for REM sleep values were mentioned earlier.

On avoidance tasks the FSL and FRL rats exhibit differences that suggest that they may be comparable to the RLA and RHA lines, respectively. Although it is difficult to compare across laboratories using different criteria for behavioral tasks, the following observations do seem pertinent. The FSL and RLA (by definition) perform poorly on an active avoidance task (Driscoll and Battig, 1982; Overstreet and Measday, 1985; Overstreet et al., 1990). In contrast, the FSL have a significantly better memory on a passive avoidance task (Overstreet, 1986) and the RLA rats perform better than the RHA rats on similar tasks (Driscoll and Battig, 1982). These results could also be related to cholinergic supersensitivity because it is well known that cholinergic stimulation can produce memory enhancement (see Overstreet, 1984, 1989). The studies of active avoidance performance of the FSL rats have not been as extensive, so direct comparisons of the two lines cannot be made because of differences in experimental design. It is not known, for example, if the performance of FSL rats is as poor as is commonly reported for RLA rats. Recent observations suggest that the FRL rats may perform as well as the SHA rats, but the FSL rats may perform somewhat better than the SLA rats (see above) (Overstreet et al., 1990).

One of the problems that these behavioral data present is the apparent lack of parallels with depressed humans. If one considers only the active avoidance data, then a good parallel exists, with the FSL and RLA rats being impaired and most studies suggesting that memory processes, when altered, are impaired in depressive disorders (McAllister, 1981). In fact, RLA rats never do acquire the avoidance response, adopting a freezing posture after the first few shocks (Driscoll and Batting, 1982), and the cognitive deficit in depressed humans appears to be an acquisition rather than a memory deficit (McAllister, 1981). Although the evidence for improved "memories" of FSL and RLA rats on the passive avoidance task appears to be at variance with the human studies, it must be remembered that human depressives do, in fact, have selectively good memories for aversive events (e.g., Bower, 1981, 1987; Breslow et al., 1981). In any case, performance of avoidance tasks is more likely a reflection of the animal's emotionality rather than its memory (Driscoll and Battig, 1982), so there really is not any disparity between the animal and human data.

Another point for consideration is the recent observation that FSL rats exhibit a greater degree of immobility than do FRL rats when exposed to mild stressors (Overstreet, 1986; Overstreet et al., 1986b, 1989). The FSL animals were significantly less active in an open field after exposure to footshock and had longer periods of immobility in the forced swim test, developed originally by Porsolt (1981). These results have been interpreted in the following manner: the mild stressors (footshock and forced swimming) lead to an activation of the cholinergic system (Janowsky and Risch, 1984). The released ACh interacts with elevated mAChRs (Overstreet et al., 1984) in the FSL rats, thereby resulting in greater immobility. Borsini and Meli (1988) have recently reviewed the forced swim test and concluded that is was useful in detecting antidepressants in rats, but not in mice. Thus, the greater immobility shown by the FSL rats may be a reflection of "depression." There are indications that human depressives may also be more sensitive to stressors (Anisman and Zacharko, 1982). There is a need to test the FSL and FRL responses to other stressors, and for further work with humans before generalizations should be made.

Reduced locomotion after footshock and impaired active avoidance performance might also be related to anxiety, rather than depression. Several workers have demonstrated, in particular, that active avoidance performance is improved by administration of anxiolytics (e.g., Boix et al., 1988; Criswell and Breese, 1989), although they do not influence the forced swim test (Borsini and Meli, 1988). To clarify this situation, we compared the baseline behavior of the FSL and FRL rats in the elevated plus-maze, a commonly used test to detect anxiolytics. There were no differences between the rats on baseline behavior; in addition, diazepam (1 mg/kg) increased the time spent in the open arms of the maze to a similar extent in the two lines (Schiller et al., 1991). Therefore, there is no evidence for a difference in anxiety-related behavior between FSL and FRL rats. Sitaram et al. (1984) have provided evidence from human studies that muscarinic supersensitivity is associated with depression and not with anxiety disorders. Once again, the rat and human data coincide.

In summary, there are a number of parallels among behavioral, pharmacological, and neurochemical variables in the FSL rats and depressed humans, as indicated in Table 12.2. All of the points listed in Table 12.2 have been observed in the FSL rats, but their occurrence in depressed humans is less clear cut. It is generally well recognized that lower body weight and reduced general activity are associated with depression and several laboratories have reported enhanced responses to cholinergic agonists. There is less consistency about the reports of elevated muscarinic receptors and enhanced responses to stress. Nevertheless, there is enough consistency between the human and rat data to indicate that further studies of the FSL rats as models for human depression are warranted.

Although the FSL rats share a considerable number of similarities with depressed humans, there is relatively little information about their response to antidepressants. Systematic investigations are being planned at the moment, but

TABLE 12.2 Some similar characteristics between FSL rats and depressed humans

1. Reduced body weight in FSL and weight loss in depressives
2. Reduced general activity in FSL and depressives
3. Increased REM sleep
4. Enhanced responses to cholinergic agonists:
adrenocortical hormones
REM sleep induction
General behavior
Associated with depression, not anxiety
5. Elevated muscarinic receptors
6. Enhanced responses to stress
7. Therapeutic response to antidepressants

there is already some information about three drugs of very different classes. Imipramine, a classical tricyclic antidepressant, was able to moderate the impaired performance of the FSL rats in an active avoidance task (Overstreet and Measday, 1985). Lithium, the prophylactic drug of choice for manic depression, has been postulated to work via cholinergic mechanisms, but workers are not agreed (Casebolt and Jope, 1989; Dilsaver, 1984; Lerer, 1985). Chronic lithium treatment did not counteract the FSL rats' greater tendency for behavioral immobility in the forced swim test, nor did it alter cholinergic depression of activity (Shiromani et al., 1990b). However, the hypothermic effect of oxotremorine was blunted (Shiromani et al., 1990a), and this result correlates with the small effect of lithium on cholinergically stimulated phosphatidyl inositol (Casebolt and Jope, 1989). Finally, rolipram, a new generation antidepressant that inhibits cyclic adenosine monophosphate (AMP) phosphodiesterase (Wachtel, 1983; Wachtel et al., 1988), was able to counteract the shock-induced suppression of activity in the FSL rats, but surprisingly further increased cholinergic sensitivity without altering receptor concentrations (Overstreet et al., 1989). These findings indicate that the FSL's tendency to become immobile is counteracted by two of the three antidepressants used. Further studies with a range of antidepressants are warranted.

DISCUSSION

Although this chapter has focused on genetic animal models of endogenous depression, it is important to keep in mind that multiple factors participate, as so eloquently reviewed by Akiskal and McKinney (1973). It is also important to realize that factors other than genes can influence cholinergic supersensitivity, as reviewed recently by Overstreet et al. (1988). They have proposed an integrative model in which genetic factors, hyperadrenocorticism, and chronic drug treatment are considered predisposing factors, and stress is considered a precipitating factor. The former may all induce a cholinergic supersensitivity. The latter may explain why stress can precipitate depression in such predisposed individuals; the stress will lead to increased release of ACh (Janowsky and Risch, 1984), acting upon supersensitive mAChRs and leading to depression.

It is unlikely, however, that cholinergic supersensitivity will provide an adequate explanation for all forms of affective disorders. Although the model is useful in accounting for some factors, it must be remembered that affective disorders have also been associated with abnormalities in other neurotransmitter systems. Another possible model is the suggestion that cholinergic supersensitivity, as produced by genetic factors and/or hormonal influences, acts as a predisposing factor toward mood disorders, whereas the actual mood alterations may come about through subsequent changes in catecholamines and/or indolamines. For example, FSL rats may be predisposed to behavioral immobility upon exposure to stress not only because they exhibit muscarinic supersensitivity but also because they exhibit GABAergic and serotonergic supersensitivity and/or dopaminergic subsensitivity (Table 12.1). More complex models of neurotransmitter interactions have been suggested recently for the Roman lines of rats (Driscoll et al., 1990). It is premature to propose such models for the Flinders lines because much less is known about the regional changes in neurotransmitter functions (see D'Angio et al., 1988; Scatton et al., 1988).

The integrative model presented above is similar to that proposed originally by Akiskal and McKinney (1973) in regard to the roles played by genetics and stress. In this chapter, however, it has been suggested that animal models of one genetic predisposing factor toward depression can be developed. These models, therefore, differ from previous animal models of depression that have emphasized various environmental factors (Katz, 1981; Maier, 1984; McKinney, 1984; Porsolt, 1981; Willner, 1984). In his critical review of these models, Willner (1984) concluded that "it is therefore possible that stress-based models might prove to be valid as models of a state of depression, even if they should turn out to be wide of the mark as models of its aetiology" (p. 11). The genetic models discussed in this chapter are more likely to be models of the trait of depression and, therefore, relevant to its etiology. Both the FSL and RLA rats are more sensitive to cholinergic agonists, as are people with a history of affective disorders. It is possible that both presynaptic and postsynaptic cholinergic mechanisms are involved in these animals (Driscoll et al., 1987; Durkin et al., 1983; Overstreet et al., 1984). Further research, which integrates stress-based models with the genetic models described in this chapter, could provide even more insights into mechanisms underlying affective disorders.

Acknowledgments. The work reported in this chapter has been supported by The Flinders University Research Budget, The Australian Research Grants Scheme, The National Health and Medical Research Council (of Australia), and the National Institutes of Health (of the USA). I wish to express my appreciation and thanks to the following coworkers: Dr. Roger W. Russell, Dr. Ann D. Crocker, Dr. Grant D. Schiller, Dr. Joe Orbach, Adam Bruzzone, John Kovaliski, Maya Penck, Kay Double, Salvatore Pepe, Robert Westphalen, and Tim Burleigh of Flinders University; Dr. Peter Shiromani, Dr. J Christian Gillin, and Rick Storch of the University of California at San Diego; Ruth Booth of UCLA; and Dr. David S.

Janowsky of the University of North Carolina. I also wish to thank Dr. Peter Driscoll and Dr. F. Robert Brush for sending papers on recent research findings on their rat lines.

REFERENCES

Akiskal HS, McKinney WT (1973): Depressive disorders: toward a unified hypothesis. *Science* 182:20–29

Anisman HA, Wahlsten D, Kokkinidis L (1975): Effects of d-amphetamine and scopolamine on activity before and after shock in three mouse strains. *Pharmacol Biochem Behav* 3:819–824

Anisman HA, Zacharko RM (1982): Depression: the predisposing influence of stress. *Behav Brain Sci* 5:89–137

Beardslee SL, Papadakis E, Altman HJ, Harrington GM, Commissaris RL (1989): Defensive burying in Maudsley reactive (MR/Har) and nonreactive (MNR/Har) rats. *Physiol Behav* 45:449–551

Berger M, Lund R, Bronisch T, von Zerssen D (1983): REM latency in neurotic and endogenous depression and the cholinergic REM induction test. *Psychiatr Res* 10:113–123

Berger M, Reimann D, Hochli D, Spiegel R (1989): The cholinergic rapid eye movement sleep induction test with RS-86. *Arch Gen Psychiatry* 46:421–428

Bignami G (1965): Selection for high rates and low rates of avoidance conditioning in the rat. *Anim Behav* 13:221–227

Blizard DA (1981): The Maudsley reactive and nonreactive rats: a North American perspective. *Behav Genet* 11:469–489

Boix R, Fernandez-Teruel A, Tobena A (1988): The anxiolytic action of benzodiazepines is not present in handling-habituated rats. *Pharmacol Biochem Behav* 31:541–546

Borsini F, Meli A (1988): Is the forced swimming test a suitable model for revealing antidepressant activity? *Psychopharmacology* 94:147–160.

Bower GH (1981): Mood and memory. *Am Psychol* 36:129–148

Bower GH (1987): Commentary on mood and memory. *Behav Res Ther* 25:443–455

Breslow R, Kocsis J, Belkin B (1981): Contribution of the depressive perspective to memory function in depression. *Am J Psychiatry* 138:227–230

Brimblecombe RW, Buxton DS, Redfern RH (1975): Drug effects on avoidance behavior in selected strains of rats. *Br J Pharmacol* 53:461P

Broadhurst PL (1975): The Maudsley reactive and nonreactive strains of rats: a survey. *Behav Genet* 5:299–319

Brush FR, Baron S, Froehlich JC, Ison JR, Pellegrino JC, Phillips DS, Sakellaris PC*Rattus norvegicus:* escape/avoidance responding, sensitivity to electric shock, discrimination learning and open-field behavior. *J Comp Psychol* 99:60–73

Brush FR, Del Paine SH, Pellegrino LJ, Rykaszewski IM, Dess NK, Collins PY (1988): CER suppression, passive-avoidance learning, and stress-induced suppression of drinking in the Syracuse High- and Low-avoidance strains of rats *(Rattus norvegicus)*. *J Comp Psychol* 102:337–349

Brush FR, Froehlich JC, Baron S (1980): Hormonal mediation of genetic difference in avoidance behavior. *Adv Physiol Sci* 13:269–272

Brush FR, Froelich JC, Sakellaris S (1979): Genetic selection for avoidance behavior in the rat. *Behav Genet* 9:309–316

Carroll BJ, Greden JF, Haskett R, Reinberg M, Abala AA, Martin FIR, Rubin RT, Heath

B, Sharp PT, McLeod WL, McLeod MF (1979): Neurotransmitter studies of neuroendocrine pathology in depression. *Acta Psychiatr Scand* 280(suppl):183–189

Casebolt TL, Jope RS (1989): Long-term lithium treatment selectively reduced receptor-coupled inositol phospholipid hydrolysis in rat brain. *Biol Psychiatry* 25:329–340

Chance WT (1980) Autoanalgesia: Opiate and non-opiate mechanisms. *Neurosci Biobehav Rev* 4:55–67

Commissaris RL, Harrington GM, Ortiz AM, Altman HJ (1986): Maudsley reactive and non-reactive rats strains—differential performance in a conflict task. *Physiol Behav* 38:291–294

Criswell HE, Breese GR (1989): A conflict procedure not requiring deprivation: evidence that chronic ethanol treatment induces tolerance to the anti-conflict action of ethanol and chlordiazepoxide. *Alcohol Clin Exp Res* 13:680–685

Crocker AD, Overstreet DH (1984): Dopamine receptor sensitivity in lines of rats bred for differences in sensitivity to anticholinesterases. Presented at C.I.N.P., Florence, Italy, June, 1984

Crocker AD, Overstreet DH (1991): Changes in dopamine sensitivity in rats selectively bred for difference in cholinergic function. *Pharmacol Biochem Behav,* 38:105–108

D'Angio M, Serrano A, Driscoll P, Scatton (1988): Stressful environmental stimuli increase extracellular DOPAC levels in the prefrontal cortex of hypoemotional (Roman high-avoidance) but not hyperemotional (Roman low-avoidance) rats. An in vivo voltammetric study. *Brain Res* 451:237–247

Davis BM, Davis KL (1980): Cholinergic mechanisms and anterior pituitary hormone secretion. *Biol Psychiatry* 15:303–310

Daws LC, Schiller GD, Overstreet DH, Orbach J: (1991) Early development of muscarinic supersensitivity in a genetic animal model of depression. *Neuropsychopharmacology,* 4:207–217

Dilsaver SC (1984): Lithium's effects on muscarinic receptor binding parameters: a relationship to therapeutic efficacy? *Biol Psychiatry* 19:1551–1566

Dilsaver SC (1986): Cholinergic mechanisms in depression. *Brain Res Rev* 11:285–316

Dilsaver SC, Greden JF (1984a): Antidepressant withdrawal phenomena. *Biol Psychiatry* 19:237–256

Dilsaver SC, Greden JF (1984b): Antidepressant withdrawal-induced activation (hypomania and mania): mechanism and theoretical significance. *Brain Res Rev* 7:29–48

Driscoll P (1988): Hypothalamic serotonin turnover in rat lines selectively bred for differences in active, two-way avoidance behavior. *Adv Biosci* 70:55–58

Driscoll P, Battig K (1982): Behavioral, emotional and neurochemical profiles of rats selected for extreme differences in active, two-way avoidance performance. In: *Genetics of the Brain,* Lieblich D, ed. Amsterdam: Elsevier, pp 95–123

Driscoll P, Claustre Y, Fage D, Scatton B (1987): Recent findings in central dopaminergic and cholinergic neurotransmission of Roman high- and low-avoidance (RHA/Verh and RLA/Verh) rats. *Behav Brain Res* 26:213

Driscoll P, Dedek J, D'Angio M, Claustre Y, Scatton B (1990): A genetically-base model for divergent stress responses: behavioral, neurochemical and hormonal aspects. In: *Farm Animals in Biomedical Research* Pliska V, Strunzinger G, eds. Berlin: Verlag Paul Parey, pp 97–107

Driscoll P, Dedek J, Fuchs A, Gentsch C (1985): Stereotypic, hypothermic and central dopaminergic effects of apomorphine in Roman high avoidance and Roman low avoidance rats. *Behav Genet* 15:591–592

Driscoll P, Dedek J, Martin JR, Battig K (1980): Regional 5-HT analysis in Roman high- and low-avoidance rats following MAO inhibition. *Eur J Pharmacol* 68:373–376

Durcan MJ, Fulker DW, Campbell IC (1983): Difference in the stereotypy response but not the hypomotility response to apomorphine in the Roman high and low avoiding strains of rats. *Psychopharmacology* 82:215–220

Durkin TP, Hashem-Zadek H, Mandel P, Kempf J, Ebel A (1983): Genotypic variation in the dopaminergic inhibitory control of striatal and hippocampal activity in mice. *Pharmacol Biochem Behav* 19:63–70

Ebel A, Durkin T, Ayad G, Mach G, Mandel P (1980): Genetically determined cholinergic involvement in morphine-induced behavioral responses in mice. *Neuropharmacology* 19:423–427

Frischknecht HR, Siegfried B, Waser PG (1988): Opioids and behavior: genetic aspects. *Experientia* 44:473–481

Fujita O, Abe L, Nakamura N (1976): Selection for high and low emotionality based on the runway test in the rat: the first seven generations of selection. *Hiroshima Forum Psychol* 3:57–62

Gentsch C, Lichtsteiner M, Feer H (1981): 3H-diazepam binding sites in Roman high- and Roman low-avoidance rats. *Experientia* 37:1315–1316

Gershon ES (1982): Genetics of the major psychoses: a review. *Res Publ Assoc Res Nerv Ment Dis* 60:121–144

Gershon ES, Berrettini W, Nurnberger JI Jr, Goldin LR (1987): Genetics of affective illness. In: *Psychopharmacology: The Third Generation of Progress,* Meltzer HY, ed. New York: Raven Press, pp 481–492

Guenaire C, Delacour J (1985): Differential acquisition of a "working memory" task by the Roman strains of rats. *Physiol Behav* 34:705–707

Ingram DK, Corfman TP (1980): An overview of neurobiological comparison in mouse strains. *Neurosci Biobehav Rev* 4:421–435

Iso H, Brush FR, Fujii m, Shimazaki M (1988): Running-wheel avoidance leaning in rats *(Rattus norvegicus):* effects of contingencies and comparisons of different strains. *J Comp Psychol* 102:350–371

Jaffard R, Ebel A, Destrade C, Durkin T, Mandel P, Cardo B (1977): Effects of hippocampal electrical stimulation on long-term memory and on cholinergic mechanisms in three inbred strains of mice. *Brain Res* 133:277–289

Janowsky DS, El-Yousef MK, David JM (1974): Acetylcholine and depression. *Psychosom Med* 36:248–257

Janowsky DS, El-Yousef MK, Davis JM, Sekerke HJ (1972): A cholinergic-adrenergic hypothesis of mania and depression. *Lancet* 2:632–635

Janowsky DS, Risch SC (1984): Cholinomimetic and anticholinergic drugs used to investigate an acetylcholine hypothesis of affective disorders and stress *Drug Dev Res* 4:125–142

Janowsky DS, Risch SC, Judd LL, Huey LY, Parker DC (1985): Brain cholinergic systems and the pathogenesis of affective disorders. In: *Central cholinergic mechanisms and adaptive dysfunction,* Singh S, ed. New York: Plenum Press, pp 309–333

Janowsky DS, Risch SC, Parker D, Huey LY, Judd LL (1980): Increased vulnerability to cholinergic stimulation in affective disorder patients. *Psychopharmacol Bull* 16:29–31

Jones D, Kelwala S, Bell J, Dube S, Jackson J, Sitaram N (1985): Cholinergic REM sleep induction: response correlation with endogenous major depressive subtype. *Psychiatr Res* 14:99–110

Katz RJ (1981): Animal models and human depressive disorders. *Neurosci Biobehav Rev* 5:231–246

Kaufman CA, Gillin JC, Hill B, O'Laughlin T, Phillips I, Gleinman JE, Wyatt RJ (1984): Muscarinic binding in suicides. *Psychiatr Res* 12:47–56

Kelsoe JR, Gillin JC, Janowsky DS, Brown JH, Risch SC, Lumkim B (1985): Failure to confirm muscarinic receptors on skin fibroblasts. *N Engl J Med* 312:861–862

Lerer B (1985): Studies on the role of brain cholinergic systems in the therapeutic mechanism and adverse effects of electroconvulsive therapy and lithium. *Biol Psychiatry* 20:20–40

Lipp H-P, Schwegler H, Heimrich B, Driscoll (1988): Infrapyramidal mossy fibers and two-way avoidance learning: developmental modification of hippocampal circuitry and adult behavior of rats and mice. *J Neurosci* 8:1905–1921

Maier SF (1984): Learned helplessness and animal models of depression. *Prog Neuro-Psychopharmacol Biol Psychiatry* 8:435–446

Marks MJ, Burch JB, Collins AC (1983): Genetics of nicotine response in four inbred strains of mice. *J Pharmacol Exp Ther* 226:291–302

Marks MJ, Miner L, Burch JB, Fulker DW, Collins AC (1984): A diallel analysis of nicotine-induced hypothermia. *Pharmacol Biochem Behav* 21:953–959

Marks MJ, Patinkin DM, Artman LD, Burch JB, Collins A (1981): Genetic influences on cholinergic drug response. *Pharmacol Biochem Behav* 15:271–279

Martin JR, Driscoll P, Gentsch C (1984): Differential response to cholinergic stimulation in psychogenetically selected rat lines. *Psychopharmacology* 83:262–267

Martin JR, Overstreet DH, Driscoll P, Battig K (1981): Effects of scopolamine, pilocarpine and oxotremorine on the exploratory behavior of two psychogenetically selected rat lines. *Psychopharmacology* 72:135–142

McAllister TW (1981): Cognitive functioning in the affective disorders. *Comp Psychiatry* 22:572–586

McKinney WT (1984): Animal models of depression: an overview. *Psychiatr Devel* 2:77–96

Meyerson LR, Wenogle LP, Abel MS, Coupet J, Lippa AS, Rough CE, Beer B (1982): Human brain receptor alterations in suicide victims. *Pharmacol Biochem Behav* 17:159–163

Miner MJ, Marks MJ, Collins AC (1984): Classical genetic analysis of nicotine-induced seizures and nicotinic receptors. *J Pharmacol Exp Ther* 231:545–554

Nadi NS, Nurnberger JI Jr, Gershon ES (1984): Muscarinic cholinergic receptors on skin fibroblasts in familial affective disorder. *N Engl J Med* 311:225–230

Nagase CS, Brush FR (1986): Genetic differences in avoidance learning covary with non-opioid stress-induced analgesia. *Ann NY Acad Sci* 467:416–418

Nagase CS, Randrich A, Brush FR (1985): Genetic differences in avoidance behavior: Cardiovascular activity, pain sensitivity and stress-induced analgesia. *Peptides* 6(suppl 1):29–35

Netherton RA, Overstreet DH (1983): Genetic and sex differences in the cholinergic modulation of thermoregulation. In: *Environment, Drugs and Thermoregulation,* Lomax P, ed. Basel: Karger, pp 74–77

Nurnberger JI Jr, Berrettini W, Mendelson W, Sack D, Gershon ES (1989): Measuring cholinergic sensitivity: I. Arecoline effects in bipolar patients. *Biol Psychiatry* 25:610–617

Nurnberger JI Jr, Gershon ES, Sitaram N, Gillin JC, Grown G, Ebert M, Gold P, Jimerson

D, Kesler L (1982): Dextroamphetamine and arecoline as pharmacogenetic probes in normal and remitted bipolar patients. *Psychopharmacol Bull* 17:80–82

Nurnberger JI Jr, Jimerson DC, Simmons-Alling S, Tamminga C, Nadi NS, Lawrence D, Sitaram N, Gillin JC, Gershon ES (1983): Behavioral, physiological and neuroendocrine response to arecoline in normal twins and well state bipolar patients. *Psychiatr Res* 9:191–200

Nurnberger JI Jr, Simon JR, Hingtgen J (1988): Differences in open field behavior in recombinant inbred strains of mice. *Soc Neurosci Abstr* 14:1073

Oliverio A, Eleftheriou BE, Bailey DW (1973): Exploratory activity: Genetic analysis of its modification by scopolamine and amphetamine. *Physiol Behav* 10:893–899

Overstreet DH (1984) Behavioral plasticity and the cholinergic system. *Prog Neuro-Psychopharmacol Biol Psychiatry* 8:133–151

Overstreet DH (1986) Selective breeding for increased cholinergic function: development of a new animal model of depression. *Biol Psychiatry* 21:49–48

Overstreet DH (1989): Memory enhancement in Animals and Humans following administration of cholinergic agents. In: *Psychobiology: Issues and Applications,* Siddle D, Bond NW, eds. Elsevier: Sydney, pp 23–33

Overstreet DH, Booth RA, Dana R, Risch SC, Janowsky DS (1986a): Enhanced elevation of corticosterone following arecoline administration to rats selectively bred for increased cholinergic function. *Psychopharmacology* 88:129–130

Overstreet DH, Double K, Schiller GD (1989): Antidepressant effects of rolipram in a genetic animal model of depression: cholinergic supersensitivity and weight gain. *Pharmacol Biochem Behav* 34:691–696

Overstreet DH, Driscoll P, Martin JR, Yamamura HI (1981): Brain muscarinic cholinergic receptor binding in Roman high- and low-avoidance rats. *Psychopharmacology* 72:143–145

Overstreet DH, Janowsky DS, Gillin JC, Shiromani PJ, Sutin EL (1986b): Stress induced immobility in rats with cholinergic supersensitivity. *Biol Psychiatry* 21:657–664

Overstreet DH, Janowsky DS, Rezvani AH (1990): Impaired active avoidance responding in rats selectively bred for increased cholinergic function. *Physiol Behav* 47:787–788

Overstreet DH, Measday M (1985): Impaired active avoidance performance in rats with cholinergic supersensitivity: Its reversal with chronic imipramine. Presented at 4th International Congress of Biological Psychiatry, Philadelphia, PA.

Overstreet DH, Russell RW (1982): Selective breeding for sensitivity to DFP. Effects of cholinergic agonists and antagonists. *Psychopharmacology* 78:150–154

Overstreet DH, Russell RW (1984): Selective breeding for differences in cholinergic function: sex differences in the genetic regulation of sensitivity to the anticholinesterase, DFP. *Behav Neural Biol* 40:227–238

Overstreet DH, Russell RW, Crocker AD, Gillin JC, Janowsky DS (1988): Genetic and pharmacological models of cholinergic supersensitivity and affective disorders. *Experientia* 44:465–472

Overstreet DH, Russell RW, Crocker AD, Schiller GD (1984): Selective breeding for differences in cholinergic function: pre- and post-synaptic mechanisms involved in sensitivity to the anticholinesterase, DFP. *Brain Res* 294:227–232

Overstreet DH, Russell RW, Hay DA, Crocker AD: (in press) Selective breeding for increase cholinergic function: biometrical genetic analysis of muscarinic responses. *Neuropsychopharmacology, (in press)*

Overstreet DH, Russell RW, Helps SC, Messenger M (1979b): Selective breeding for sensitivity to the anticholinesterase, DFP. *Psychopharmacology* 65:15–20

Overstreet DH, Russell RW, Helps AC, Runge P, Prescott AM (1979a): Sex differences following manipulation of the cholinergic system by DFP and pilocarpine. *Psychopharmacology* 61:49–58

Pepe S, Overstreet DH, Crocker AD (1988): Enhanced benzodiazepine responsiveness in rats with increased cholinergic function. *Pharmacol Biochem Behav* 31:15–20

Porsolt RD (1981): Behavioral despair. In: *Antidepressants: Neurochemical, Behavioral and Clinical Perspectives,* Enna SJ, ed. New York: Raven Press, pp 121–139

Pradhan SN, Bose S (1978): Interaction among central transmitters. In: *Psychopharmacology: A Generation of Progress,* Lipton M, ed. New York: Raven Press, pp 271–282

Reinstein DK, BeBoissiere T, Robinson N, Wurtman RJ (1983): Radial maze performance in three strains of mice: Role of the fimbria fornix *Brain Res* 263:162–176

Risch SC, Janowsky DS, Gillin JC (1981a): Muscarinic supersensitivity of anterior pituitary ACTH and beta endorphine release in major depressive illness. *Peptides* 4:789–792

Risch SC, Kalin NH, Janowsky DS (1981b): Cholinergic challenge in affective illness: Behavioral and neuroendocrine correlates. *J Clin Psychopharmacol* 1:186–192

Risch SC, Siever LH, Gillin JC, Janowsky DS, Sitaram N, Weker J, Cohen RM, Murphy DL (1983): Differential mood effects of arecoline in depressed patients and normal volunteers. *Psychopharmacol Bull* 19:696–698

Russell RW, Overstreet DH (1987): Mechanisms underlying sensitivity to organophosphorus anticholinesterase compounds. *Prog Neurobiol* 28:97–129

Russell RW, Overstreet DH, Messenger M, Helps SC (1982): Selective breeding for sensitivity to DFP. Generalization of effects beyond criterion variables. *Pharmacol Biochem Behav* 17:885–891

Scatton B, D'Angio M, Driscoll P, Serrano A (1988): An in vivo voltammetric study of the response of mesocortical and meso accumbens dopaminergic neurons to environmental stimuli in strains of rats with differing levels of emotionality. *Ann NY Acad Sci* 537:124–137

Schiller GD, Byrne C, Orbach J, Overstreet DH (1989): Regulation of muscarinic receptors in two lines of rats genetically selected for differential cholinergic sensitivity. Presented at Australian Neuroscience Society, Melbourne, Australia, February, 1989

Schiller GD, Daws LC, Orbach J, Overstreet DH (1991): Absence of anxiety in a genetic animal model of depression with cholinergic supersensitivity. *Brain Res,* Bull 26:443–447

Schiller GD, Orbach J, Overstreet DH (1988): Effects of intracerebroventricular administration of site selective muscarinic drugs in rats genetically selected for differing cholinergic sensitivity. Presented at the Australasian Society for Clinical and Experimental Pharmacology, Adelaide, Australia, December, 1988

Shiromani PJ, Gillin JC, Hendrickson P (1987): Acetylcholine and the regulation of REM sleep—basic mechanisms and clinical implications for affective illness and narcolepsy. *Annu Rev Pharmacol Toxicol* 27:137–156

Shiromani PJ, Overstreet DH, Levy D, Goodrich CA, Campbell SS, Gillin JC (1988): Increased REM sleep in rats selectively bred for cholinergic hyperactivity. *Neuropharmacology* 1:127–133

Shiromani PJ, Overstreet DH, Lucero S (1990b): Failure of dietary lithium to alter immobility in an animal model of depression. *Lithium* 1:241–244

Shiromani PJ, Overstreet DH, Lucero S, Double KL, Jeong D-O (1990a): Dietary lithium blunts oxotremorine-induced hypothermia in a genetic animal model of depression. *Lithium* 1:186–190

Shiromani PJ, Velazquez-Moctezuma J, Overstreet DH, Shalauta M, Lucero S, Floyd C (1991): Effects of sleep deprivation on sleepiness and increased REM sleep in rats selectively bred for cholinergic hyperactivity. *Sleep,* 14:116–120.

Sihotang K, Overstreet DH (1983): Studies on the possible relationship of brain proteins to behavioral sensitivity to DFP. *Life Sci* 32:413–420

Sitaram N, Dube S, Jones D, Pohl R, Gershon S (1984): Acetylcholine and alpha$_1$-adrenergic sensitivity in the separation of depression and anxiety. *Psychopathol* 17(Suppl. 3):24–39

Sitaram N, Jones D, Dube S, Keshavan M, Bell J, Davies A, Reynal P (1987): The association of supersensitive cholinergic REM-induction and affective illness within pedigrees. *J Psychiatr Res* 21:487–497

Sitaram N, Nurnberger JI Jr, Gershon E, Gillin JC (1980): Faster cholinergic REM sleep induction in euthymic patients with primary affective illness. *Science* 208:200–202

Sitaram N, Nurnberger JI Jr, Gershon E, Gillin JC (1982): Cholinergic regulation of mood and REM sleep: a potential model and marker for vulnerability to depression. *Am J Psychiatry* 139:571–576

Smolen A, Smolen TN, Wehner JM, Collins AC (1985): Genetically determined differences in acute responses to diisopropylfluorophosphate. *Pharmacol Biochem Behav* 22:623–630

van Abeelen JHF, Gilissen L, Hannsen T, Lenders A (1972): Effects of intrahippocampal injections of methscopolamine and neostigmine upon exploratory behavior in two inbred mouse strains. *Psychopharmacologia* 24:470–475

van Abeelen JHF, Smits AJM, Raaijma, Kers WGM (1971): Central Location of a genotype-dependent cholinergic mechanism controlling exploratory behavior in mice. *Psychopharmacologia* 19:324–328

van Abeelen JHF, Strijbosch H (1969): Genotype-dependent effects of scopolamine and eserine on exploratory behavior in mice. *Psychopharmacologia* 16:81–88

Wachtel H (1983): Potential antidepressant activity of rolipram and other selective cyclic 3′,5′-monophosphate phosphodiesterase inhibitors. *Neuropharmacology* 22:367–372

Wachtel H, Loschmann PA, Pietzuch P (1988): Absence of anticholinergic sensitivity of rolipram, an antidepressant with a novel mechanism of action, in three different animal models in vivo. *Pharmacopsychiatry* 21:218–221

Wallis EM, Overstreet DH, Crocker AD (1988): Selective breeding for increased cholinergic function: Psychopharmacological evidence for altered serotonergic sensitivity. *Pharmacol Biochem Behav* 31:345–350

Weissman MM, Gammon GD, John K, Merikangas KR, Warner V, Prusoff BL, Sholonski D (1987): Children of depressed parents. *Arch Gen Psychiatry* 44:847–853

Willner P (1984): The validity of animal models of depression. *Psychopharmacology* 83:1–16

Wolff EA III, Putnam FW, Post RM (1985): Motor activity and affective illness. The relationship of amplitude and temporal distribution to changes in affective state. *Arch Gen Psychiatry* 42:288–294

Individual Differences in the Activity of the Hypothalamus–Pituitary–Adrenocortical System After Stressors: Use of Psychogenetically Selected Rat Lines as a Model

CLAIRE-DOMINIQUE WALKER, MICHEL L. AUBERT, MICHAEL J. MEANEY, AND PETER DRISCOLL

The ability of an organism to cope with stressful situations represents one of the major adaptive processes required to adapt successfully to changing environmental conditions. The "coping" processes involve complex interactions among cognitive mechanisms, central neurochemical changes, and neuroendocrine responses, all processes aimed at improving readily available energy and at triggering adequate behavioral responses. Considerable variations among individuals in response to various stressors can occur at virtually any level during the course of the response. One approach to this complex problem has been to use psychogenetically selected rat strains (inbred) or lines (outbred) that have been selected according to behavioral parameters, and in which selection traits have been maintained over generations.

In this chapter we will describe the various processes involved in the physiological and behavioral activation of the organism in response to stressors. We will then discuss these mechanisms in light of studies conducted with various strains or lines of rats, which pertain to the notion of individual differences in the stress response. Furthermore, we will discuss one particular aspect of this response, namely, the activation of the hypothalamus–pituitary–adrenal (HPA) axis, in two lines of rats, the Wistar-derived Roman High- (RHA) and Low- (RLA) Avoidance rats, originated in the early 1960s (Bignami, 1965). Finally, early expression of phenotypic differences between these two rat lines during the neonatal period will be examined.

STRESS RESPONSE: DISSECTING THE PROCESSES?

Neuroendocrine, Metabolic, and Behavioral Responses

After exposure to an aversive stimulus, various central pathways are activated (Feldman, 1985) and converge to stimulate the production of corticotropin-releasing factor (CRF) (Vale et al., 1981) from the neurons of the paraventricular nucleus (PVN) of the hypothalamus (Swanson et al., 1983). At the level of the median eminence, CRF is released from nerve endings into the hypophyseal portal system supplying the anterior pituitary (Bruhn et al., 1984; Rivier and Plotsky, 1986). Depending on the type of stressor applied, another weaker adrenocorticotropin hormone (ACTH) secretagogues such as vasopressin (AVP) (Plotsky et al., 1985a,b), oxytocin (OT) (Gibbs, 1984, 1986), or angiotensin II (AII) (Spinedi and Negro-Vilar, 1983) are released from the hypothalamus, in addition to CRF. Corticotropin-releasing factor stimulates the production of beta-endorphins and ACTH from the anterior pituitary gland (Rivier et al., 1982; Vale et al., 1983) and this latter peptide triggers the production of glucocorticoids from the adrenal cortex (Normand et al., 1980). Corticosterone in the rat and mouse, or cortisol in the human, dog, or monkey act via a negative feedback mechanism through specific receptors (Reul and De Kloet, 1985; De Kloet and Reul, 1987) at the pituitary, hypothalamic, and suprahypothalamic level to regulate the activity of the axis (Keller-Wood and Dallman, 1984; Dallman et al., 1987). Application of a stressor will also activate the sympathetic nervous system to release catecholamines centrally [epinephrine (EPI) and norepinephrine (NEPI)] and in the peripheral circulation via the release of NEPI at nerve endings and EPI from the adrenal medulla (Axelrod and Reisine, 1984).

In addition to this well known scenario, some types of stress reaction may concomitantly trigger the secretion of AVP and/or OT from the posterior pituitary (Lang et al., 1983).

The effects of activation of the HPA axis and, thus, discharge of significant amounts of ACTH, glucocorticoids, and catecholamines into the peripheral circulation can be appreciated both at the metabolic and behavioral levels. The rapid increase in substrate availability represents a major aspect of the metabolic response and is primarily mediated by increased levels of circulating glucocorticoids. Indeed, these will increase lipolysis, gluconeogenesis, and cellular glucose uptake while decreasing protein synthesis (Dallman et al., 1989). Elevated catecholamine concentrations will cooperate to elevate metabolic fuel and oxygen supply to the cells by increasing heart rate and mean arterial blood pressure (Fischer et al., 1982), as well as lipolysis and glycogenolysis. Apart from the metabolic effects, activation of the HPA axis also results in several disturbances related to other endocrine systems, in particular the reproductive system (Rivier et al., 1986), as well as in suppression of the immune system (McEwen, 1987).

The physiological stress response will not be complete without inducing marked changes in the arousal level (Hennessy and Levine, 1979) of the animal and altering the general behavioral state. Changes in coping behavior during stress have been generally classified into various categories depending on whether a fight/flight or a freezing response is the overall result. All three patterns have been associated with activation of various brain areas and have been related to differential hormonal patterns (Henry, 1986). These patterns have been well studied in rodents and are particularly interesting when individual differences between rat lines are considered. For instance, RLA rats are known to exhibit freezing behavior repeatedly after receiving foot shocks (making them poor avoiders) whereas their RHA counterparts are more active avoiders and will thus tend to exhibit the flight pattern. Relative to human behavior, an elegant study has shown, for instance, that differentiation between low-anxious, high-anxious, and repressive personalities may result in different coping styles following a phrase association task (Weinberger et al., 1979). Moreover, these different coping styles were associated with marked differences in elevations of heart rate and skin resistance (an indirect index of sympathetic activation of sweat glands), the high-anxious and repressor types demonstrating higher increases after being stressed than the low-anxious type of personality.

A direct effect of activation of the HPA axis on behavioral parameters has been clearly demonstrated for CRF, ACTH, and glucocorticoids. Intracerebroventricular (ICV) injection of CRF has been shown to increase locomotor activity and learning ability in a dose-dependent manner (Koob et al., 1982). Of interest in the context of individual differences in emotionality is the observation that CRF apparently increased the "stressfulness" of a novel situation: rats pretreated with icv CRF exhibited higher emotionality than vehicle-treated rats in an open field (Koob and Bloom, 1985). However, it is likely that the central effects of CRF on behavior can be separated from the action of CRF on the pituitary–adrenal axis. Alternatively, ACTH was demonstrated to have beneficial effects on acquisition of a conditioned avoidance response (CAR) in a type of rats called "superior avoiders" but not in the "inferior" avoiders (Bohus et al., 1968), suggesting that the effects of ACTH on the acquisition of CAR may vary among individuals or rat lines. Similarly, avoidance responses were increased following median eminence lesions (Levine and Soliday, 1960). Avoidance extinction was delayed by ACTH independently of glucocorticoid production (Bohus and Lissak, 1968; Levine, 1968; Guth et al., 1971). By contrast, corticosterone (Bohus et al., 1968; Conner and Levine, 1969), but not dexamethasone (Kasper-Pandi et al., 1970), was shown to have opposite effects on both acquisition and extinction of an avoidance response. The interaction of CRF, ACTH, and glucocorticoids with behavioral parameters is important when considering basal HPA axis activity and the avoidance score of some rat lines such as the Roman or the Syracuse High- and Low-Avoidance rats. In both cases, selection for high versus low avoidance scores may indeed have selected for high versus low levels of circulating glucocorticoids or for low versus high levels of ACTH in the periphery and CRF in the central nervous system. We will return to this issue later in the chapter.

Individual differences related to the physiological responses to stressors may lie in many of the central mechanisms triggering such neuroendocrine responses, including cognitive processes (Davidson and Levine, 1972) and neurochemical activation.

Cognitive Processes

Intuitively, it is clear that the magnitude of the physiological and behavioral response to an aversive stimulus or to a novel situation depends on how the stressor is perceived by the organism. This cognitive "perception" is influenced by a number of factors, including (1) prior experience or conditioning (Hennessy and Leving, 1977), (2) the degree of expectancy, and (3) the ability to "cope" with the situation and to engage in a certain type of behavior, as well as the dominant/subordinate relationship.

Prior exposure to a certain type of stressor can decrease further behavioral and endocrine activation to this particular type of stressor (De Souza and Van Loon, 1982) while leaving untouched, or exaggerating, the magnitude of the response to a different type of stressor (Dallman and Jones, 1973; Chalmers et al., 1974; Sakellaris and Vernikos-Danellis, 1975). In the case of a shock stressor applied to induce an avoidance response, plasma corticosterone levels decrease as avoidance training proceeds, and are much lower in the stable phase of the response than during the training/acquisition phase (Coover et al., 1973). It has also been demonstrated that individual differences in basal, pretest concentrations of corticosterone were absent in naive animals but developed during the training phase (Wertheim et al., 1969). The response will be modified, therefore, depending on how novel the situation is perceived.

Another aspect of prior exposure is reflected in the effects of early handling or exposure to stressor(s) during the neonatal period. Such early stimulation has been shown to alter some parameters of the adult stress response, including magnitude and duration of the stress-induced corticosterone secretion (Levine, 1967; Meaney et al., 1989), hippocampal glucocorticoid receptor number, and age-related cognitive impairments associated with spatial memory tasks (Meaney et al., 1988).

Bound to the notion of prior exposure to a certain situation is the concept of expectancy. A series of elegant studies have shown that frustration (induced by a lack of reward) and its opposite, elation (induced by a shift to higher reinforcement densities) can specifically alter the direction of the activation of the pituitary–adrenal system, in that the former stimulates production of corticosterone while the latter tends to suppress it (Levine et al., 1972). Thus, the shift from a predictable to an unpredictable event is sufficient to cause an affective state that is reflected in the activation of the HPA axis.

Finally, the magnitude of the activation of the HPA axis depends on the ability of the organism to cope with the aversive situation or to have some kind of control over it (ability to terminate the stressor). Such a phenomenon has been demonstrated in many species, including humans (Breier et al., 1987), monkeys (Hanson et al., 1976), and rodents (Davis et al., 1977; Swenson and Vogel, 1983). Furthermore, it has been demonstrated that the availability of outlets for frustration, that is, the ability to engage in either a consummatory behavior (Levine et al., 1979) or a fight, appears to reduce the stress-induced activation of the HPA axis. Rats able to fight, or at least to assume the fighting posture, have a diminished ACTH and corticosterone production after electroshocks when compared to animals shocked alone (Conner et al., 1971; Weinberg et al., 1980). Similarly, the level of dominance/subordination also influences the magnitude of the endocrine response to stress. Subordinate animals have been repeatedly found to exhibit higher levels of glucocorticoids after stress than their dominant counterparts, and this phenomenon has been well documented in several species including baboons (Sapolsky, 1982, 1988, 1989), mice (Schuhr, 1987), and rats (Popova and

Naumenko, 1972). Thus, the importance of behavioral control and the psychological and environmental setting in which stress occurs should be added to the actual "physical" scale of stressor intensity in determining the magnitude of the neuroendocrine, metabolic, and behavioral stress response.

Neurochemical "Sensors"

In addition to cognitive perception of an aversive situation, neurochemical transmission relays the stressful signal to the neuroendocrine component of the system. This part of the system can be first viewed as a whole array of chemoreceptors and baroreceptors mainly located at the periphery, which send their visceral and somatosensory information to central catecholaminergic cell groups (noradrenergic A1,A2,A6 and adrenergic C1,C2,C3). Direct projections of these cell groups have been demonstrated to the PVN, where CRF cell bodies are located, and to the supraoptic nucleus where AVP cell bodies are found (Sawchenko and Swanson, 1982). Following central changes in catecholamine output, CRF and ACTH release are modified, but the direction of these changes still remain somewhat controversial (Plotsky et al., 1990). Other neurotransmitters like dopamine, serotonin, and gamma-aminobutyric acid (GABA) also affect, directly or indirectly, the activity of the HPA axis. Although studies on central catecholamine and other neurotransmitter metabolism may simply reflect various degrees of arousal or emotionality, they can still provide useful information on which pathways are involved in the neuroendocrine stress response and how these pathways can differ among individuals or among selected lines. For instance, under stress conditions, hypothalamic serotonin turnover was increased in RLA/Verh compared to RHA/Verh rats (Driscoll and Bättig, 1982; Driscoll et al., 1983) whereas dopamine metabolism in the prefrontal cortex was increased in RHA/Verh compared to RLA/Verh rats (D'Angio et al., 1988; Driscoll et al., 1990).

PSYCHOGENETICALLY SELECTED STRAINS AND LINES: MODELS TO STUDY INDIVIDUAL DIFFERENCES

As implied above, individual differences in the response to aversive stimuli can occur at multiple levels that may be difficult to apprehend as a whole. The need to understand possible physiological mechanisms underlying or accompanying individual differences in behavior among humans and animals was a good reason for the selection of various strains of rodents (Treiman and Levine, 1969), often according to behavioral tests reflecting differences in the degree of emotionality. Emotional reactivity had traditionally been measured by the degree of locomotor activity, exploratory behavior, and by the defecation score in an open field (Broadhurst, 1969; Denenberg, 1969). It was also demonstrated, however, that defecation score in the open field and emotionality were best correlated with increases in plasma corticosterone, whereas measurement of activity scores or exploratory behavior were, under certain circumstances, dissociable from emotionality level and corticosterone production (Denenberg, 1969; Stern et al., 1973).

Some of the rat strains and lines that have been directly or indirectly selected for various parameters of emotionality have been reviewed by Broadhurst (1975) and by Gentsch et al. (1988).

Maudsley Reactive and Nonreactive Rats

Originated by Broadhurst (1957), the Maudsley Reactive (MR) and Maudsley Nonreactive (MNR) rat strains have been selected according to differences in their defecation score in an open field, in which the MR rats defecate more than the MNR rats, suggesting a higher level of emotionality for the MR rats (Gray et al., 1965). Among other psychological and physiological parameters, MR animals were found to be poorer avoiders in a two-way shuttle box, to ambulate less, and to have a higher body and adrenal weight than their MNR counterparts (Broadhurst, 1975). Under resting conditions, MR rats had a higher blood pressure and heart rate than did MNR rats (Blizard et al., 1980), and they also exhibited greater increases in heart rate following the onset of visual and acoustic stimuli (Blizard, 1971). Most of the subsequent studies with these rats have been concerned with central and peripheral noradrenergic mechanisms, however (Blizard, 1988), and little work has been done on defining the activity of the HPA axis or of other neurotransmitter systems.

Syracuse High- and Low-Avoidance Rats

More recently, Long–Evans rats were selectively bred for their avoidance behavior in a two-way shuttle box. This selection for the Syracuse High-Avoidance (SHA) and Syracuse Low-Avoidance (SLA) strains demonstrated, in addition, that avoidance behavior could be selected independently on the general activity level (Brush et al., 1979). Defecation score in the open field was higher in SLA than in SHA rats, indicating a higher emotionality in the SLA line (Brush et al., 1988). Differences in plasma corticosterone levels and adrenal weight have been observed between the two strains in that SLA rats exhibit lower circulating glucocorticoids but have higher adrenal weights than SHA rats (Iso et al., 1988). Some of the psychological differences observed between SLA and SHA rats, including shuttle box (active) avoidance, passive avoidance, and open field defecation score, are comparable to those observed in other psychogenetically selected lines and strains, but studies on the HPA axis have been somewhat at odds, suggesting that endocrine correlates may not be tightly linked to avoidance genotype (Brush, 1991).

Roman High- and Low-Avoidance Rats

The selection for the original Roman High-Avoidance (RHA) and Roman Low-Avoidance (RLA) lines from Wistar stock was accomplished using similar criteria (avoidance behavior in a two-way shuttle box) as those used for the selection of SLA and SHA rats (Bignami, 1965; Broadhurst and Bignami, 1965). Psychological and neurochemical differences between RLA/Verh and RHA/Verh rats, selected and bred in Switzerland since 1972 (see Driscoll, 1986), have been extensively described (e.g., Driscoll and Bättig, 1982; Driscoll et al., 1983, 1990). For instance, RHA/Verh rats are more resistant to pentobarbital administration (Driscoll and Käsermann, 1977), but exhibit faster sleep induction times after ether (Käsermann and Driscoll, 1980) than RLA/Verh rats. Additionally, it was shown that stress-induced increases in plasma ACTH, corticosterone, prolactin, and defecation were higher in RLA/Verh compared to RHA/Verh rats when mild, psychological types of stressors were applied. Stronger types of stress like exposure to ether, footshock, and immobilization were less able to induce differential endocrine and metabolic responses between the rat lines under the conditions (e.g., only one time latency, 10 min, measured) used in that experiment (Gentsch et al., 1982). Finally, stress-induced increases in heart rate were higher in magnitude and duration in RLA/Verh as compared to RHA/Verh rats (D'Angio et al., 1988). Thus, important differences in HPA axis activity between RLA/Verh and RHA/Verh rats had been observed in a number of studies but the locations at which these differences were expressed still remained unclear. We therefore undertook some studies on the HPA axis activity in RLA/Verh and RHA/Verh rats, focusing on differences in basal- and stress-stimulated endocrine responses, on target tissue sensitivity, and on the feedback effects on corticosteroids.

HPA AXIS ACTIVITY IN RLA/VERH AND RHA/VERH RATS

Adult Rats

Measurement of pituitary and adrenal responses to two different types of stressors (exposure to ether vapors and open field) have shown that RLA/Verh secreted more corticosterone than RHA/Verh rats under mildly stressful conditions (open field), but similar amounts of corticosterone and ACTH after ether presentation. Under basal conditions, however, RLA/Verh rats were found to have lower plasma ACTH levels and higher corticosterone levels than RHA/Verh rats (Walker et al., 1989). It should be mentioned here that such a relationship between normal to low circulating levels of ACTH and high levels of corticosteroids has been described for depressive and anorexic patients (Gold et al., 1986a,b). In those patients, pituitary and adrenal responses to exogenous CRF were diminished, whereas the adrenal cortex was hyperresponsive to ACTH stimulation. Are similar modifications of pituitary and adrenal sensitivity observable in behaviorally distinct rat lines?

Our results concerning *in vivo* and *in vitro* pituitary secretion of ACTH after exogenous CRF challenge, in RLA/Verh and RHA/Verh rats, are shown in Figure 13.1. In both methods, pituitary output was greater in RLA/Verh than RHA/Verh rats, suggesting a higher sensitivity of the corticotropes of RLA/Verh rats to exogenous CRF. As it is possible that, *in vivo,* the magnitude of the CRF-induced ACTH response reflected modifications in the time course of the response, in addition to changes in pituitary sensitivity to CRF, this possibility was explored in freely moving cannulated rats, with the results being shown in Figure 13.2. No differences in the time course of CRF-induced ACTH secretion were observed, but the magnitude of the response remained greater in RLA/Verh than in RHA/Verh rats (integrated area under the curve: RLA/Verh ± 65.7 ± 9.9 ng/ml/180 min, RHA/Verh = 32.8 ± 5.6 ng/ml/180 min).

Thus, lower basal levels of ACTH in the hyperemotional RLA/Verh line, together with an increased pituitary sensitivity to CRF, suggest that the hypothalamic discharge of CRF may be impaired in RLA/Verh as compared to RHA/Verh rats. It is possible to explain such a hypothesis in terms of increased glucocorticoid negative feedback effect on hypothalamic and pituitary secretions? Despite the fact that both hippocampal and pituitary glucocorticoid receptor number are decreased in RLA/Verh as compared to RHA/Verh rats (Walker et al., 1989), it is possible that the elevated levels of corticosterone found in RLA/Verh versus RHA/Verh rats will exert a stronger inhibitory influence on CRF and/or ACTH release.

Alternatively, one may consider that high corticosterone output in the presence of relatively low circulating levels of ACTH in RLA/Verh rats would be indicative of changes in adrenal sensitivity to ACTH between the two rat lines. We tested thishypothesis by challenging RLA/Verh and RHa/Verh male rats with exogenous ACTH according to a protocol described earlier (Dallman et al., 1978). As shown in Figure 13.3, basal corticosterone levels were significantly higher in RLA/Verh than in RHA/Verh rats at both the circadian peak (PM) and nadir (AM) of the corticosterone rhythm. By contrast, ACTH-induced corticosterone release was similar in both rat lines in the morning (AM), suggesting an absence of adrenal hypersensitivity in RLA/Verh as compared to RHA/Verh rats. It is possible, however, that the ACTH dose used in this study might have maximally stimulated the adrenal, thus obviating the distinction of differential sensitivity between the two rat lines.

In conclusion, we have shown that differences in the basal and stimulated activity of the HPA axis exist between RLA/Verh and RHA/Verh rats, and that at least some of these differences can be explained in terms of altered sensitivity of pituitary cells to CRF. In addition, it is suggested that changes in glucocorticoid feedback potency or corticosterone clearance rate (via changes in concentrations of its carrier protein, CBG) may participate in this phenomenon. When our results on adrenal secretion in the Roman lines are compared to results obtained in other rat lines, however, it appears that basal circulating levels of corticosterone are not always positively correlated to the degree of emotionality (emotionality: MR > MNR, SLA > SHA, RLA > RHA; basal corticosterone levels: MR < MNR, SLA < SHA, RLA > RHA) (Broadhurst, 1975; Gentsch et al., 1982; Brush, 1991).

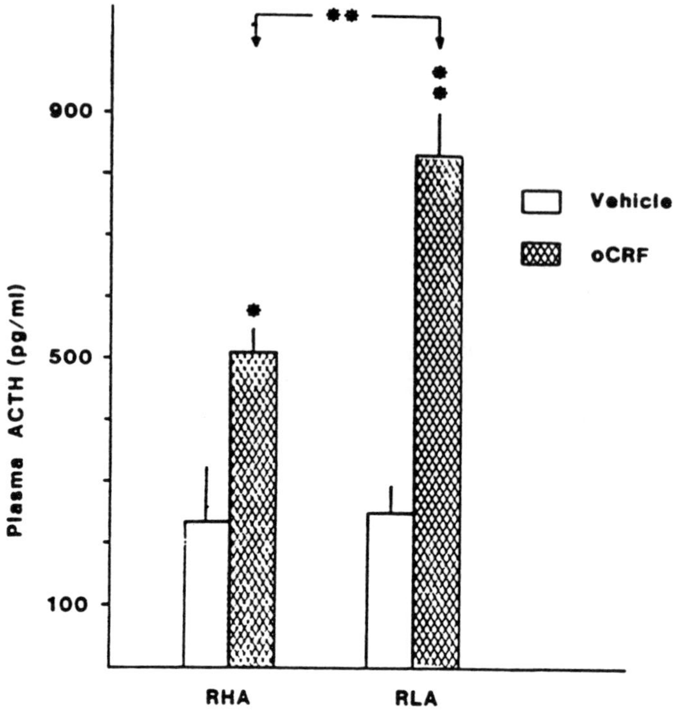
**
*
**
900
500
100
Plasma ACTH (pg/ml)
Vehicle
oCRF
RHA
RLA

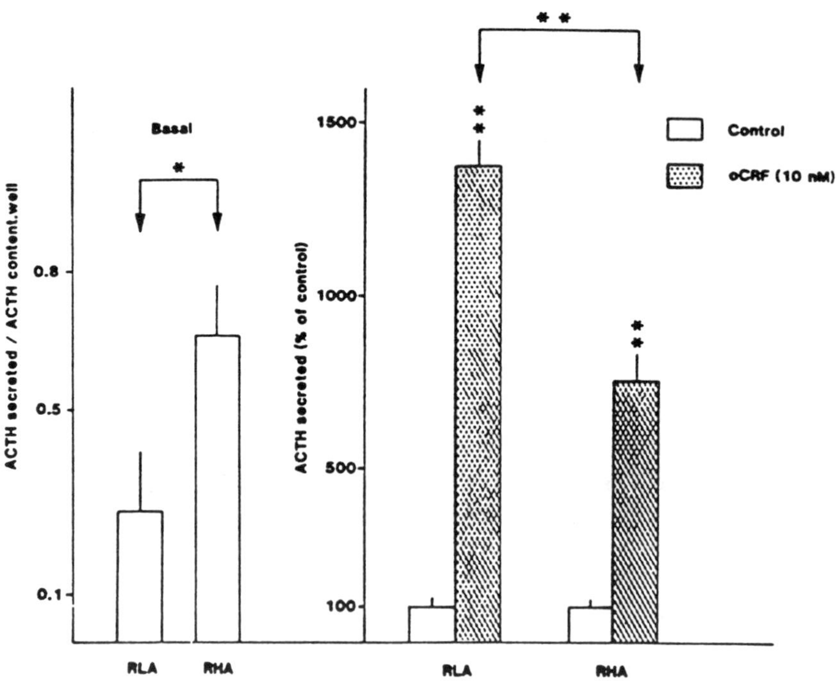
Basal
*
0.8
0.5
0.1
ACTH secreted / ACTH content.well
RLA
RHA
**
1500
1000
500
100
ACTH secreted (% of control)
Control
oCRF (10 nM)
RLA
RHA

Neonatal Rats

Line differences between adult RLA/Verh and RHA/Verh rats have been demonstrated at the pituitary and adrenal levels, but it has not been determined whether these differences are linked to genetic determinants of early developmental patterns or result from genetically influenced changes related to learning and environmental stimulation during the neonatal, preweaning, and postweaning periods. The fact that behavioral differences in the avoidance response and emotionality (including differences in corticosterone secretion) have been maintained over many generations favors the first hypothesis, although we cannot preclude early experiences modulating the importance of such genetic determinants on physiological responses to external stimuli. It has been shown that RLA/Lu and RHA/Lu pups (the Canadian stock) exhibited similar acquisition rates of avoidance response and comparable defecation scores in an open field at weaning age, thus suggesting that line-related differences appeared later in life (days 35–70) (Satinder, 1981). Similarly, the higher levels of locomotor activity regularly observed in adult RHA/Verh compared to RLA/Verh rats (Bättig et al., 1976; Gentsch et al., 1981; Driscoll and Bättig, 1982) were observed in neonates only after weaning age (day 24) (Fümm, 1982). We were therefore interested in determining whether differences in pituitary–adrenocortical activity and pituitary sensitivity to CRF, as observed in adult rats, could appear prior to weaning age and be dissociated from the rather nondifferential behavioral traits of very young RLA/Verh and RHA/Verh rats.

Pituitary ACTH responses to two stressful stimuli (exposure to an open field for 10 min and to ether vapors for 3 min) were determined as a function of postnatal age (Fig. 13.4). Basal levels of ACTH were not different between RLA/Verh and RHA/Verh neonates at any of the ages tested, although by day 21 they tended to be higher in RHA/Verh rats. In agreement with previous studies using a 10-min time-point determination after stress, preweanlings were hyporesponsive to stressors at very early ages (Walker et al., 1986) and the emergence of significant elevations of ACTH with stress occurred earlier following exposure to ether than exposure to an open field, emphasizing the lack of equivalence of various stressors

FIGURE 13.1. *In vivo (top)* and *in vitro (bottom)* response of pituitary corticotropes of RLA/Verh (RLA) and RHA/Verh (RHA) adult male rats to an exogenous CRF challenge. *In vivo,* oCRF was injected at a dose of 10 μg/kg i.p., 30 min prior to sacrifice. *In vitro,* dispersed pituitary cells were maintained in culture for 3 days and challenged on day 4 with oCRF (10nM/well). Basal ACTH secretion was measured after a 90-min preincubation period without the peptide. Media were collected 60 min after addition of the treatment and assayed for ACTH. Pituitary cells were lysed and removed after media collection for determination of total ACTH content/well. For a detailed protocol, see Walker et al. (1989). * = $p < 0.05$, ** = $p < 0.01$. Reproduced with permission of *Journal of Endocinology* from Walker C-D, et al. (1989). Differential activation of the pituitary-adrenocortical axis after stress in the rat: use of two genetically selected lines (RLA and RHA) as a model. *J Endocinol* 123:477–485.

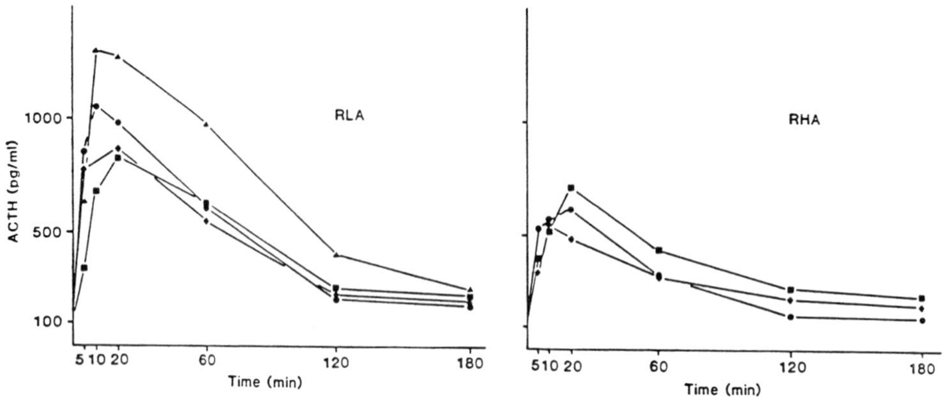

FIGURE 13.2. Time course of ACTH secretion after oCRF (5 μg/kg, i.v.) administration in adult male RLA/Verh (RLA) and RHA/Verh (RHA), cannulated rats. Ovine CRF was injected in phosphate-buffered saline (pH 7.4) containing BSA (0.1%) and ascorbic acid (0.01%) at time zero. Four RLA and three RHA rats are shown, the integrated surface under the curve being calculated over the entire 180 min. Peak concentrations of ACTH were reached 10 to 20 min after administration of the peptide.

in stimulating HPA axis activity, even during the neonatal period. Also demonstrated in these experiments is the absence of line-related differences in stimulated levels of ACTH prior to weaning age. Similar conclusions could be drawn when corticosterone responses to stress were examined (Table 13.1). Although some differences in resting corticosterone levels were observed between RLA/Verh and RHA/Verh rats on day 5 postnatally, no line-related changes in stimulated concentrations were observed.

At this point, it is worth wondering whether differences in maternal care and behavior (Driscoll et al., 1979; Fümm and Driscoll, 1981) could influence the development of differential responses to stress. The answer appears to be negative, since cross-fostering of RLA/Verh and RHA/Verh neonates from birth onward did not change the magnitude of the ACTH and corticosterone responses to ether exposure at weaning age, and basal levels of ACTH or corticosterone were not altered by this procedure (data not shown). Furthermore, cross-fostering did not alter adult shuttle box behavior (Driscoll and Bättig, 1982).

Adult RLA/Verh and RHA/Verh rats differ not only in basal and stimulated HPA axis activity but also in pituitary sensitivity to CRF, as well as in their pituitary and hippocampal glucocorticoid receptor number (Walker et al., 1989). Measurements of the pituitary response to CRF challenge during the neonatal period failed to demonstrate any significant differences in the magnitude of the response between the two rat lines (Fig. 13.5). By contrast, pituitary glucocorti-

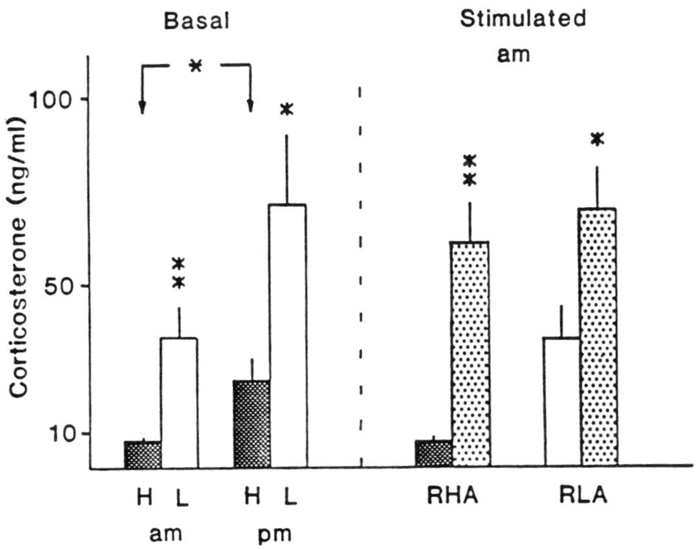

FIGURE 13.3. Basal- and ACTH-stimulated corticosterone concentrations in RLA/Verh (RLA,L) and RHA/Verh (RHA,H) adult male rats. Rats were injected with dexamethasone (300 μg/100 g, i.p.) at the onset of lights on (AM) or 2 hr prior to lights off (PM). Two hr later, they were anesthetized with pentobarbital (50 mg/kg, i.p.) and the jugular vein was exposed. A first blood sample was withdrawn 30 min later for zero time point, just prior to the administration of ACTH (20 ng/kg, i.v.). Subsequent blood samples were collected 15 and 30 min after the treatment. Plasma corticosterone concentrations generally peaked 15 min after the ACTH injection and, therefore, the 15-min time point was chosen to represent maximal adrenal stimulation. Each value represents the mean ± SEM of 9 to 12 animals. * = $p < 0.05$, ** = $p < 0.01$ vs. RHA/Verh rats *(left)* and vs. basal values *(right)*.

coid receptor number was already elevated in RHA/Verh compared to RLA/Verh rats by day 10 of life (Table 13.2). This last observation is in agreement with the increased receptor number found in the pituitary of adult RHA/Verh compared to RLA/Verh rats (Walker et al., 1989).

In summary, whereas adult RHA/Verh and RLA/Verh rats exhibited marked differences in their resting and stimulated HPA axis activity, as well as in their pituitary sensitivity to CRF, no similar variations between the two rat lines were observed during the neonatal period. It is likely, therefore, that the expression of differential HPA axis activity only appears after weaning age, possibly concomitant with other differences in behavioral traits. In support of this hypothesis is the recent report of Castanon et al. (1992) showing that the differences in the activity of the adrenocortical axis between the RLA and RHA lines only appear around 20 weeks of age and remain absent on week 14. However, when differences in prolactin

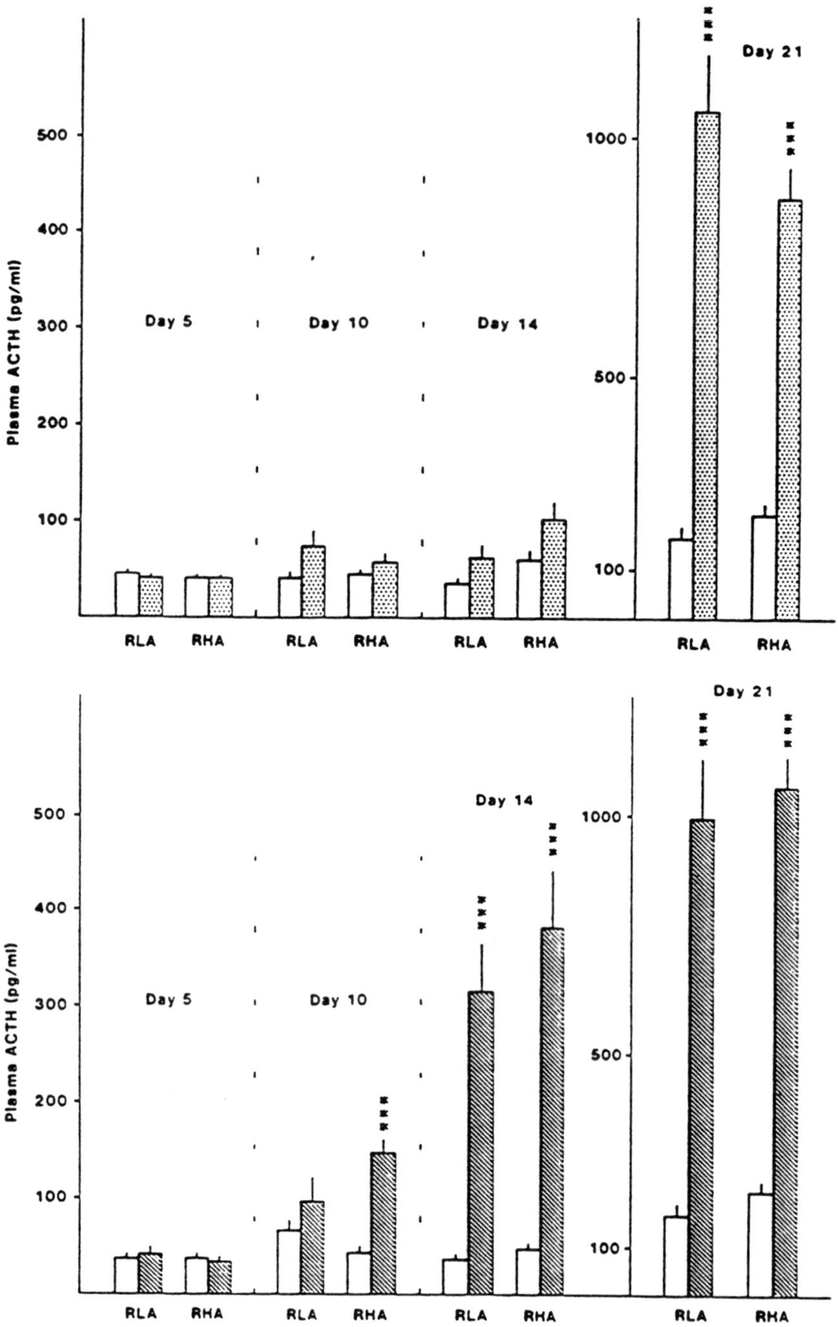

FIGURE 13.4. Age-dependent changes in plasma ACTH concentrations in RLA/Verh (RLA) and RHA/Verh (RHA) neonates under resting conditions *(open bars)* and after a 10-min exposure to an openfield *(top panel, dotted bars)*, or after a 3-min exposure to ether vapors *(bottom panel, hatched bars)*. The rats were sacrificed 10 min after the onset of the stressors. Values represent the mean ± SEM of 6 to 16 animals per treatment. *** = $p < 0.001$ vs. unstressed rats.

TABLE 13.1 Age-related changes in basal or stressor-induced corticosterone concentrations in RLA/Verh and RHA/Verh neonates

Age (days)	Treatment	RLA/Verh	RHA/Verh	p
5	Control (O)	3.7 ± 0.6	1.8 ± 0.1	0.02
	Control (E)	11.5 ± 3.3	3.9 ± 0.7	0.03
	Openfield	6.9 ± 1.1[a]	8.1 ± 2.3[a]	ns
	Ether	9.9 ± 1.0	5.6 ± 1.3	ns
10	Control (O)	2.7 ± 0.4	1.8 ± 0.2	ns
	Control (E)	3.4 ± 0.4	3.0 ± 0.3	ns
	Openfield	4.0 ± 0.7	2.8 ± 0.8	ns
	Ether	6.6 ± 0.4[b]	8.0 ± 0.9[b]	ns
14	Control (O)	12.0 ± 2.2	11.5 ± 3.1	ns
	Control (E)	10.7 ± 1.2	6.8 ± 0.8	0.01
	Openfield	12.5 ± 1.5	12.6 ± 2.4	ns
	Ether	30.1 ± 1.6[b]	25.8 ± 2.3[b]	ns
21	Control (O)	98.1 ± 14	97.7 ± 7.3	ns
	Openfield	160.4 ± 8.4[b]	168.8 ± 5.9[b]	ns
	Ether	152.9 ± 9.9[b]	149.3 ± 8.2[b]	ns

Each control value refers to a particular experiment. O, openfield; E, ether. Experimental protocol is described in Fig. 13.4. Values are mean ± SEM of at least six animals per group.
[a]$p < 0.05$, [b]$p < 0.01$ vs. control group.

secretion were examined, higher prolactin secretion was found in RLA compared to RHA rats after stress from 6 weeks on.

CONCLUSIONS

Individual differences in the endocrine and behavioral responses to stressors are observed in many species and are related to a complex network of sensory, cognitive, neurochemical, and endocrine "signals." The quality and magnitude of these signals vary among individuals, thus resulting in differential physiological and behavioral responses. The observation that HPA axis activity is markedly altered in some defective psychological states make it clear that the relationships between cognitive perception, emotionality, and HPA axis activity need to be investigated further. Several lines of evidence now suggest that individual differences in humans are also associated with changes in central neurotransmitter secretion and turnover rate. The use of psychogenetically selected rat lines is proving to be a useful model in pursuing these lines of research (e.g., Driscoll et al., 1983, 1990; Walker et al., 1989; see also the chapter by Overstreet in this volume for another type of genetic selection), as (even for the Swiss sublines of the Roman lines alone) relationships have already been established between emotion-

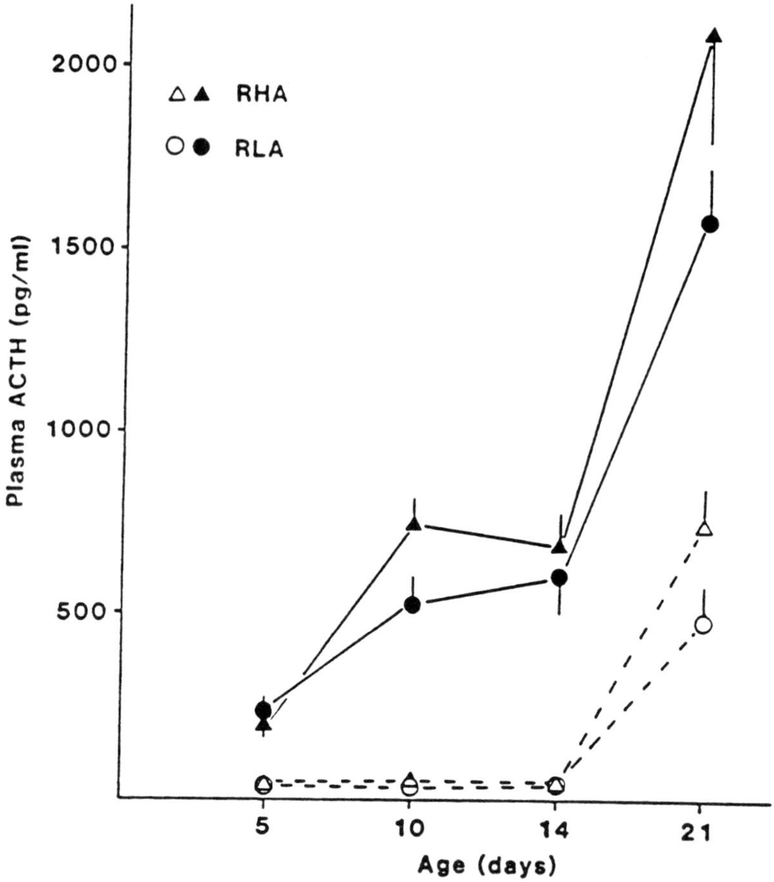

FIGURE 13.5. Effect of vehicle (○, △) or oCRF (●, ▲) (10 μg/kg, i.p.) injection in RLA/Verh (RLA) and RHA/Verh (RHA) neonates on plasma ACTH concentrations. Rats were sacrificed 30 min after the treatment. Values are mean ± SEM of 6 to 10 animals per determination. Significant differences ($p < 0.01$) between vehicle and oCRF-injected rats were observed at all ages for both lines of rats. No line-related differences were observed in either basal or stimulated ACTH levels.

ality level, neurotransmitter systems, basal HPA axis activity, and target tissue sensitivity, and even between the brain and immunological system (Sandi et al., 1991). It is to be hoped that in the future some of these changes can be used as markers of certain psychological states. A better understanding of the factors controlling neuroendocrine and behavioral responses to external aversive stimuli will certainly contribute to a better diagnosis and treatment of affected emotional states in humans.

TABLE 13.2 Pituitary glucocorticoid receptor (type II) population in RLA/Verh and RHA/Verh rats as a function of age.

Age (days)	3H-RU 28362 bound (fmol/mg prot) RHA/Verh	RLA/Verh	$p<$
5	249.9 ± 62 (3)	183.5 ± 29 (6)	ns
10	209.3 ± 48 (6)	124.0 ± 19 (7)	0.05
21	279.5 ± 62 (6)	150.5 ± 21 (7)	0.05
100	245.6 ± 14 (13)	173.0 ± 18 (10)	0.01

Receptor concentration was determined with binding of tritiated-RU28362, a type II agonist, according to a method described previously (see Walker et al. 1989). No changes in the affinity of the receptor for its ligand (Kd) were observed at any age tested. Values represent the mean ± SEM of 3–13 determinations per age group and per rat type.

REFERENCES

Axelrod J, Reisine RD (1984): Stress hormones: their interaction and regulation. *Science* 224:452–459

Bättig K, Driscoll P, Schlatter J, Uster HJ (1976): Effects of nicotine on the exploratory locomotion patterns of female Roman high- and low-avoidance rats. *Pharmacol Biochem Behav* 4:435–439

Bignami G (1965): Selection for high rates and low rates of avoidance conditioning in the rat. *Anim Behav* 13:221–227

Blizard DA (1971): Autonomic reactivity in the rat: effects of genetic selection for emotionality. *J Comp Physiol Psychol* 76:282–289

Blizard DA (1988): The locus ceruleus: a possible neural focus for genetic differences in emotionality. *Experientia* 44:491–495

Blizard DA, Liang B, Emmel DK (1980): Blood pressure, heart rate, and plasma catecholamines under resting conditions in rat strains selectively bred for differences in response to stress. *Behav Neural Biol* 29:487–492

Bohus B, Lissak K (1968): Adrenocortical hormones and avoidance behaviour of rats. *Int J Neuropharmacol* 7:301–306

Bohus B, Nyakas C, Endroczi E (1968): Effects of adrenocorticotropic hormone on avoidance behaviour of intact and adrenalectomized rats. *Int J Neuropharmacol* 7:307–314

Breier A, Albus M, Pickar D, Zahn TP, Wolkowitz OM, Paul SM (1987): Controllable and uncontrollable stress in humans: alterations in mood and neuroendocrine and psychophysiological function. *Am J Psychiatry* 144:1419–1424

Broadhurst PL (1957): Determinants of emotionality in the rat. Situational factors. *Br J Psychol* 48:1–12

Broadhurst PL (1969): Psychogenetics of emotionality in the rat. *Ann NY Acad Sci* 159:806–824

Broadhurst PL (1975): The Maudsley Reactive and Nonreactive strains of rats: a survey. *Behav Genet* 5:299–319

Broadhurst PL, Bignami G (1965): Correlative effects of psychogenetic selection: a study of the Roman High and Low Avoidance strains of rats. *Behav Res Ther* 2:273–280

Bruhn TO, Plotsky PM, Vale W (1984): Effect of paraventricular lesions on corticotropin-releasing factor like immunoreactivity in the stalk median eminence: studies on the adrenocorticotropin response to ether stress and exogenous CRF. *Endocrinology* 114:57–62

Brush FR (1991): Genetic determinants of individual differences in avoidance learning: behavioral and endocrine characteristics. *Experientia* 47, 1039–1050

Brush FR, Del Paine SN, Pellegrino LJ, Rykaszewski IM, Dess NK, Collins PY (1988): CER suppression, passive-avoidance learning, and stress-induced suppression of drinking in the Syracuse High- and Low-Avoidance strains of rats (rattus norvegicus). *J Comp Psychol* 102:337–349

Brush FR, Froehlich JC, Sakellaris PC (1979): Genetic selection for avoidance behavior in the rat. *Behav Genet* 9:309–316

Castanon N, Dulluc J, LeMoal M, Mormede P (1992): "Prolactin as a link between behavioral and immune differences between the Roman rat lines." Physiology and Behaviour (in press).

Chalmers DV, Hohf JC, Levine S (1974): The effects of prior aversive stimulation on the behavioral and physiological responses to intense acoustic stimuli in the rat. *Physiol Behav* 12:711–717

Conner RL, Levine S (1969): The effects of adrenal hormones on the acquisition of signaled avoidance behavior. *Horm Behav* 2:127–138

Conner RL, Vernikos-Danellis J, Levine S (1971): Stress, fighting and neuroendocrine function. *Nature* 234:564–566

Coover DG, Ursin H, Levine S (1973): Plasma corticosterone levels during active-avoidance learning in rats. *J Comp Physiol Psychol* 82:170–174

Dallman MF, Akana SF, Cascio CS, Darlington DN, Jacobson L, Levin N (1987): Regulation of ACTH secretion: variations on a theme of B. In: *Recent Progress in Hormone Research 13*. New York: Academic Press

Dallman MF, Darlington DN, Suemaru S, Cascio CS, Levin N (1989): Corticosteroids in homeostasis. *Acta Physiol Scand* 136(suppl 583):27–34

Dallman MF, England WC, Rose JC, Wilkinson CW, Shinsako J, Siedenburg F (1978): Nycthemeral rhythm in adrenal responsiveness to ACTH. *Am J Physiol* 235:R210–R218

Dallman MF, Jones MT (1973): Corticosteroid feedback control of ACTH secretion: effect of stress-induced corticosterone secretion on subsequent stress response in the rat. *Endocrinology* 92:1367–1375

D'Angio M, Serrano A, Driscoll P, Scatton B (1988): Stressful environmental stimuli increase extracellular DOPAC levels in the prefrontal cortex of hypoemotional (Roman high-avoidance) but not hyperemotional (Roman low-avoidance) rats. An in vivo voltammetric study. *Brain Res* 451:237–247

Davidson JM, Levine S (1972): Endocrine regulation of behavior. *Annu Rev Physiol* 34:375–408

Davis H, Porter JW, Livingstone J, Herrmann T, McFadden L, Levine S (1977): Pituitary-adrenal activity and lever press, shock escape behavior. *Physiol Psychol* 5:280–284

De Kloet ER, Reul JMHM (1987): Feedback action and tonic influence of corticosteroids on brain function: a concept arising from the heterogeneity of brain receptor systems. *Psychoneuroendocrinology* 12:83–105

Denenberg VH (1969): Open-field behavior in the rat: what does it mean? *Ann NY Acad Sci* 159:852–859

De Souza EB, Van Loon GR (1982): Stress-induced inhibition of the plasma corticosterone response to a subsequent stress in rats: a nonadrenocorticotropin mediated mechanism. *Endocrinology* 110:23–33

Driscoll P (1986): Roman High- and Low-Avoidance rats: present status of the Swiss sublines, RHA/Verh and RLA/Verh, and effects of amphetamine on shuttle-box performance. *Behav Genet* 16:355–364

Driscoll P, Bättig K (1982): Behavioral, emotional and neurochemical profiles of rats selected for extreme differences in active, two-way avoidance performance. In: *Genetics of the Brain,* Lieblich I, ed. Amsterdam: Elsevier Biomedical Press, pp 95–123.

Driscoll P, Dedek J, D'Angio M, Claustre Y, Scatton B (1990): A genetically-based model for divergent stress responses: behavioral, neurochemical and hormonal aspects. *Adv Anim Breeding Genet* 5(suppl):97–107

Driscoll P, Dedek J, Martin JR, Zivkovic B (1983): Two-way avoidance and acute shock stress induced alterations of regional noradrenergic, dopaminergic and serotonergic activity in Roman High- and Low-Avoidance rats. *Life Sci* 33:1719–1725

Driscoll P, Fümm H, Bättig K (1979): Maternal behavior in two rat lines selected for differences in the acquisition of two-way avoidance. *Experientia* 35:786–788

Driscoll P, Käsermann HP (1977): Differences in the response to pentobarbital sodium of Roman High- and Low-Avoidance rats. *Arzneim Forsch/Drug Res* 27:1582–1584

Feldman S (1985) Neural pathways mediating adrenocortical responses. *Fed Proc* 44:169–175

Fischer LA, Rivier J, Rivier C, Spiess J, Vale W, Brown MR (1982): Corticotropin-releasing factor (CRF): central effects on mean arterial pressure and heart rate in rats. *Endocrinology* 110:2222–2223

Fümm H (1982): *Die Ontogenese des Explorationsverhaltens bei RHA/Verh und RLA/Verh Ratten*. Ph.D. Thesis, ETH Zürich, no. 6994

Fümm H, Driscoll P (1981): Litter size manipulations do not alter maternal behavior traits in selected lines of rats. *Anim Behav* 29:1267–1269

Gentsch C, Lichtsteiner M, Driscoll P, Feer H (1982): Differential hormonal and physiological responses to stress in Roman High- and Low-Avoidance rats. *physiol Behav* 28:259–263

Gentsch C, Lichtsteiner M, Feer H (1981): Locomotor activity, defecation score and corticosterone levels during an openfield exposure: a comparison among individually and group-housed rats, and genetically selected rat lines. *Physiol Behav* 27:183–186

Gentsch C, Lichtsteiner M, Feer H (1988): Genetic and environmental influences on behavioral and neurochemical aspects of emotionality in rats. *Experientia* 44:482–489

Gibbs DM (1984): Dissociation of oxyocin, vasopressin and corticotropin secretion during different types of stress. *Life Sci* 35:487–491

Gibbs DM (1986): Vasopressin and oxytocin: hypothalamic modulators of the stress response: a review. *Psychoneuroendocrinology* 11:131–140

Gold PW, Gwirtsman H, Avgerinos P, Nieman LK, Gallucci WT, Kayne W, Jimerson D, Ebert M, Rittmaster R, Loriaus L, Chrousos GP (1986b): Abnormal hypothalamic-pituitary-adrenal function in anorexia nervosa. *N Engl J Med* 314:1335–1342

Gold PW, Loriaux L, Roy A, Kling MA, Calabrese JR, Kellner CH, Nieman LK, Post RM, Pickar D, Gallucci W, Avgerinos P, Paul S, Oldfield EH, Cutler GB, Chrousos GP (1986a): Responses to corticotropin-releasing hormone in the hypercortisolism of depression and Cushing's disease. *N Engl J Med* 314:1329–1335

Gray JA, Levine S, Broadhurst PL (1965): Gonadal hormone injections in infancy and adult emotional behavior. *Anim Behav* 13:33–45
Guth S, Seward JP, Levine S (1971): Differential manipulation of passive avoidance by exogenous ACTH. *Horm Behav* 2:127–138
Hanson JD, Larson ME, Snowdon CT (1976): The effects of control over high intensity noise on plasma cortisol levels in rhesus monkeys. *Behav Biol* 16:333–340
Hennessy MB, Levine S (1977): Effects of various habituation procedures on pituitary-adrenal responsiveness in the mouse. *Physiol Behav* 18:799–802
Hennessy JW, Levine S (1979): Stress, arousal and the pituitary-adrenal system: a psychoendocrine hypothesis. In: *Progress in Psychobiology and Physiological Psychology,* Sprague JM, Epstein AN, eds. 8:133–178
Henry JP (1986): Neuroendocrine patterns of emotional response. In: *Emotion: Theory, Research, and Experience*. Academic Press, New York 3:37–60
Iso H, Brush FR, Fujii M, Shimazaki M (1988): Running-wheel avoidance learning in rats (rattus norvegicus): effects of contingencies and comparison of different strains. *J Comp Psychol* 102:350–371
Käsermann HP, Driscoll P (1980): Differential sensitivity to ether anesthesia in RHA/Verh and RLA/Verh rats. In: *Animal Quality and Models in Biomedical Research,* SpiegelA, Erichsen S, Solleveld HA, eds. Stuttgart: Gustav Fischer Verlag, pp 377–381
Kasper-Pandi P, Hansing R, Usher D (1970): The effect of dexamethasone blockade of ACTH release on avoidance learning. *Physiol Behav* 5:361–363
Keller-Wood ME, Dallman MF (1984): Corticosteroid inhibition of ACTH secretion. *Endocr Rev* 5:1–24
Koob GF, Bloom FE (1985): Corticotropin releasing factor and behavior. *Fed Proc* 44:259–263
Koob GF, Le Moal M, Bloom FE, Sutton RE, Rivier J, Vale W (1982): Corticotropin-releasing factor produces behavioral activation and improves learning in rats. *Soc Neurosci Abstr* 8:145
Lang RE, Heil JWE, Ganten D, Hermann K, Unger T, Rascher W (1983): Oxytocin, unlike vasopressin, is a stress hormone in the rat. *Neuroendocrinology* 37:314–316
Levine S (1967): Maternal and environmental influences on the adrenocortical response to stress in weanling rats. *Science* 156:258–260
Levine S (1968): *Hormones and Conditioning*. Nebraska Symp. Motivation, Lincoln, Nebraska, pp 85–102
Levine S, Goldman L, Coover GD (1972): Expectancy and the pituitary-adrenal system. In: *Physiology, Emotion and Psychosomatic Illness,* Ciba Foundation Symposium 8:281–296
Levine S, Soliday S (1960): The effects of hypothalamic lesions on conditioned avoidance learning. *J Comp Physiol Psychol* 53:497–501
Levine S, Weinberg J, Brett LP (1979): Inhibition of pituitary-adrenal activity as a consequence of consummatory behavior. *Psychoneuroendocrinology* 4:275–286
McEwen BS (1987): Influences of hormones and neuroactive substances on immune function. In: *The Neuro-Immune-Endocrine Connection,* Cotman CW, ed. New York: Raven Press
Meaney MJ, Aitken DH, Van Berkel C, Bhatnager S, Sapolsky RM (1988): Effect of neonatal handling on age-related impairments associated with the hippocampus. *Science* 239:766–768
Meaney MJ, Aitken DH, Viau V, Sharma S, Sarrieau A (1989): Neonatal handling of rats

alters adrenocortical negative feedback sensitivity: a model for the study of individual differences in the pituitary-adrenal response to stress. *Neuroendocrinology* 50:587–604

Normand M, Lalonde J, Lavoie M, Barden N (1980): Adrenocortical responses to adrenocorticotropin in the rat. *Can J Physiol Pharmacol* 58:1279–1285

Plotsky PM, Bruhn TO, Vale W (1985a): Evidence for multifactor regulation of the adrenocorticotropin secretory response to hemodynamic stimuli. *Endocrinology* 116:633–639

Plotsky PM, Bruhn TO, Vale W (1985b): Hypophysiotropic regulation of adrenocorticotropin secretion in response to insulin-induced hypoglycemia. *Endocrinology* 117:323–329

Plotsky PM, Cunningham ET, Widmaier EP (1990): Catecholaminergic modulation of corticotropin-releasing factor and adrenocorticotropin secretion. *Endocr Rev* 10:437–458

Popova NK, Naumenko EV (1972): Dominance relations and the pituitary-adrenal system in rats. *Anim Behav* 20:108–111

Reul JMHM, De Kloet ER (1985): Two receptor systems for corticosterone in the rat brain: microdistribution and differential occupation. *Endocrinology* 117:2505–2511

Rivier CL, Brownstein M, Spiess J, Rivier J, Vale W (1982): In vivo corticotropin-releasing factor-induced secretion of adrenocorticotropin, beta-endorphin, and corticosterone. *Endocrinology* 110:272–278

Rivier CL, Plotsky PM (1986): Mediation by corticotropin-releasing factor (CRF) of adenohypophyseal hormone secretion. *Annu Rev Physiol* 48:475–494

Rivier CL, Rivier J, Vale W (1986): Stress-induced inhibition of reproductive functions: role of endogenous corticotropin-releasing factor. *Science* 231:607–609

Sakellaris PC, Vernikos-Danellis J (1975): Increased rate of response of the pituitary-adrenal system in rats adapted to chronic stress. *Endocrinology* 97:597–602

Sandi C, Castanon N, Vitiello S, Neveu PJ, Mormède P (1991): Different responsiveness of spleen lymphocytes from two lines of psychogenetically selected rats (Roman high and low avoidance). *J Neuroimmunol* 31:27–33

Sapolsky RM (1982): The endocrine stress response and social status in the wild baboon. *Horm Behav* 15:279–292

Sapolsky RM (1988): Individual differences and the stress response: studies of a wild primate. *Adv Exp Med Biol* 245:399–411

Sapolsky RM (1989): Hypercortisolism among socially-subordinate wild baboons originates at the CNS level. *Arch Gen Psychiatry* 46:1047–1051

Satinder KP (1981): Ontogeny and interdependence of genetically selected behaviors in rats: avoidance response and open-field. *J Comp Physiol Psychol* 95:175–187

Sawchenko P, Swanson L (1982): The organization of noradrenergic pathways from the brain stem to the paraventricular and supraoptic nuclei in the rat. *Brain Res Rev* 4:275–325

Schuhr B (1987): Social structure and plasma corticosterone levels in female albino mice. *Physiol Behav* 40:689–693

Spinedi E, Negro-Vilar A (1983): Angiotensin II and ACTH release: site of action and potency relative to corticotropin releasing factor and vasopressin. *Neuroendocrinology* 37:446–453

Stern JM, Erskine MS, Levine S (1973): Dissociation of open-field behavior and pituitary-adrenal function. *Horm Behav* 4:149–162

Swanson LW, Sawchenko PE, Rivier J, Vale W (1983): Organization of ovine corticotropin-releasing factor immunoreactive cells and fibers in the rat brain: an immunohistochemical study. *Neuroendocrinology* 36:165–186

Swenson RM, Vogel WH (1983): Plasma catecholamine and corticosterone as well as brain catecholamine changes during coping in rats exposed to stressful footshock. *Pharmacol Biochem Behav* 18:689–693

Treiman DM, Levins E (1969): Plasma corticosteroid response to stress in four species of wild mice. *Endocrinology* 84:676–680

Vale W, Spiess J, Rivier C, Rivier J (1981): Characterization of a 41-residue ovine hypothalamic peptide that stimulates the secretion of corticotropin and beta-endorphin. *Science* 213:1394–1397

Vale W, Vaughan J, Smith M, Yamamoto G, Rivier J, Rivier C (1983): Effects of synthetic ovine corticotropin-releasing factor, glucocorticoids, catecholamines, neurohypophyseal peptides, and other substances on cultured corticotropic cells. *Endocrinology* 113:1121–1131

Walker CD, Perrin M, Vale W, Rivier C (1986): Ontogeny of the stress response in the rat: role of the pituitary and the hypothalamus. *Endocrinology* 118:1445–1451

Walker CD, Rivest RW, Meaney MJ, Aubert ML (1989): Differential activation of the pituitary-adrenocortical axis after stress in the rat: use of two genetically selected lines (RLA and RHA) as a model. *J Endocrinol* 123:477–485

Weinberg J, Erskine MS, Levine S (1980): Shock-induced fighting attenuates the effects of prior shock experience in rats. *Physiol Behav* 25:9–16

Weinberger DA, Schwartz GE, Davidson RJ (1979): Low-anxious, high-anxious, and repressive coping styles: psychometric patterns and behavioral and physiological responses to stress. *J Abnormal Psychol* 88:369–380

Wertheim GA, Conner RL, Levine S (1969): Avoidance conditioning and adrenocortical function in the rat. *Physiol Behav* 4:41–44

Index